NANOELECTRONICS, NANOPHOTONICS, QUANTUM AND EMERGING TECHNOLOGIES

SELECTED TOPICS IN ELECTRONICS AND SYSTEMS

Editor-in-Chief: **M. S. Shur**   ISSN: 1793-1274

Selected Topics in Electronics and Systems – Vol. 68

NANOELECTRONICS, NANOPHOTONICS, QUANTUM AND EMERGING TECHNOLOGIES

Editors

F Jain
University of Connecticut, USA

C Broadbridge
Southern Connecticut State University, USA

M Gherasimova
University of Bridgeport, USA

H Tang
Yale University, USA

World Scientific

NEW JERSEY · LONDON · SINGAPORE · BEIJING · SHANGHAI · HONG KONG · TAIPEI · CHENNAI

Published by

World Scientific Publishing Co. Pte. Ltd.

5 Toh Tuck Link, Singapore 596224

USA office: 27 Warren Street, Suite 401-402, Hackensack, NJ 07601

UK office: 57 Shelton Street, Covent Garden, London WC2H 9HE

Library of Congress Control Number: 2024044630

British Library Cataloguing-in-Publication Data
A catalogue record for this book is available from the British Library.

Selected Topics in Electronics and Systems — Vol. 68
NANOELECTRONICS, NANOPHOTONICS, QUANTUM AND EMERGING TECHNOLOGIES

ISBN 978-981-12-9741-0 (hardcover)
ISBN 978-981-12-9742-7 (ebook for institutions)
ISBN 978-981-12-9743-4 (ebook for individuals)

For any available supplementary material, please visit
https://www.worldscientific.com/worldscibooks/10.1142/13966#t=suppl

Desk Editors: Soundararajan Raghuraman/Steven Patt

Typeset by Stallion Press
Email: enquiries@stallionpress.com

Preface

This special issue on *Nanoelectronics, Nanophononics, Quantum and Emerging Technologies* contains the proceedings of the 32nd annual Symposium of the Connecticut Microelectronics and Optoelectronics Consortium (CMOC), held virtually on February 28, 2024 and hosted by Information Technology Staff, University of Connecticut (Storrs Campus).

Organized by a team of seven academic institutions and thirteen companies across the United States, this symposium attracted speakers from both academia and industry with topics representative of CMOC's dynamic and relevant mission.

This year's Symposium program featured an increased emphasis on quantum materials and devices, including a keynote presentation on improving the coherence in superconducting quantum circuits. The keynote talk was complemented by 22 invited and contributed oral presentations and 30 contributed posters displayed in the virtual gallery. This volume contains 28 papers expanded for inclusion in the proceedings.

The topics discussed at the Symposium ranged from recent advances in microelectronics, such as gate-all-around field effect transistors (GAA FETs) to quantum computing, featured in a special session with three invited presentations, to topics in additive manufacturing.

Phase change memory cells and plasmonic nanogap antennas for sub-bandgap photodetection were discussed in the quantum materials session. Presentation topics also included in-memory computing and applications in neural networks, quantum dot gate FETs (QDG FETs) and optical logic.

Microfluidic systems used to evaluate electrochemical sensors and GAA FETs for biosensing applications were some of the topics of discussion in sensor applications. Explosives detection utilizing inkjet-printed devices and flexible 5G electronics for smart city applications were presented as examples of work in additive manufacturing.

In summary, the papers collected in this special issue broadly illustrate relevant aspects of high-performance materials and emerging quantum and nanoscale devices for implementing high-speed electronic systems. We would like to take this opportunity to express our thanks to the authors, presenters, participants and reviewers for their contributions to the vibrant symposium and the proceedings presented here. We are grateful to the staff of World Scientific for their careful attention to the preparation of this volume.

Guest Editors:
F. Jain (*University of Connecticut*)
C. Broadbridge (*Southern Connecticut State University*)
M. Gherasimova (*University of Bridgeport*)
H. Tang (*Yale University*)

Contents

Interconnected Plasmonic Nanogap Antennas for Sub-Bandgap Photodetection via Hot Carrier Injection#

John Grasso ©*, Rahul Raman ©† and Brian G. Willis ©‡

Chemical and Biomolecular Engineering,
University of Connecticut, 191 Auditorium Road,
Storrs, CT 06269, USA
*John.grasso@uconn.edu
†Rahul.raman@uconn.edu
‡Brian.willis@uconn.edu

Modern integrated circuits have active components on the order of nanometers. However, optical devices are often limited by diffraction effects with dimensions measured in wavelengths. Nanoscale photodetectors capable of converting light into electrical signals are necessary for the miniaturization of optoelectronic applications. Strong coupling of light and free electrons in plasmonic nanostructures overcomes these limitations by confining light into sub-wavelength volumes with intense local electric fields. Localized electric fields are intensified at nanorod ends and in nanogap regions between nanostructures. Hot carriers generated within these high-field regions from nonradiative decay of surface plasmons can be injected into the conduction band of adjacent semiconductors, enabling sub-bandgap photodetection. The optical properties of these plasmonic photodetectors can be tuned by modifying antenna materials and geometric parameters like size, thickness, and shape. Electrical interconnects provide connectivity to convert light into electrical signals. In this work, interconnected nanogap antennas fabricated with 35 nm gaps are encapsulated with ALD-deposited TiO_2, enabling photodetection via Schottky barrier junctions. Photodetectors with high responsivity ($12\,\mu A/mW$) are presented for wavelengths below the bandgap of TiO_2 ($3.2\,eV$). These plasmonic nanogap antennas are sub-wavelength, tunable photodetectors with sub-bandgap responsivity for a broad spectral range.

Keywords: Plasmonic dimers; nanofabrication; Schottky; photodetector; localized surface plasmon resonance.

1. Introduction

Modern nanophotonics has flourished due to plasmonics, yielding tremendous developments in various interdisciplinary fields by harnessing favorable light-matter

*Corresponding author.
#This chapter appeared previously on the International Journal of High Speed Electronics and Systems. To cite this chapter, please cite the original article as the following: J. Grasso, R. Raman, and B. G. Willis, *Int. J. High Speed Electron. Syst.*, **33**, 2440052 (2024), doi:10.1142/S0129156424400524.

interactions [1]. Plasmonic metal (Ag, Au, Cu) nanostructures can produce localized surface plasmons (LSPs) upon illumination with light from electrons reaching higher energy states via photoexcitation [2]. The primary effects of the generation of LSPs are local electromagnetic (EM) enhancements and hot carrier production [2]. Hot carrier generation (high-energy electrons/holes emanating from electronic excitation decay) is proportional to EM enhancements and has been successfully used in photocatalytic applications as a method to reduce activation energy and alter selectivity toward favorable products [3, 4]. Plasmonic nanostructures may exhibit resonances that can be tuned by altering composition, geometry, and spatial configurations resulting in control of wavelengths where maximum plasmonic effects occur [2]. Further EM field enhancement occurs when plasmonic nanostructures are arranged as dimers that create hotspots for hot carrier generation. Localized electric fields are intensified in nanogap regions between nanostructures where enhancements can reach over 1000 [2]. These nanogap regions enhance hot carrier formation and there is an inverse relationship between nanogap size and hot carrier generation rates [2]. Another advantage of using dimer pairs is that they have large extinction cross-sections and act as antennas to increase light collection. Plasmonic dimers can be electrically connected for electro-optic applications such as photodetectors [5].

Photodetection is possible using internal photoemission at Schottky barriers where photoexcited carriers transfer from metal to a semiconductor material through an interface [6, 7]. The Schottky barrier may be lower in energy compared to the semiconductor material bandgap, which enables sub-bandgap photodetection due to the lower energy required to promote the charge of carriers into the conduction band. By integrating Schottky diodes with plasmonic nanostructures, photodetection can occur over a large wavelength range by collecting hot carriers [8]. Pertsch *et al.* recently measured the responsivity of a pair of Au plasmonic nanostructures contacting TiO_2 semiconductor layers with a Au/TiO_2 Schottky barrier around 1 Ev [9]. The structures were fabricated as single devices through focused ion beam milling which has limitations for large array fabrication due to lower throughput compared to electron-beam lithography. Knight *et al.* measured the responsivity of gold nanorod arrays on Si and measured a Schottky barrier of 0.5 eV, which is half of the barrier reported for TiO_2 [8]. Kos *et al.* studied photocurrent detection through quantum tunneling with gold nanospheres and observed responsivities in the range of 0.01 $\mu A/mW$ [10]. In this study, we evaluate the responsivity and detectivity of plasmonic photodetectors formed using arrays of interconnected dimers with Au/TiO_2 Schottky barriers. We measure photocurrents at different bias voltages, optical wavelengths, polarization directions, and analyze optical extinction data pre- and post-TiO_2 deposition.

2. Experimental

Nanostructures were fabricated using an Elionix BODEN 150 electron beam writer on a fused silica wafer with poly (methyl methacrylate) (PMMA) photoresist and a layer of E-spacer. Samples were developed using a methyl isobutyl ketone (MIBK) and isopropanol (IPA) mixture followed by an O_2 plasma descum treatment of 75 W for 30 seconds. Metal was deposited in an electron beam evaporator with 4 nm Ti and 45 nm Au. Remover PG was used for lift-off processing. A second lithography step was used to add bonding pads and electrical connections to nanostructures. Shipley 1805 and LOR 3A resists were applied followed by exposure with a Heidelberg MLA-150 maskless aligner tool. Metal deposition was performed again to deposit 10 nm Ti and 200 nm Au contacts followed by lift-off processing with Remover PG. A third lithography step was used with Shipley 1805 and LOR 3A resists and the MLA-150 tool to create openings around the photodetectors for atomic layer deposition (ALD) of the semiconductor layers. A Savanah ALD system was used to deposit Al_2O_3 and TiO_2 on samples at 150° C with water co-reactant and trimethyl aluminum (TMA) and tetrakis(dimethylamino) titanium (TDMAT) used as precursors, respectively. Another lift-off step was used to remove the ALD coated photoresist leaving TiO_2 only on the device regions. Optical extinction curves were measured using an ellipsometer set in polarized transmission mode to measure LSP intensity. Scanning electron microscopy (SEM) was used to image nanostructures using a 4 nm layer of a sputter-deposited Au/Pd target to assist in charge dissipation. PROSEM software (Genisys) was used to analyze high-resolution SEM images to measure nanostructure dimensions.

A Keithley 2612B SMU was used to measure photocurrents and device I-V characteristics in the dark. Photodetection experiments were performed at various bias voltages, spanning 1–3.5 V. The device was blocked from light with a shutter for 30 seconds followed by cycles of exposure to light for 5 seconds and blocking for 10 seconds. Edmund Optics 50 nm bandpass filters and an Edmund Optics wire grid VIS-NIR polarizing film were used to measure wavelength and polarization dependence, respectively. The power from the light source for each stage of testing was measured with a Newport 843-R power meter. Dark currents are subtracted using a baseline correction algorithm to record photocurrents as changes in current upon illumination.

3. Results and Discussion

3.1. *Plasmonic nanostructure design*

Figure 1 illustrates the design of interconnected plasmonic nanogap photodetectors. Device areas are $25 \times 25 \, \mu m^2$ regions with arrays of nanostructures contacted by Au lines. Photodetectors consist of nanorod dimers with bisecting interconnect lines encapsulated in ALD TiO_2. Prior work with interconnected nanogap antennas guided geometric parameters to minimize perturbations to dimer resonance, as well as tuning plasmonic resonances around 800 nm [5, 11, 12]. Twelve cycles of ALD Al_2O_3 were deposited prior to TiO_2 layers as thin insulating layers to reduce dark currents and improve rectifying characteristics [9]. Titania (TiO_2) film thickness is measured as 45.7 nm via spectroscopic ellipsometry using a flat Si wafer

(a)   (b)

(c)

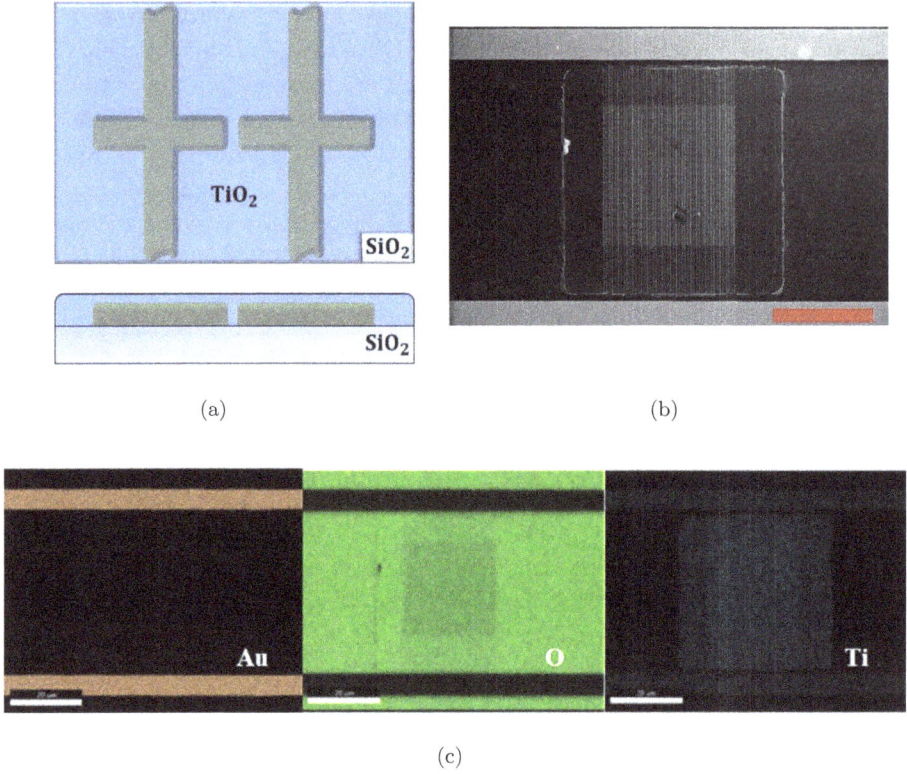

Fig. 1. Interconnected plasmonic dimers are coated with TiO_2 via ALD at 150° C. (a) Simplified schematic of plasmonic photodetectors within the unit cell; top and side views. (b) SEM image of plasmonic array post-ALD. The scale bar is 30 μm. (c) EDAX elemental mapping analysis images for Au, O, and Ti.

included in the ALD run. Previous work suggests that as-deposited TiO_2 at 150° C from TDMAT and water results in amorphous thin films, verified by grazing incidence X-ray diffraction (GIXRD) [13]. Energy dispersive X-ray analysis (EDAX) was used to perform elemental mapping of photodetectors. Figure 1(c) confirms the selective deposition of TiO_2, which covers the nanostructure regions. Faint traces of titanium are also detected on the Au contact lines due to the Ti adhesion layers applied during metal deposition, but they are unrelated to the ALD TiO_2.

Optical extinction spectra were investigated for samples before and after ALD TiO_2, as shown in Fig. 2. The uncoated antennas show a strong dipole resonance at 800 nm (blue line). There is a smaller peak near 530 nm assigned to the interconnect lines [14]. After ALD TiO_2, the surface plasmon resonance peak redshifts from 800 to beyond 1000 nm. This shift is expected due to the increased refractive index (2.39 @ 623.8 nm) for the surrounding medium (TiO_2) and is well understood in the literature [2, 15]. High-resolution SEM images were used to extract geometric parameters of fabricated nanostructures and verify conformal ALD coatings.

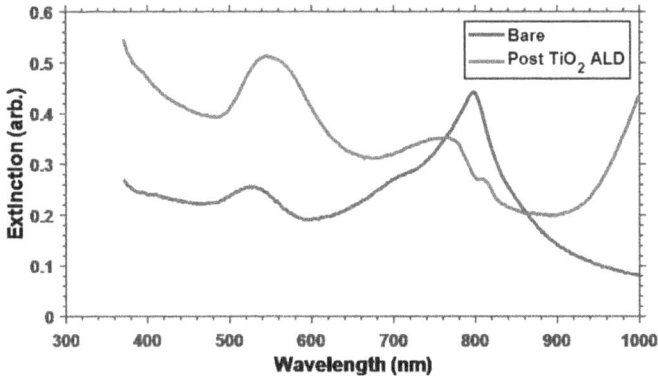

Fig. 2. Experimental extinction spectra pre- and post-TiO$_2$ ALD at 150°C.

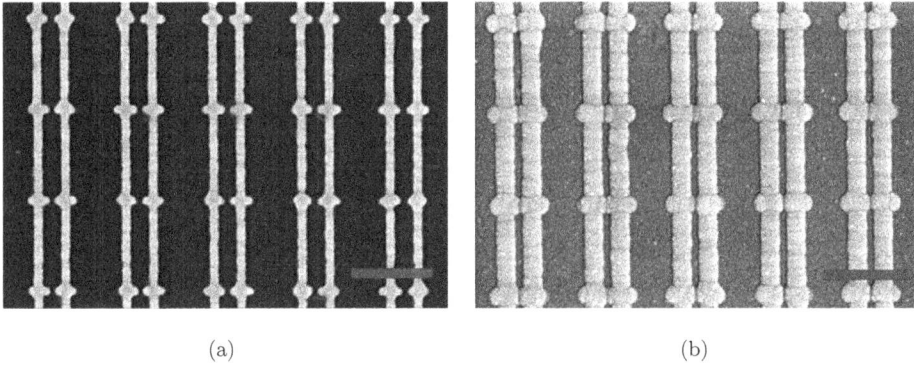

(a)   (b)

Fig. 3. SEM images of plasmonic nanostructures. (a) as-fabricated, before ALD. (b) post-ALD of TiO$_2$ at 150°C. Scale bar is 500 nm.

Figure 3(a) depicts bare nanostructures before ALD. Average nanorod lengths and widths were 134.2 ± 9.5 nm and 53.6 ± 0.8 nm, respectively. The average nanogap distance was 33.6 ± 6.7 nm. Nanostructures were successfully encapsulated in TiO$_2$, as shown in Fig. 3(b). The TiO$_2$ layers fill in the nanogaps to create Au/TiO$_2$/Au plasmonic junctions. Further investigation is required to characterize how TiO$_2$ deposition conditions and film thickness affect plasmon resonances and subsequent photoresponses.

3.2. *Electrical characteristics*

Current-voltage (I-V) characteristics of Au/TiO$_2$ Schottky barrier junctions at several temperatures ranging from 10°C to 30°C are given in Fig. 4(a). The metal-semiconductor-metal (MSM) structures feature two Schottky barriers connected in series. Under voltage bias conditions, one diode is forward biased and the other is

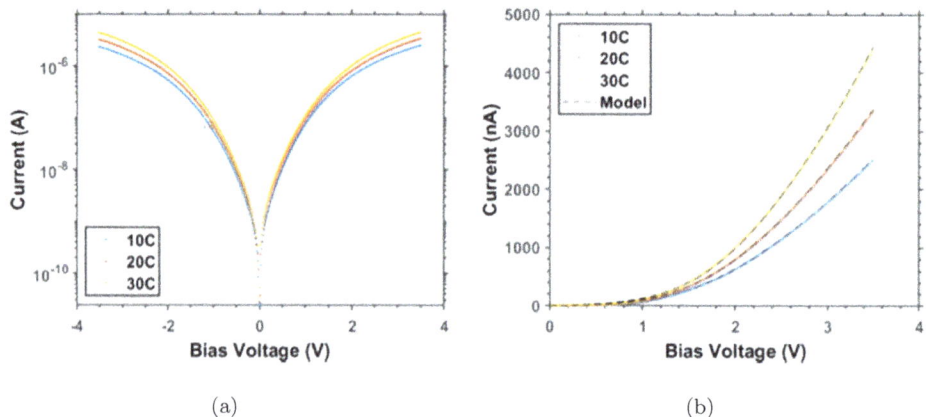

Fig. 4. I-V current characteristics of plasmonic photodetector at several temperature ranges. (a) ln(I) for positive and negative applied bias. (b) thermionic emission current modeled with extracted Schottky diode parameters from Cheung method.

reverse biased. Thus, the photodetectors exhibit symmetric I-V behavior for both bias conditions. At low to moderate bias voltages, current flow is governed by the Schottky contact under reverse bias. Image force lowering, or the Schottky effect, and tunneling currents contribute to current in these regimes [16]. For sufficiently large bias voltages, the forward biased Schottky contact in the MSM structure dominates current flow and can be modeled according to thermionic emission theory, as follows [16–20]:

$$I = AA^*T^2 \exp\left(-\frac{\Phi_B}{k_B T}\right)\left[\exp\left(\frac{q(V - IR_s)}{nk_B T}\right) - 1\right] \qquad (1)$$

where A is the effective diode area (cm^2), A* is the effective Richardson constant for TiO$_2$ (1200 A/cm^2/K^2) [21], k$_B$ is the Boltzmann constant, q is electronic charge, V is applied bias, Φ_B is the Schottky barrier height (eV), n is an ideality factor, and R$_s$ is series resistance in ohms. Electronic properties of Schottky contacts are characterized by barrier height and ideality factor. Other properties, such as series resistance, also play an important role in determining device performance, thus accurate parameters are crucial for effective device design. Methods which rely on extrapolation of the linear region of ln (I) − V curves may introduce uncertainty in parameter estimation due to considerable deviation from linearity at sufficiently large applied voltages, caused by series resistance [18]. Alternatively, barrier height, ideality factor, and series resistance can be determined by a method pioneered by Cheung and Cheune [17]. Figure 4(b) depicts the current modeled according to parameters calculated from the Cheung method. The model is in good agreement with the experimental I-V characteristics, especially at large applied voltages, suggesting thermionic emission is the dominant transport mechanism in this region.

Using the Cheung method, barrier height, ideality factor, and series resistance are estimated as $0.75\,\mathrm{eV}$, 15.6, and $256\,\mathrm{k\Omega}$ at $30^\circ\,\mathrm{C}$, respectively. The estimated barrier height is lower than the theoretical barrier ($1.0\,\mathrm{eV}$) for Au/TiO_2 and may be a result of the TiO_2 film quality and Au interface. Amorphous TiO_2 exhibits increased conductivity attributed to its electronically defective nature, which would lead to increased leakage current and non-ideality [22]. Additionally, trap states at the interface could shift the fermi level toward the conduction band, resulting in lowered barrier heights [19]. An ideality factor greater than 1 suggests a deviation from the ideal thermionic emission current. Several factors can contribute to n being greater than unity, including the presence of an interfacial layer [23], trap states in the band gap, barrier height inhomogeneity [21, 24, 25] and increased contribution from other current mechanisms, such as Fowler-Nordheim tunneling and generation-recombination currents within the depletion region [19]. Series resistance is related to the bulk resistivity of TiO_2 and also increases with insulating layers between the metal and semiconductor, such as Al_2O_3. Large deviations may additionally be a result of impurities in the TiO_2 film. Post-deposition annealing to promote amorphous-to-crystalline phase transition may improve Schottky diode characteristics.

3.3. *Photodetection*

Photocurrent measurements were performed with excitation wavelengths from $300\,\mathrm{nm}$ to $2800\,\mathrm{nm}$ using a fiber-coupled tungsten-halogen light source (ThorLabs, SLS201L). The fiber source is collimated and focused onto a sample with a spot diameter of $0.72\,\mathrm{mm}$ and light power of $2.4\,\mathrm{mW}$. Photocurrents were measured while varying applied bias, spanning from $1.5\,\mathrm{V}$ to $3.5\,\mathrm{V}$. Responsivity (R) and specific detectivity (D^*) were calculated as photodetector metrics, according to the following equations [9]:

$$R = \frac{I_{ph}}{P}\left(\frac{A_{beam}}{A_{eff}}\right) \tag{2}$$

$$D^* = \frac{A_{eff}R}{\sqrt{2qI_D}} \tag{3}$$

where I_{ph} is the photocurrent, A_{beam} is the area of the light beam, A_{eff} is the effective Au/TiO_2 contact area of the photodetector ($6.8\mathrm{e}\text{-}10\,\mathrm{m}^2$), P is light power, and I_D is dark current. Figure 5(a) illustrates time-dependent photo response data, demonstrating square wave responses upon illumination. Figure 5(b) shows an increase in both responsivity and specific detectivity as applied bias increases. Responsivity of $10^{-3}\,\mathrm{A/W}$ and specific detectivity of 10^5 Jones for the plasmonic nanogap arrays show three and two orders of magnitude enhancement compared to single nanogap antennas of similar design, respectively [9]. Similarly, there is a $1000\times$ improvement in responsivity compared to nanorod arrays on n-Si [8]. Dark currents

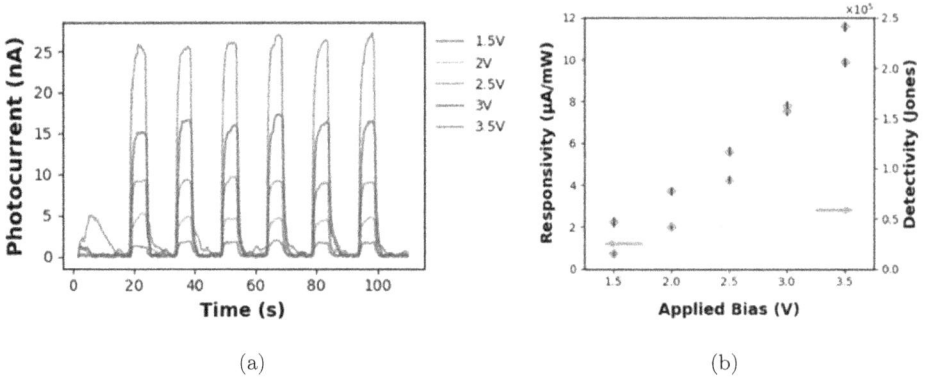

(a)   (b)

Fig. 5. Comparison of photocurrent as a function of applied bias. (a) Time dependence of the photocurrent at 1.5–3.5 V voltage bias. (b) Responsivity; left axis, and detectivity; right axis, photodetector metrics. Error bars correspond to standard error of 6 repeated measurements.

increase with larger bias voltages; however, decreased detectivity is not observed due to the offsetting increase in responsivity, see Eq. (3).

Previous studies of plasmonic junctions used monochromatic light from a white light laser, but quartz tungsten halogen sources contain a small proportion of photons with energy sufficient to transverse the intrinsic bandgap of TiO_2 (3.2–3.4 eV) [13], which may contribute to photocurrent via the photoelectric effect. To ensure photocurrents are produced via internal photoemission, the wavelength dependency of responsivity was investigated. Wavelength dependence was studied with the use of 50 nm bandpass filters ranging from 700 nm to 1100 nm. Power readings were measured for each filter. Figure 6(a) confirms sub-bandgap photodetection was achieved and responsivity values are consistently greater than $8\,\mu A/mW$. The photon energy of NIR light is sufficiently below the expected band gap of TiO_2 to support a mechanism involving photocurrent generation via photoemission of

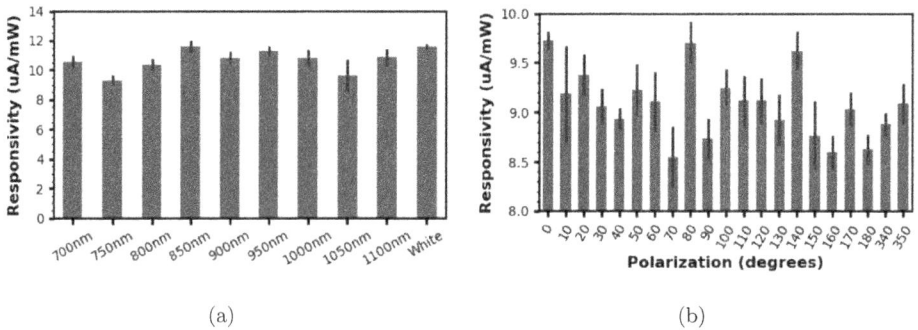

(a)   (b)

Fig. 6. (a) Wavelength and (b) polarization responsivity dependence at 3.5 V applied bias. Error bars represent standard error of 6 repeated measurements.

electrons from the Au contacts into the TiO$_2$ semiconductor layer. Similar responsivity was observed with filters compared to the white light spectrum. External quantum efficiency (EQE) is calculated (EQE (%) = [R$_\lambda$E$_\lambda$/q] × 100) based on responsivity and photon energy (E$_\lambda$) at the central wavelength (λ) for each optical filter. Quantum efficiency is between 1.1% and 1.8% for all wavelengths. This is significantly larger than previous reports of single nanogap antenna pairs with photodetectors exhibiting 0.01% EQE [9].

Photocurrents generated from hot carriers are expected to increase at resonance, but this effect is not clearly evident in Fig. 6(a). One possible explanation that a stronger wavelength dependence is not observed is that the resonance wavelength is shifted beyond the range of this investigation. Figure 2 shows that the extinction is increasing beyond 1000 nm. Additionally, the large bandwidths of optical filters (50 nm) may smear the response and be too broad to detect any wavelength dependency on photoresponses. Further study with different nanostructure designs is needed to elucidate wavelength-dependent photocurrents corresponding to plasmonic resonances.

In addition to wavelength effects, light polarization is also expected to influence hot carrier generation through the polarization dependence of the plasmon resonances. Photocurrent experiments were conducted using white light and a NIR polarizing filter to investigate how polarization affects responsivity, as depicted in Fig. 6(b). For nanorods, theory indicates the greatest EM field enhancement occurs for light polarized parallel to the longitudinal dimensions of the rods in the direction of the nanogaps. Maximum responsivity, 9.7 μA/mW, is observed for a polarization angle of 0 degrees relative to the longitudinal axis. Increasing the angle toward the transverse direction is accompanied by a decrease in responsivity, which is consistent with expectations. The data are scattered and the trend is not smooth, but the variation in photocurrents with polarization axis hints at plasmonic influences on photoemission of electrons from the metal into the semiconductor.

4. Conclusions

Arrays of nanofabricated plasmonic dimers with interconnects were coated with TiO$_2$ by ALD to study sub-bandgap photodetection with Au/TiO$_2$/Au Schottky barrier nanojunctions. Non-linear I-V characteristics support the formation of Schottky barriers and a photocurrent mechanism where photo-excited electrons are injected into the semiconductor layer. Estimation of Schottky barrier heights and ideality factors suggest non-ideal behavior, especially at low voltage biases, which may be attributed to TiO$_2$ trap states, barrier height inhomogeneity, insulating layers, and/or amorphous crystallinity. Nonetheless, high responsivities of 12 μA/mW and detectivities of 2.1e5 Jones demonstrate three orders of magnitude improvement compared to similar plasmonic photodetectors. Further optimization of device designs, materials, and process engineering may lead to a new class of nanoscale photodetectors based on plasmonic nanojunctions.

Acknowledgments

The authors acknowledge the National Science Foundation (NSF) (Grant No. 2150158) and the Office of Naval Research (Grant No. N00014-22-1-2567). This work was performed in part at the Harvard University Center for Nanoscale Systems (CNS); a member of the National Nanotechnology Coordinated Infrastructure Network (NNCI), which is supported by the National Science Foundation under NSF award no. ECCS-2025158. Electron microscopy was performed at the UConn/Thermo Fisher Scientific Center for Advanced Microscopy and Materials Analysis (CAMMA).

ORCID

John Grasso ◉ https://orcid.org/0000-0003-4466-6894

Rahul Raman ◉ https://orcid.org/0009-0009-0382-1123

Brian G. Willis ◉ https://orcid.org/0000-0002-1720-4451

References

1. J. Tang, Q. Guo, Y. Wu, J. Ge, S. Zhang and H. Xu, *ACSNano* **18**, 4 (2024).
2. G. Baffou, *Thermoplasmonics: Heating metal nanoparticles using light* (Cambridge University Press, Cambridge, 2018).
3. J. Fojt, T. P. Rossi, P. V. Kumar and P. Erhart, *ACSNano* **18**, 8 (2024).
4. Y. Yuan, L. Zhou, H. Robatjazi, J. L. Bao, J. Zhou, L. Yuan, M. Lou, M. Lou, S. Khatiwada, E. A. Carter, P. Nordlander and N. J. Halas, *Science* **378**, 6622 (2022).
5. R. Raman, J. Grasso and B. G. Willis, *Int. J. High Speed Electron. Syst.* **32**, (2n04) (2023).
6. R. T. Tung, *Appl. Phys. Rev.* **1**, 011304 (2014).
7. Y. Lao and A. Perera, *Adv. OptoElectron.* **16**, 105304 (2015).
8. M. Knight, H. Sobhani, P. Nordlander and N. Halas, *Science* **332**, 6030 (2011).
9. P. Pertsch, R. Kullock, V. Gabriel, L. Zurak, M. Emmerling and B. Hecht, *Nano Lett.* **22**, 17 (2022).
10. D. Kos, D. R. Assumpcao, C. Guo and J. J. Baumberg, *ACSNano* **15**, 9 (2021).
11. C. Zhang, T. Gao, D. Sheets, J. N. Hancock, J. Tresback and B. Willis, *J. Vac. Sci. Technol. B* **39**, 5 (2021).
12. D. T. Zimmerman, B. D. Borst, C. J. Carrick, J. M. Lent, R. A. Wambold, G. J. Weisel and B. G. Willis, *J. Appl. Phys.* **123**, 063101 (2018).
13. R. Khan, H. Ali-Loytty, J. Saari, M. Valden, A. Tukianinen, K. Lahtonen and N. V. Tkachenko, *Nanomaterials* **10**, 8 (2020).
14. B. G. Willis, J. Grasso, C. Zhang and R. Raman, *ACS Appl. Nano Mater.* **6**, 23 (2023).
15. P. K. Jain and M. A. El-Sayed, *J. Phys. Chem. C* **11**, 47 (2007).
16. J. Fu, M. Hua, S. Ding, X. Chen, R. Wu, S. Liu, J. Han, C. Wang, H. Du, Y. Yang and J. Yang, *Sci. Rep.* **6**, 35630 (2016).
17. S. K. Cheung and N. W. Cheune, *Appl. Phys. Lett.* **49**, 2 (1986).
18. G. Guler, O. Gullu, S. Karatas and O. F. Bakkaloglu, *J. Phys.: Conf. Ser.* **153**, 012054 (2009).
19. M. A. Kadaoui, W. B. Bouiadjra, A. Saidane, S. Belahsene and A. Ramdane, *Superlattices Microstruct.* **82**, 1502–1503 (2015).

20. C. D. Lien, F. C. T. So and M. A. Nicolet, *IEEE Trans. Electron. Devices* **31**, 10 (1984).
21. G. Rawat, H. Kumar, Y. Kumar, C. Kumar and D. Somvanshi, *IEEE Electron Device Lett.* **38**, 5 (2017).
22. M. Hannula, H. Ali-Loytty, K. Lahtonen, E. Sarlin, J. Saari and M. Valden, *Chem. Mater.* **30**, 4 (2018).
23. S. Chaliha, M. N. Borah, P. C. Sarmah and A. Rahman, *Int. J. Thermophys.* **31**, 2030–2039 (2010).
24. N. A. Al-Ahmadi, *Heliyon* **6**, 04852 (2020).
25. H. Altuntas, A. Bengi, U. Aydemir, T. Asar, S. S. Cetin, I. Kars, S. Altindal and S. Ozcelik, *Mater. Sci. Semicond. Process.* **12**, 6 (2009).

Recent Advances and Applications of Semiconductor Optical Amplifiers*

Niloy K. Dutta ©

Department of Physics, University of Connecticut,
Storrs, CT 06269, USA
niloy.dutta@uconn.edu

This paper describes the recent advances in device designs and optical transmission applications of semiconductor optical amplifiers (SOA). The device advances described are quantum-dot-based SOA and photonic-integrated circuits using SOA. The use of nonlinear properties of SOAs in high-speed optical transmission is discussed.

Keywords: Semiconductor optical amplifiers; quantum dot; photonic integration; optical transmission; optical phase conjugation.

1. Introduction

A semiconductor optical amplifier (SOA) is a device very similar to a semiconductor laser. Optical amplifiers amplify light through stimulated emission which is responsible for gain. SOA has been fabricated with a regular double heterostructure, multi-quantum well (MQW), and quantum dot (QD) light amplifying region. QD-based SOAs are currently being investigated for many applications [1]. Semiconductor amplifiers are important for photonics integrated circuits, optical time division multiplexing, and optical phase conjugation (OPC). This paper describes the current status in these areas.

2. Quantum Dot (QD) SOA

The quantum dot semiconductor optical amplifiers (QD-SOA) have some advantages over conventional bulk or quantum well devices because of higher saturation power, faster gain recovery and large amplification bandwidth [1–7]. The quantum dot active region (i.e. quantum dots and the embedding layers), for SOA, is often surrounded by a set of separate confinement layers that have n-type and p-type cladding layers on either side similar to that for a regular laser. There has

---

*This chapter appeared previously on the International Journal of High Speed Electronics and Systems. To cite this chapter, please cite the original article as the following: N. K. Dutta, *Int. J. High Speed Electron. Syst.*, **33**, 2440053 (2024), doi:10.1142/S0129156424400536.

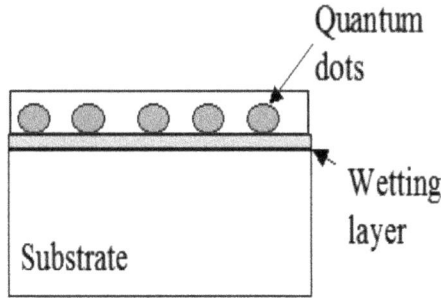

Fig. 1. Schematic of quantum dot formation. The top figure shows the planar substrate. The lower figure shows quantum dots on the wetting layer. The dots are covered in a high band gap material for laser or amplifier devices.

been considerable work on the fabrication of semiconductor nanostructures using the phenomenon of island formation during strained layer epitaxy, a process called Stranski-Krastanow growth mode. A formation of clusters is observed during the epitaxial growth of a semiconductor layer (e.g. InGaAs) on top of another (e.g. GaAs) that has a lattice constant several (3 to 5) percent smaller. For the first few layers, the atoms arrange themselves in a planar layer called the wetting layer. As the epitaxial growth proceeds, the atoms bunch up to form clusters as shown in Fig. 1. Cluster growth is energetically favorable since it relaxes the strain and thus reduces the strain energy. Since the QDs appear spontaneously, the process is often called a self-assembly process. Although this process may produce dots of different sizes, lasers and amplifiers with QD active regions have been fabricated. The self-assembly process is observed in several material systems. For example, GaAs and InAs to form InGaAs quantum dots, InP and GaP to form InGaP quantum dots and Ge and Si which form GeSi dots. Although the advantages of low dimensional structures (such as QD and Q-wire) were determined early on [2–7], the experimental challenge has been to fabricate QDs ($\sim$10 nm in dimension) of uniform size in a reproducible way.

The lateral confinement in this heterostructure could be either strongly index guided (buried heterostructure) or weakly index guided (e.g. ridge waveguide type) i.e. the lateral index and current confinement follow the same methods as that for a laser.

The structure of a quantum-dot SOA along with the scanning electron photomicrographs of the QD active region is shown in Fig. 2 [8]. The quantum-dot layers consist of InAs QDs (grown by molecular beam epitaxy) embedded in $In_{0.17}Ga_{0.83}As$. The QD material in this case has a ground state emission peak of $\sim$1300 nm. The embedding layer affects the emission wavelength [1]. In order to obtain sufficient gain from the quantum dots, the quantum-dot layers are generally repeated a few times (10 in Fig. 2). Generally, close stacking of quantum-dot layers causes the vertical lining of dots (Fig. 2(a)), which leads to strain accumulation. The researchers

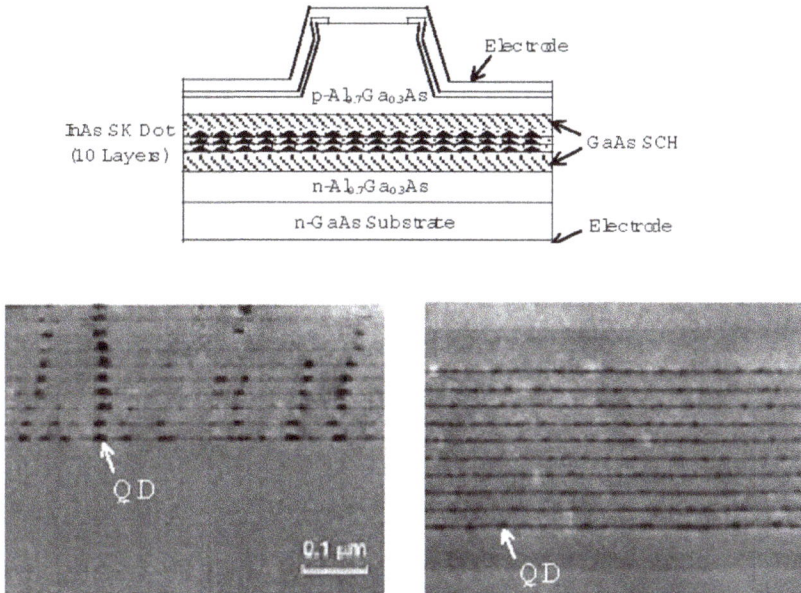

Fig. 2. Structure of ridge waveguide type quantum dot (QD) SOA. GaAs separate confinement heterostructure (SCH) is also shown (a) aligned quantum dots (b) nonaligned QDs [8].

in [8, 9] developed a technique to eliminate lining up (Fig. 2(b)) and hence strain accumulation.

The quantum dots have a discrete set of energy levels in the conduction band and valence band. Depending on the size of the dots and the bandgap of the wetting layer (the layer in which the dots are embedded), one or more of these discrete

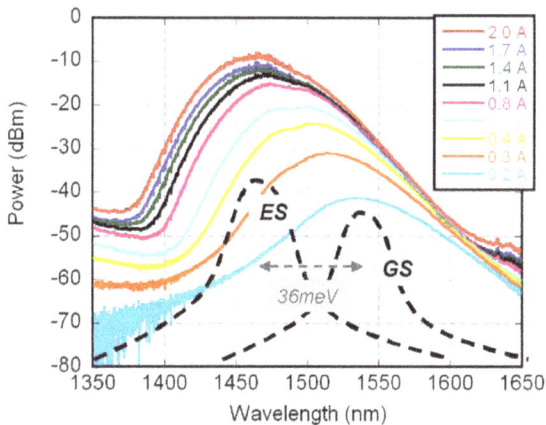

Fig. 3. The amplified spontaneous emission (ASE) spectra of a quantum dot (QD) amplifier at different bias currents. Also shown in the figure are the extrapolated QD ground and excited state emission profiles. This spectrum is for the device shown in [9].

energy levels are distinct i.e. separated from the wetting layer. The lowest energy level is called ground state (GS) here and the higher energy level is called excited state (ES) level here. Optical transitions take place from both levels. The higher energy levels merge with the wetting layer. These differences are evident in the amplified spontaneous emission (ASE) spectrum of the QD SOA.

The amplified spontaneous emission (ASE) from an InAs/InGaAs/InP QD SOA is shown in Fig. 3 [9].

At low currents, the conduction band ground state of the QD is populated, and the spectrum peak is near 1540 nm. The broadening of the spectrum is believed to be due to distribution in the dimension of the dots, i.e. not all dots are of the same size. At high currents, a second peak in the spectrum (near 1470 nm) emerges due to transitions from the excited state in the conduction band to the valence band.

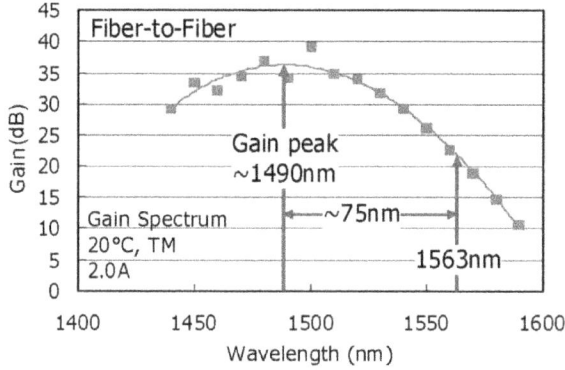

(a)

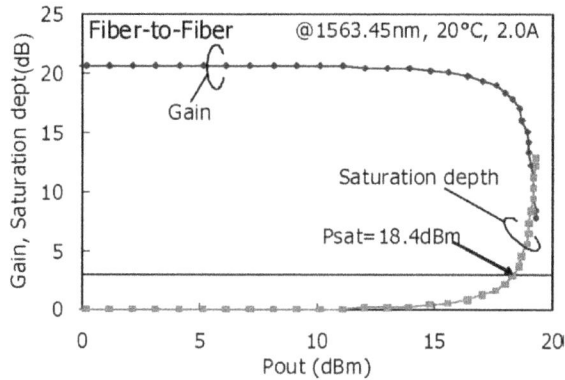

(b)

Fig. 4. (a) The measured gain spectra of a quantum dot (QD) amplifier at 20C. The device was put in a packaged module. (b) The measured gain is a function of output power [9].

In Fig. 3, the extrapolated QD excited state (ES) and the ground state (GS) ASE profile are also shown. The ∼70 nm difference between the peaks of the spectrum (ES and GS transitions) corresponds to an energy spacing of ∼36 meV, which agrees well with the calculation using the dot dimensions. The measured gain as a function of the wavelength of the QD-SOA in Fig. 3 is shown in Fig. 4. Note that a fiber-to-fiber gain of >30 dB is feasible. The 3-dB gain saturation power is ∼17 dB.

3. Photonic Integration with Amplifiers

Integrated distributed feedback (DFB) laser and amplifier structures are important because they have low spectral widths under modulation [10, 11]. The schematic of such a structure is shown in Fig. 5. The DFB laser and the amplifier have an MQW active region. Semi-insulating Fe-doped InP layers are used for both the current confinement and isolation of the DFB laser and the amplifier. The DFB laser region has a grating etched under the active region which provides frequency selective feedback.

For low chirp operation, the laser is modulated with low current (∼5 mA) which produces a small chirp and the amplifier is used to amplify the output to the required power levels (∼10 mW to 30 mW).

Photonic integrated circuits (PICs), for multichannel wavelength division multiplexed (WDM) transmission systems, both as transmitter and receiver ICs have been reported [12, 13]. The schematic architecture of a 10-channel transmitter PIC is shown in Fig. 6. It has ten tunable lasers which can be set to a 200 GHz spacing required by the International Telecommunication Union (ITU) for a WDM system. Each laser is followed by an electroabsorption (EA) modulator, a semiconductor optical amplifier (SOA), a back facet power monitor for the laser, and a variable optical attenuator (VOA) on each laser path. The latter is used to equalize the power output for each wavelength signal out of the AWG. The AWG multiplexes all modulated light signals onto a single fiber. The entire PIC is built using InP/InGaAsP device fabrication technology. The layers are grown using the MOCVD growth process.

The transmitter generates 10 wavelengths using tunable DFB lasers in the ITU grid. Good error free performance of the integrated transmitter has been reported

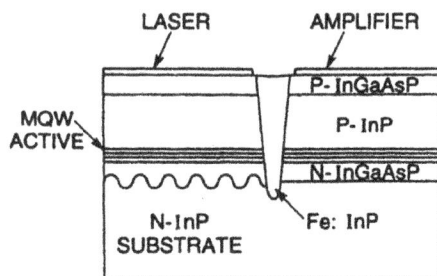

Fig. 5. Schematic of a distributed feedback (DFB) laser and amplifier structure [10].

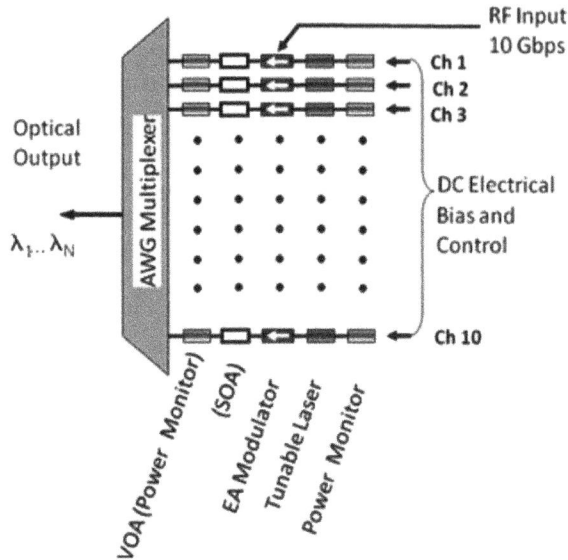

Fig. 6. Schematic architecture of a 10 channel WDM transmitter PIC [12].

[12]. Also, the reliability of the PICs is good for system applications. Transmitter and Receiver PICs with higher total transmission capacity have been reported using polarization multiplexing and QPSK modulation scheme [13].

4. Optical Transmission System Applications

In an optical time division multiplexing (OTDM) system, the data rates are high enough that the process of multiplexing (generation of high data rate signals from several low data rate signals) and demultiplexing (the process of conversion of high data rate signals to several low data rate signals) is carried out optically. The demultiplexing is necessary since the electronics for error rate determination is not available at high (>100 Gb/s) data rates. Several papers on various aspects of OTDM systems and all optical processing using optical amplifiers has been published [14–16]. For the demultiplexing process, an optical clock (a timing signal) needs to be extracted from the incoming high speed (>100 Gb/s optical data). This clock recovery process has been carried out using cross-gain modulation, cross-phase modulation and four-wave mixing [14–17].

A schematic of optical demultiplexing using four-wave mixing (FWM) is shown in Fig. 7. This technique allows any regularly spaced set of bits to be simultaneously extracted from the incoming data stream using a semiconductor optical amplifier (SOA) [14,15].

Consider the high-speed data (at a data rate B) at frequency ω_1 and an optical clock (at a clock rate B/N) at frequency ω_2 injected into an optical amplifier. The output of the amplifier will include four-wave mixing signals at frequencies $2\omega_1 - \omega_2$

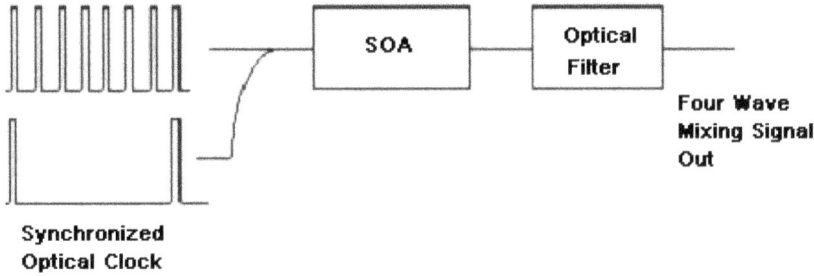

Fig. 7. Schematic of a four-wave mixing-based demultiplexer. The four-wave mixing takes place in the semiconductor amplifier (SOA) only when there is a temporal overlap between the signal and the clock pulse.

and $2\omega_2 - \omega_1$. One of the FWM signals (typically $2\omega_2 - \omega_1$) is filtered out using an optical filter at the exit port of the nonlinear element and represents the data signal before further processing. Thus, a signal representing the data every Nth bit time slot (where the mixing occurs with the clock signal) is generated by the four-wave mixing process (Fig. 7). The clock signal can then be delayed by 1 additional bit each time to retrieve the original data in successive time slots or a bit-delayed parallel system (similar to Fig. 7) could be set up for simultaneous extraction of N data streams each at B/N data rate. A complete system would need N SOA elements in order to extract N data bits simultaneously. The FWM process can have modulation bandwidths as high as 1.5 Tb/s [18]. Thus in principle demultiplexing from Tb/s optical data is feasible using the FWM process.

Among the semiconductor devices suitable for high-speed all-optical demultiplexing is an SOA-based Mach-Zehnder interferometer (MZI) [14–16]. These devices operate on the principle of a phase change caused by an incident optical clock signal (cross-phase modulation). This phase change can be adjusted so that the interferometer only yields output when a data signal overlaps a clock signal for demultiplexing applications. The schematic of a MZI device operating as a demultiplexer is shown in Fig. 8. The input signal beam is split into two beams by an input y-junction, which propagate through the semiconductor optical amplifiers positioned in the upper and lower arms of the interferometer. These beams then merge and interfere at an output y-junction. One of the amplifiers has the clock pulse incident on it. The clock frequency is 1/N time the signal bit rate. The clock pulse is absorbed by the amplifier changing its carrier density, which induces an additional phase shift for the signal traveling through that amplifier arm. This changes the output of the device for every Nth signal bit. The demultiplexed signal appears at the output at the clock frequency.

The arms of the MZI can be configured with built-in phase shifters (or SOA bias current can be used as a phase shifter) so that the phase difference for the high data rate signal propagating through the two arms is π (before the arrival of the clock pulse). Under this condition, the signals propagating through the two arms interfere

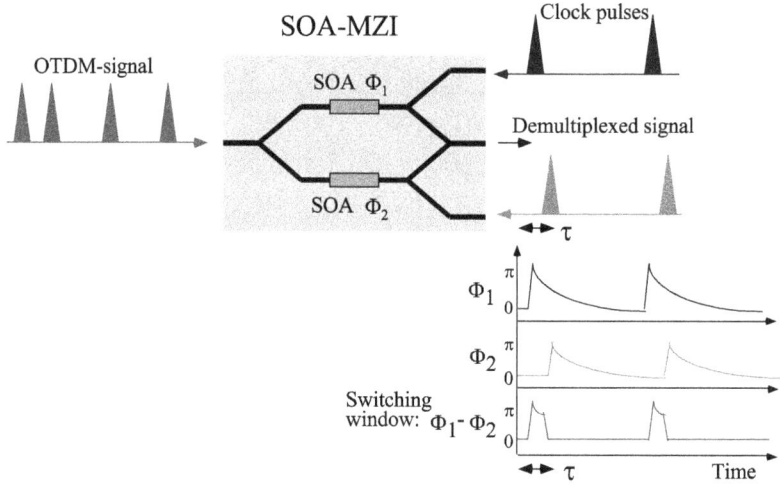

Fig. 8. Schematic of an optical amplifier-based Mach-Zehnder demultiplexer which uses a differential phase shift scheme. The figure shows the principle of operation. ϕ_1 and ϕ_2 represent the phases produced by the two-clock pulse sets one of which is delayed with respect to the other in the two arms. The phase difference $(\phi_1 - \phi_2)$ between the two arms occurs for a time of duration τ. This increases the speed capability of the demultiplexer [14, 15].

destructively, resulting in no output. In the presence of the clock pulse, an additional π phase shift is introduced in one arm of the interferometer, which results in a nonzero output for every Nth bit. Similar to the case of FWM, a complete system would need N MZI elements in order to extract N data bits simultaneously. The response speed i.e. the signal speed which the MZI can demultiplex is determined by the phase response time in the SOA. QD based SOA which has a fast phase response is important for this application.

Figure 9 shows the experimental results of demultiplexing an 80 Gb/s signal into 8 channels at 10 Gb/s each [15]. This experiment utilized a Michelsen interferometer which operates on the same principle as Mach-Zehnder interferometer.

As stated earlier a timing clock needs to be recovered from a high-data rate signal. The schematic of an experimental setup used for clock recovery from high-speed data signals is shown in Fig. 10. The input optical signal and the local oscillator optical signal produce a four-wave mixing (FWM) signal at the output of the SOA.

In a transmission system, the timing signal (clock) extracted from the input optical data is used to demultiplex the data to lower data rate channels, the error rate of which are then measured using traditional electronic circuits. The clock recovery process could be carried out using an optical phase-locked loop in conjunction with an electrical low-frequency phase-locked loop. This is shown schematically in Fig. 10. In this process, the high data rate optical signals are converted to a low data rate electrical signal using a four-wave mixing process followed by a photodetector. The phase and frequency of this electrical signal are locked to that of a

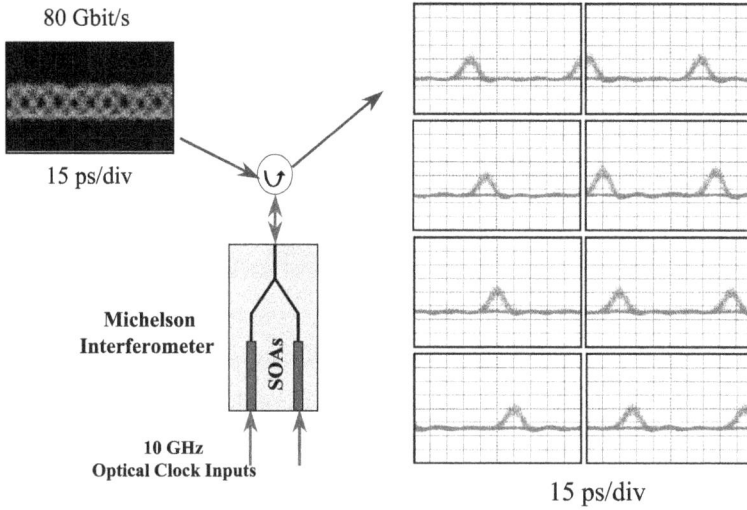

Fig. 9. The figure shows the demultiplexing of a 80 Gb/s signal to eight 10 Gb/s channels using a Michelson demultiplexer. Scale = 15 ps/div in all traces [15].

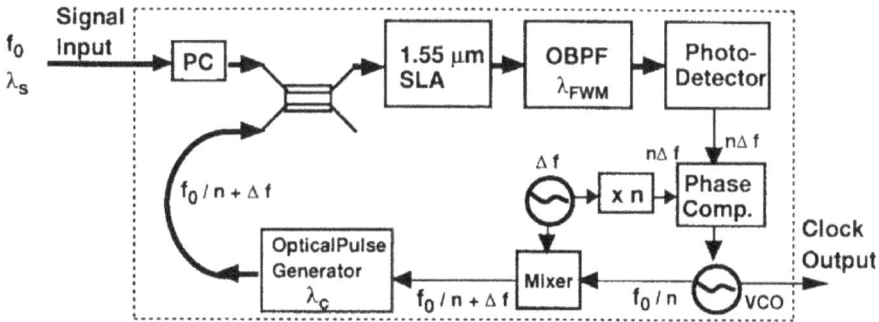

Fig. 10. Schematic of a clock recovery diagram used for recovering the clock from a high-speed optical signal using FWM in SOA (labeled SLA in the figure). The high-speed optical input (at >100 Gb/s) is down-converted to a low frequency (at ~1 MHz) electrical signal using an optical-to-electrical nonlinear down converter. This low-frequency signal is phase-locked using a conventional electrical phase-locked loop [17].

voltage-controlled local oscillator (VCO) using conventional electrical phase-locked loop techniques.

The key to the operation of this optical phase-locked loop is the four-wave mixing optical-to-electrical converter. It is a semiconductor optical amplifier which has two optical signal inputs at frequencies ω_1 and ω_2. The optical frequency ω_1 is the input wavelength carrying data and the frequency ω_2 is the optical clock generated by a laser at the demultiplexer. The wavelength ω_1 carries the data at B Gb/s and the wavelength ω_2 is the optical clock at a frequency F ~ B/N GHz +

f kHz. The four-wave mixing process produces an optical signal at a frequency of $2\omega_1 - \omega_2$. This optical signal has several electrical components including a low-frequency component close to Nf kHz produced by mixing the Nth harmonic of the clock (NF $\sim$ B + Nf) and the incoming signal at B Gb/s. This low-frequency component in the optical signal is converted to an electrical signal using an avalanche photodiode, filtered, and amplified. This low-frequency electrical signal and the Nf kHz signal from a signal generator are used in a conventional phase-locked loop configuration to lock the frequency F to B/N. The conventional phase-locked loop has a phase detector (electrical mixer) and a voltage-controlled oscillator. A stable optical clock is generated using a low-frequency electrical phase locked loop which would provide the needed stability, and an optical front end (semiconductor optical amplifier) which would convert the high data rate signal to a low-frequency optical signal.

The schematic of an optical time division multiplexed (OTDM) transmission setup is shown in Fig. 11. A stabilized mode-locked fiber laser generates short pulses at a 10 GHz repetition rate. The 3.5 ps wide pulses are compressed to 0.98 ps pulses and are coded with data by transmission through a LiNbO$_3$ modulator. It then goes through a time division multiplexer (TDM) which converts the data stream to 400 Gb/s data. The TDM consists of a suitably designed sequence of optical delay lines. The 400 Gb/s data are transmitted through various lengths of fiber. The clock is extracted from the transmitted data using a four-wave mixing (FWM) SOA-based clock recovery process. A mode-locked fiber laser operating at 10 GHz, the exact frequency of which is recovered by phase locking to the received 400 Gb/s data, provides the optical clock. The demultiplexing in this experiment is carried out using also a FWM process.

The error rate performance of the demultiplexed lower data rate channels are shown in Fig. 12. A Bit-error-rate of $<10^{-9}$ is obtained suggesting good performance. Very high-speed OTDM transmission systems are now being developed

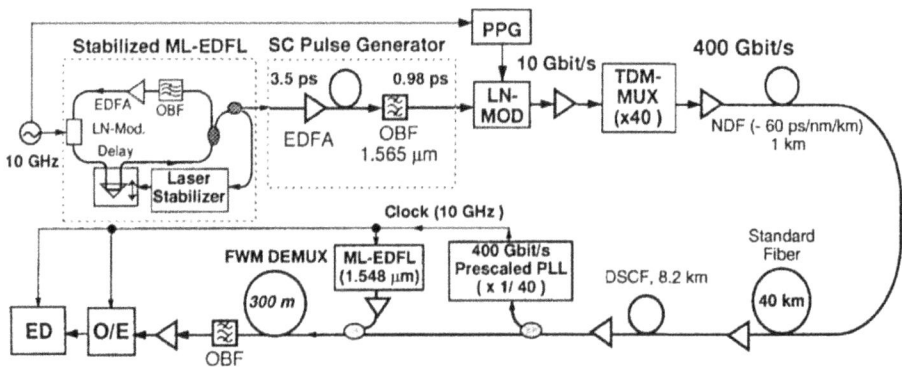

Fig. 11. Schematic of a 400 Gb/s optical transmission experiment set up [17].

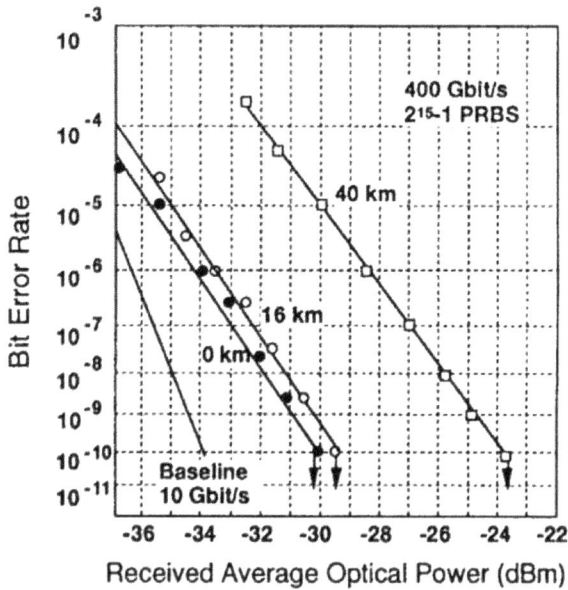

Fig. 12. Measured Bit-error-rate (BER) results at 10 Gb/s after demultiplexing of the 400 Gb/s data into forty 10 Gb/s data channels. The baseline is the BER data after the modulator. The three other lines are the data after transmission through ~10 m (labeled 0 km in the figure), 16 km and 40 km of fiber [17].

in several research laboratories for future installment in commercial traffic. Semiconductor optical amplifiers play a key role in the development of these systems. They can be used for both optical demultiplexing and optical clock recovery using either the FWM process or Mach-Zehnder-type devices. A detailed review is given in Refs. [8, 19, 1].

Lightwave transmission systems using advanced modulation formats such as QPSK and 16 QAM systems are being studied. These systems employ phase shifts in addition to amplitude shifts for modulation. For example, QPSK has four phase shifts $0, \pi/4, 2\pi/4, 3\pi/4$. The system using 16QAM has in addition four amplitude levels. Thus, reducing the accumulation of phase noise in optical fiber transmission is important for such systems. One way of reducing phase noise is optical phase conjugation (OPC) using SOA [20]. In the four-wave mixing, the signal generated has a conjugate phase from that of the incoming signal. This is shown in Fig. 13.

The principle of optical phase conjugation using four-wave mixing in SOA is shown in Fig. 14. The phase conjugator is placed mid-span. The accumulated phase fluctuation gets reversed, after going through the SOA which after transmission to an equal length of span compensates for the phase variance. This is very similar to dispersion compensation which involves putting a spool of fiber of opposite dispersion periodically throughout the transmission link.

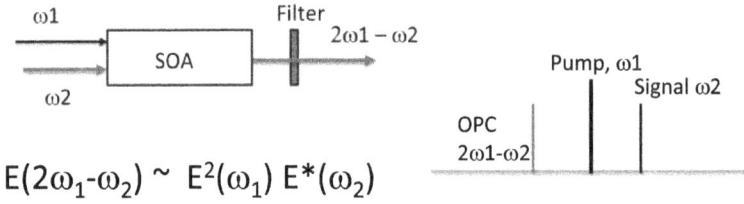

$$E(2\omega_1-\omega_2) \sim E^2(\omega_1) \, E^*(\omega_2)$$

Fig. 13. Schematic of four-wave mixing. E (ω_1) is the input electric field of the pump which is CW. E (ω_2) is the electric field of the input signal. The electric field of the FWM (noted as OPC signal) signal at ($2\omega_1 - \omega_2$) is also shown. Note E $* (\omega_2)$ i.e. phase is reversed.

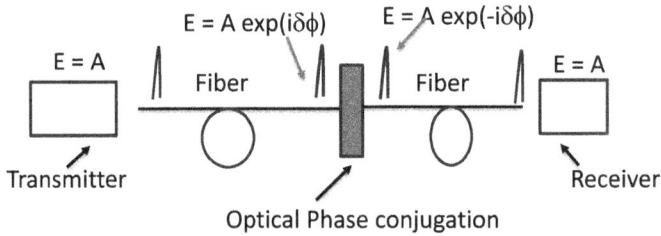

Fig. 14. A phase compensator is inserted midspan to reverse accumulated phase variance, which after transmission to an equal length of span compensates the phase variance.

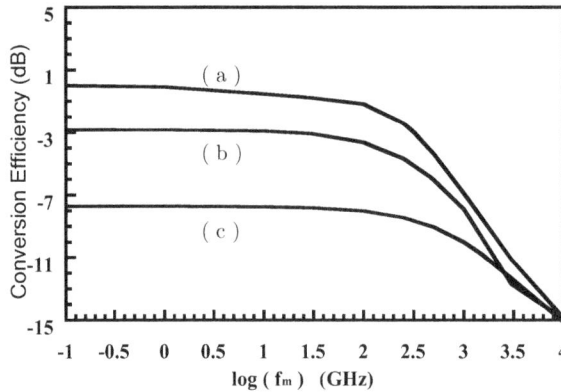

Fig. 15. Calculated frequency response of FWM conversion efficiency when $g_0 L = 7$. (a) wavelength difference between the CW pump and the input signal $\lambda_d = 0.5$ nm, FWM bandwidth = 320 GHz, (b) $\lambda_d = 2.5$ nm, FWM bandwidth = 460 GHz, (c) $\lambda_d = 10$ nm, FWM bandwidth = 1500 GHz. The x-axis is the log (base 10) of the modulation frequency f_m in GHz [18].

Experiments using OPC and SOA are being investigated in several laboratories. An important requirement is that the FWM process operate at high speed. The calculated bandwidth of the FWM process is shown in Fig. 15 [18]. The figure shows the FWM process can operate at high frequencies. The parameters used are suitable for a quantum dot semiconductor optical amplifier.

5. Summary

The paper describes the advances in device design and optical transmission application of semiconductor optical amplifiers (SOAs). Fabrication and performance of amplifiers with quantum dot active region are described. These devices have lower noise and high saturation power compared to SOAs with multi-quantum wells and regular double heterostructure active regions. SOAs are important for optical demultiplexing, and clock recovery in high-speed (400 Gb/s or higher) optical time division multiplexed transmission. They can also be used for phase distortion reduction in phase shift-based optical transmission system. Four-wave mixing in SOA plays an important role in the above systems.

ORCID

Niloy K. Dutta ⊕ https://orcid.org/0009-0003-3859-0747

References

1. N. K. Dutta, *Semiconductor Optical Amplifiers* (2nd edn.) (World Scientific, 2007).
2. V. M. Ustinov, A. E. Zhukov, A. Yu. Egorov and N. A. Maleev, *Quantum Dot Lasers* (Oxford University Press, 2003).
3. T. Akiyama, H. Kuwatsuka, T. Simoyama, Y. Nakata, K. Mukai, M. Sugawara, O. Wada and H. Ishikawa, J. Quantum. Electron. **37**, (2001) 1059.
4. P. Borri, W. Langbein, J. Hvam, F. Heinrichsdorff, M. Mao and D. Bimberg, Phys. Stat. Sol. **224**, (2001) 419.
5. P. Borri, W. Langbein, J. M. Hvam, F. Heinrichsdorff, M.-H. Mao and D. Bimberg, IEEE Photon. Technol. Lett. **12**, (2000) 594.
6. M. Sugawara, N. Hatori, T. Akiyama, Y. Nakata and H. Ishikawa, Jpn. J. Appl. Phys. **40**, (2001) L488.
7. M. Sugawara, H. Ebe, N. Hatori, M. Ishida, Y. Arakawa, T. Akiyama, K. Otsubo and Y. Nakata, Phys. Rev. B **69**, (2004) 235332.
8. M. Sugawara, T. Akiyama, N. Hatori, Y. Nakata, Y. Otsubo and H. Ebe, Proc. SPIE **4905**, (2002) 135.
9. G. Contestabile, A. Maruta and K. Kitayama, IEEE Photonic Tech. Lett. **22**, 987 (2010).
10. N. K. Dutta, J. Lopata, R. Logan and T. Tanbun-Ek, Appl. Phys. Lett. **51**, (1991) 1676.
11. U. Koren, B. I. Miller, G. Raybon, M. Oron, M. G. Young and T. L. Koch, Appl. Phys. Lett. **57**, (1990) 1375.
12. F. A. Kish *et al.*, IEEE J. Sel. Top. Quantum Electron. **17**, (2011) 1470.
13. R. Nagarajan, M. Kato, D. Lambert, P. Evans, S. Corzine, V. Lal, J. Rahn, A. Nilsson, M. Fisher, M. Kuntz, J. Pleumeekers, A. Dentai, H. Tsai, D. Krause, H. Sun, K. Wu, M. Ziari, T. Butrie, M. Reffle, M. Mitchell, F. Kish and D. Welch Semicond. Sci. Technol. **27**, (2012) 94003.
14. A. Piccirilli, H. Presby and N. K. Dutta, *MillCom*, Vol. 99 (Atlantic City, NJ, 1999).
15. I. Kang *et al.*, IEEE J. Sel. Areas Quantum Electron. **14**, (2008) 258.
16. S. Kawanishi and M. Saruwatari, IEEE J. Lightwave Technol. **11**, (1993) 2123.

17. S. Kawanishi, IEEE J. Quantum Electron. **QE-34**, (1998) 2064.
18. C. Wu and N. K. Dutta, J. Appl. Phys. **87**, (2000) 2076.
19. H. Sotobayashi and T. Ozeki, Chapter 2, in *WDM Technologies — Optical Networks*, A. K. Dutta, N. K. Dutta and M. Fujiwara (eds.) (Elsevier Science, 2004).
20. A. Sobhanan *et al.*, Adv. Opt. Photonics **14**, (2022) 571.

https://doi.org/10.1142/9789811297427_0003

ZSM-5-Based Catalyst Structure Design and Selectivity Control for Liquid Hydrocarbon Formation: A Mini Review#

Zichen Wang ⊚, Chunxiang Zhu ⊚, Binchao Zhao ⊚, Chenxin Deng ⊚, Fangyuan Liu ⊚ and Pu-Xian Gao ⊚*

Department of Materials Science and Engineering & Institute of Materials Science, University of Connecticut, Storrs, CT 06269, USA
*puxian.gao@uconn.edu

The selectivity limitation has long posed a significant challenge in the conversion of CO or CO_2 into liquid hydrocarbon-based sustainable fuels via Fischer–Tropsch synthesis (FTS) and other related processing pathways due to the Anderson–Schulz–Flory (ASF) distribution. The unique pore structure, thermal stability, and acidic sites of ZSM-5 have enabled its widespread applications in various catalytic processes of value-added chemicals and sustainable fuels. In the reaction pathways from CO/CO_2 to liquid HCs of interest, incorporating ZSM-5 into metal catalysts provides new opportunities to enhance product selectivity and catalytic activity, and even lower the reaction energetics. This review highlights recent advancements over the past five years in the structural design of ZSM-5-based catalysts aimed at improving the conversion selectivity of advanced liquid biofuels such as renewable diesel and sustainable aviation fuel. It explores innovative strategies for optimizing catalyst composition, the acidity, mesoporosity, and structures of ZSM-5 to design more efficient, selective, and robust catalysts. Additionally, the review addresses the direct hydrogenation of CO and CO_2 into C5+ liquid hydrocarbons, focusing on catalyst selection, acidic site optimization, and the structural configuration of ZSM-5. The mechanisms involved in these processes are also surveyed.

Keywords: ZSM-5; structure design; selectivity; hydrocarbons; sustainable biofuel.

#This chapter appeared previously on the International Journal of High Speed Electronics and Systems. To cite this chapter, please cite the original article as the following: Z. Wang, C. Zhu, B. Zhao, C. Deng, F. Liu and P.-X. Gao, *Int. J. High Speed Electron. Syst.*, **33**, 2440054 (2024), doi:10.1142/S0129156424400548.

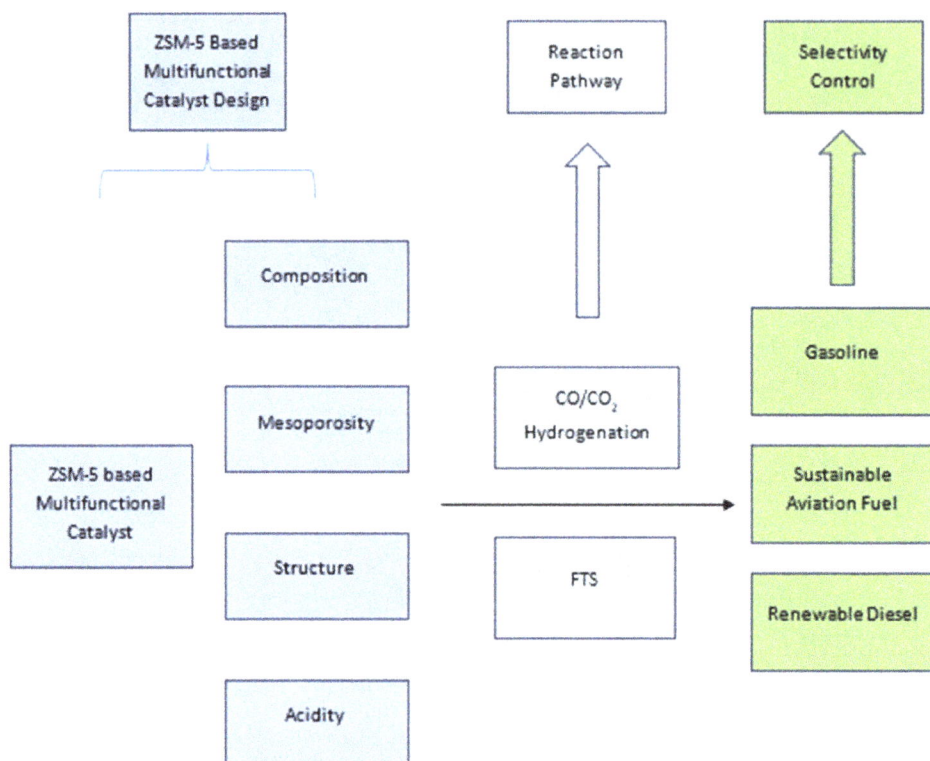

Scheme 1. Organization flow of this review on the selectivity control of ZSM-5-based multifunctional catalysts toward sustainable liquid hydrocarbon production.

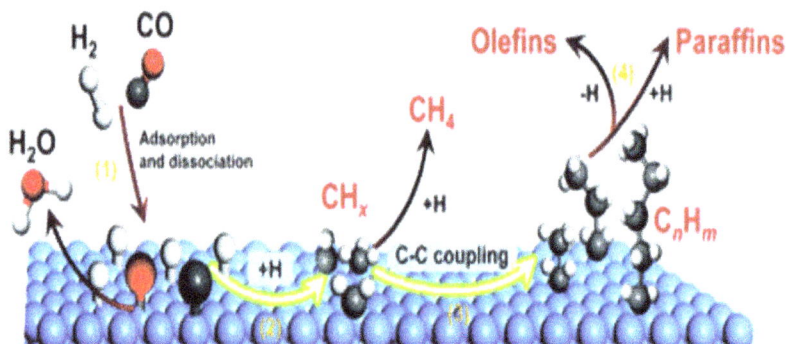

Scheme 2. Schematic illustration of key steps in Fischer–Tropsch synthesis [1].

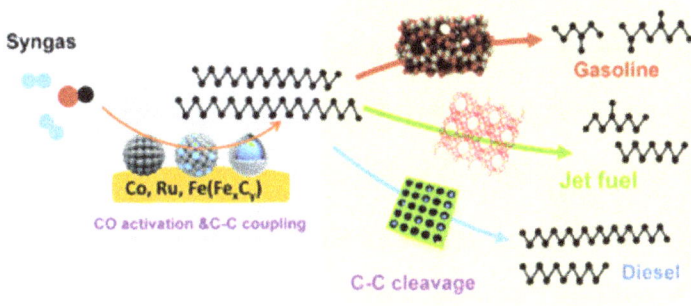

Scheme 3. Bifunctional catalysts composed of FT metal or metal carbide nanoparticles and a component for C–C cleavage for the direct conversion of syngas into liquid hydrocarbon of interest including gasoline, aviation fuel, and diesel fuel [1].

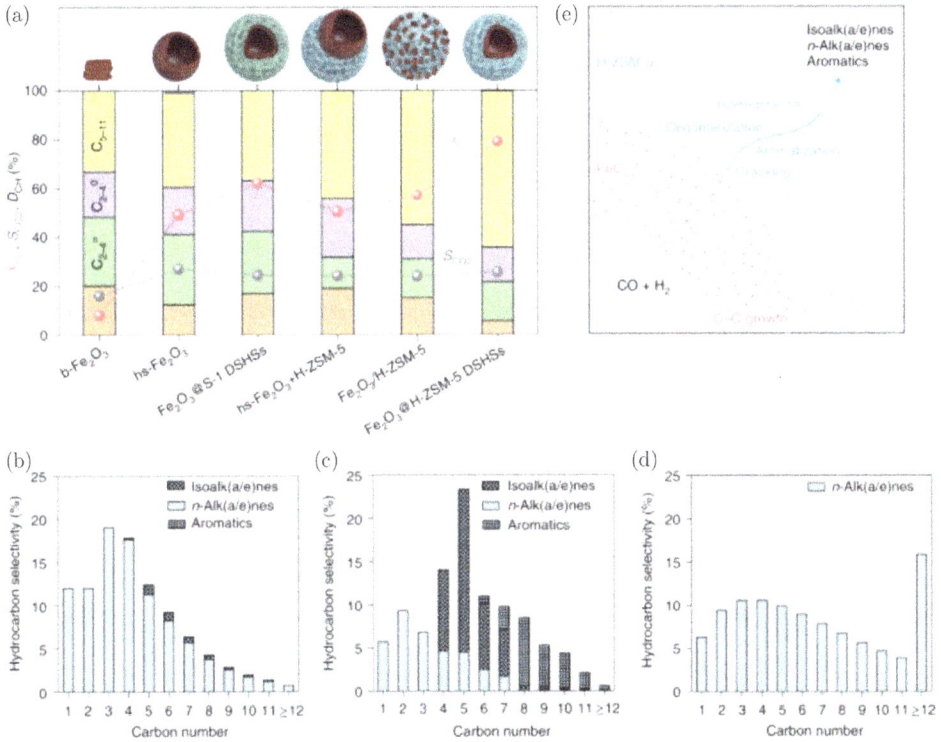

Fig. 1. Catalytic performances of different catalyst materials in syngas conversion. (a) CO conversion (X_{CO}), CO_2 selectivity (S_{CO2}) and distribution of hydrocarbon production (D_{CH}) in syngas conversion catalyzed by b-Fe_2O_3, hs-Fe_2O_3, Fe_2O_3@S-1 DSHSs, hs-Fe_2O_3+H-ZSM-5, Fe_2O_3/H-ZSM-5, and Fe_2O_3@H-ZSM-5 DSHSs. The top images show the structural model of each catalyst. (b), (c) Detailed hydrocarbon distributions in syngas conversion over hs-Fe_2O_3 (b) and Fe_2O_3@H-ZSM-5 DSHSs (c). (d) Theoretical ASF distribution with an alpha value of 0.75. (e) Schematic illustration of the C_5–C_{11} hydrocarbon formation pathway on Fe_2O_3@H-ZSM-5 DSHSs [2].

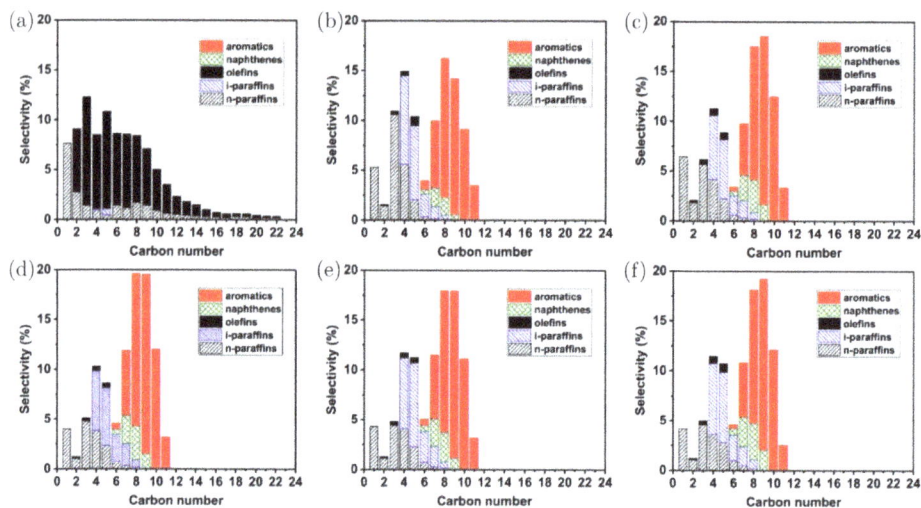

Fig. 2. Detailed hydrocarbon distribution obtained over hollow zeolites after alkaline treatment mixed with $Fe_3O_4@MnO_2$. None (a), Hol-Z(27) (b), Hol-Z(27)-1.00 (c), Hol-Z(27)-0.75 (d), Hol-Z(27)-0.25 (e), and Hol-Z(27)-0.00 (f). Hol-Z(27)-y means that the rate of SiO_2/Al_2O_3 (SAR) is 27 and y represents the molar ratio of TPAOH in the mixed alkaline solution [3].

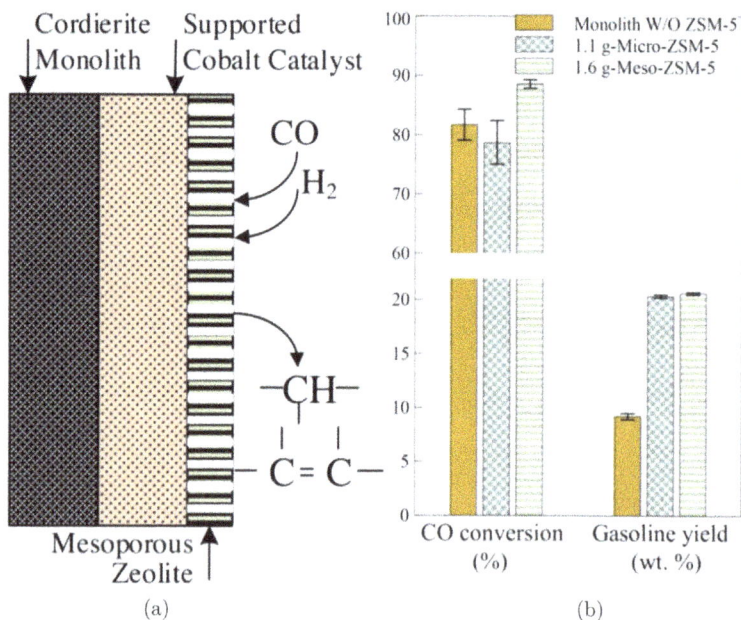

Fig. 3. (a) Layered bifunctional structured catalyst. (b) Gasoline weight percent comparison between monolith W/O ZSM-5, 1.1 g-Micro-ZSM-5, and 1.6 g-Meso-ZSM-5. [4].

ORCID

Zichen Wang ⊕ https://orcid.org/0009-0003-5510-7489

Chunxiang Zhu ⊕ https://orcid.org/0000-0003-3364-0509

Binchao Zhao ⊕ https://orcid.org/0000-0003-0772-8873

Chenxin Deng ⊕ https://orcid.org/0000-0002-6932-6514

Fangyuan Liu ⊕ https://orcid.org/0000-0002-3755-7705

Pu-Xian Gao ⊕ https://orcid.org/0000-0002-2132-4392

References

1. W. Zhou *et al.*, New horizon in C1 chemistry: breaking the selectivity limitation in transformation of syngas and hydrogenation of CO 2 into hydrocarbon chemicals and fuels, *Chem. Soc. Rev.*, 48, 12, 3193–3228, 2019, doi: 10.1039/C8CS00502H.
2. J. Xiao *et al.*, Tandem catalysis with double-shelled hollow spheres, *Nat. Mater.*, 21, 5, 572–579, 2022, doi: 10.1038/s41563-021-01183-0.
3. Y. Xu, J. Wang, G. Ma, J. Lin and M. Ding, Designing of hollow ZSM-5 with controlled mesopore sizes to boost gasoline production from syngas, *ACS Sustain. Chem. Eng.*, 7, 21, 18125–18132, 2019, doi: 10.1021/acssuschemeng.9b05217.
4. C. Zhu, D. P. Gamliel, J. A. Valla and G. M. Bollas, Fischer-Tropsch synthesis in monolith catalysts coated with hierarchical ZSM-5, *Appl. Catal. B Environ.*, 284, 119719, 2021, doi: 10.1016/j.apcatb.2020.119719.

© 2025 World Scientific Publishing Company
https://doi.org/10.1142/9789811297427_0004

In-Memory Computing Using Dot-Product
via Multi-Bit QD-NVRAMs[#]

R. H. Gudlavalleti ⊚[*],[†], J. Chandy ⊚[*], E. Heller[‡] and F. Jain ⊚[*],[§]

[*]University of Connecticut, CT, USA

[†]Biorasis Inc., Storrs, CT, USA

[‡]Synopysis Inc., Ossining, NY, USA
[§]faquir.jain@uconn.edu.

This paper presents in-memory computing using fast write/erase quantum dot (QD) nonvolatile random access memory (NVRAM). In comparison to NVMs, multi-state NVRAMs offer enhanced Compute-In-Memory capability for applications in deep neural network architecture. Dot product is the methodology that enables an array structure for multiply and accumulate (MAC) operation. We show an approach to dot product computation using multi-state quantum dot channel (QDC) FETs and QD-NVRAM.

Keywords: Quantum dot FETs; multi-state FETs; QD-NVRAMs; 16-state FETs.

1. Introduction

In-memory computing has been demonstrated using $HfZrO_2$ ferroelectric floating gate memory cell [1] where the gate voltage is modulated by charging/discharging using corresponding pull-up pFET and pull-down nFET. In general, resistive random access memory (RRAM) and various other types of random access memory devices have been proposed [2] for in-memory computing.

Compute-in-memory (CIM) technology is one of the emerging solutions towards efficient data transfer between logic processor and memory by computing within the memory macro [2]. Static random access memory (SRAM) based CIM has attracted for its access speed, write energy and compatibility with CMOS technologies. The efficiency of the in-memory computation can be increased by handling multiple bits in the same clock cycle. SWS-FETs based SRAM for CIM applications was reported [3]. Quantum dot (QD)-NVRAM has been demonstrated showing faster write/erase time but data can be retained without power. Integrating QD-NVRAMs with SWS-QDC/SWS FETs have been suggested for multi-bit in-memory computing [4]. Our

---

[#]This chapter appeared previously on the International Journal of High Speed Electronics and Systems. To cite this chapter, please cite the original article as the following: R. H. Gudlavalleti, J. Chandy, E. Heller and F. Jain, *Int. J. High Speed Electron. Syst.*, **33**, 2440055 (2024), doi:10.1142/S012915642440055X.

approach in this paper is to increase memory density and computation ability using QD-NVRAM and multi-state QDC-QDG-FET/SWS-FET cells.

2. Multi-State QDC-QDG FETs and QD-NVRAM

2.1. *Multi-state QDC-QDG FETs*

In this section, schematic of a QDC-QDG-FET structure is shown in Fig. 1(a). QDC-QDG-FET builds on experimentally reported QDC-QDG FET with two Ge QD layers in the gate region. Figure 1(b) presents the quantum simulations. We envision four Ge QD layers on Si QD layers in the gate region, which will permit additional states.

Simulation of the energy bands, such as those in Fig. 1(b), was performed by solving the Poisson equation (1) and the Schrödinger equation (3), self-consistently [5].

$$\nabla \cdot (\varepsilon \nabla \varphi) = q(n_{QM} + n - N_D^+ + N_A^- - p), \tag{1}$$

where ε is the permittivity, φ is the electrostatic potential, and n_{QM} is the 2D electron gas in the quantum well layers. Also, n and p are the 3D electron and hole concentrations, respectively. Lastly, N_D^+ and N_A^- are the ionized donor and acceptor concentrations, respectively. The 2D carrier concentration, n_{QM}, is expressed by Eq. (2), where E_F is the Fermi level, E_n are the eigen energies of the bound states, and ψ_n are the corresponding wavefunctions. Finally, Θ is the Heaviside Step function.

$$n_{QM} = \sum_n \frac{m*}{\pi \hbar^2} kBT\Theta(E_F - E_n) \ln \left[1 + \exp \left(\frac{E_f - E_n}{kT} \right) \right] |\psi_n|^2. \tag{2}$$

The Schrödinger equation (Eq. (3)) provides E_n and ψ_n for a given potential energy profile V, which is a function of position along the heterostructure. V, is

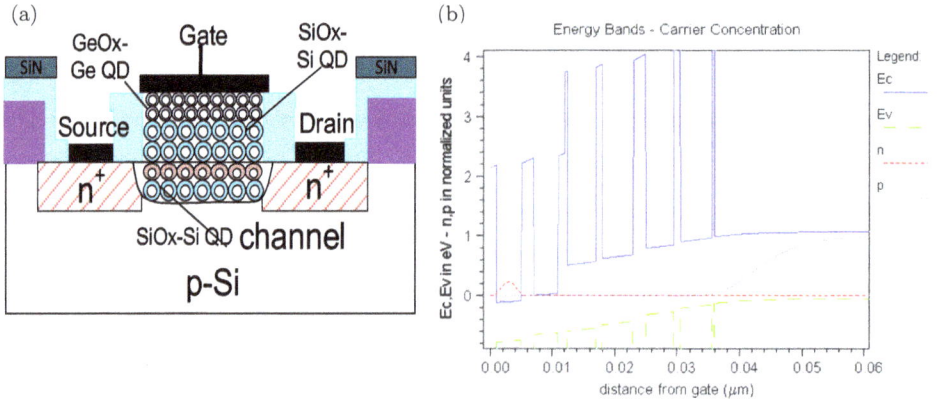

Fig. 1. (a) Asymmetric SiOx-cladded Si quantum dot channel with Si and Ge QD layers in the gate region. (b) Simulation results indicate the wavefunction is bound to the top Ge QD layers.

defined by the individual material properties of each layer, such as energy gap and electron affinity, plus the electrostatic potential, φ, from the Poisson equation. These equations are solved iteratively until a self-consistent solution is determined.

$$\frac{\hbar^2}{2}\nabla \cdot \left(\frac{1}{m*}\nabla\psi_n\right) + (E_n - V)\psi_n = 0. \tag{3}$$

The QDSL band structure and density of states (DOS) were determined by approximating the quantum dots as 3D cuboids, and solving the Schrödinger equation (3), by applying Kronig Penny (K-P) model. In this case, V (Eq. (4)) is the potential profile due to differences between the QD core and cladding properties, and each, $E(q)$ (Eq. (5)) is the energy solution of the K-P characteristic equation, for each direction, at wavevector q [6].

$$V(r) = V_X(x) + V_Y(y) + V_Z(z), \tag{4}$$

$$E(q) = E_X(q_X) + E_Y(q_Y) + E_Z(q_Z). \tag{5}$$

Finally, the DOS was determined by integrating the band structure over the 1st Brillouin Zone.

2.2. *QD-NVRAM with asymmetric SiOx-Si QDC for additional states*

Figure 2 is QD-NVRAM with asymmetric SiOx-cladded Si quantum dot channel and Si QD floating gate and Ge quantum dot access channel (QDAC). This builds on QD-NVRAM using Si Quantum dot layers in the gate region [6].

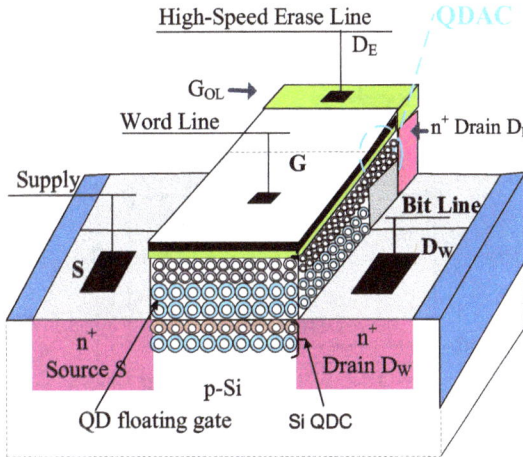

Fig. 2. QD-NVRAM with asymmetric SiOx-cladded Si quantum dot channel and Si QD floating gate and Ge quantum dot access channel (QDAC).

3. In-Memory Dot-Matrix Computation Using QD-NVRAM Crossbar Arrays

3.1. *QD-NVRAM Write, Read, and Erase mechanism*

During the Write operation, as the control gate (VG) increases above the threshold voltage of the bottom QD layer, the electrons from the inversion channel fill the bottom quantum dot (QD) layer (adjacent to the tunnel oxide) near drain D1. With further increase in the gate voltage (VG), the electrons start filling the upper QD layers. This results in a change in the threshold voltage (VTH) and manifests multi-bit storage. The transport of electrons within each QD layer is due to the formation of mini-energy bands [7], which form quantum dot superlattice (QDSL) layers when the array of QDs is formed with thin cladding/barrier (SiOx) layer.

The Write mechanism of a QD-NVRAM is similar to that of a conventional flash memory. The additional drain D2 of a QD-NVRAM is not used and biased during the Write operation.

The charges stored in the quantum dot floating gate layer of a QD-NVRAM can be removed using conventional Erase mechanism via control and tunnel gate dielectric layers by biasing the gate and source. In addition, two high-speed erasing mechanisms are available by biasing secondary drain D2. Positive D2 bias removes the stored charges in the floating gate by attracting the electrons towards D2, whereas the negative D2 bias injects the electrons to the source by repelling the electrons away from D2.

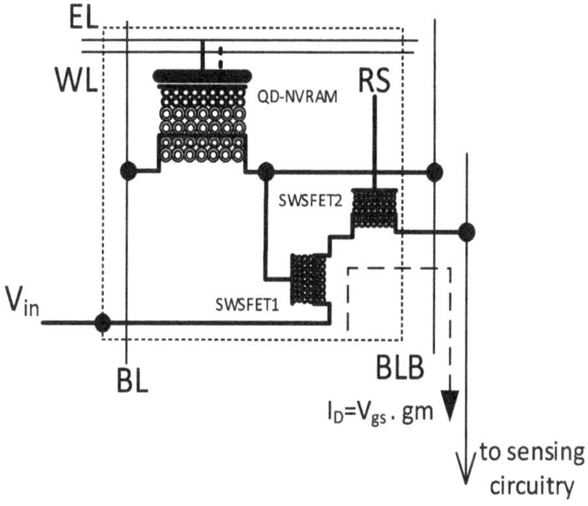

Fig. 3. QD-NVRAM integrated with SWS-FET-based storage and in-memory computation.

3.2. *QD-NVRAM and Multi-state SWS-FET Dot-matrix computation*

Figure 2 shows the QD-NVRAM bit-cell schematic and the biasing conditions of the word-line (WL), the bit-line (BL), and Erase-line (EL) for different operations. In this configuration, the QD-NVRAM stores the multibits depending on the pulse voltage applied on the WL and data can be sensed through the BLB line.

The in-memory analog dot-product computation is accomplished using QD-NVRAM (as storage) and two multi-state transistors (SWS-QDC FET1 and SWS-QDC-FET2), as shown in Fig. 3. The conductance of *SWSFET*1 depends on the state level stored in the QD-NVRAM, and the conductance of SWSFET2 is controlled by row-select (RS). Due to multiple states associated with SWSFET1 and SWSFET2, the output current (shown as dashed arrow) to the sensing circuitry can

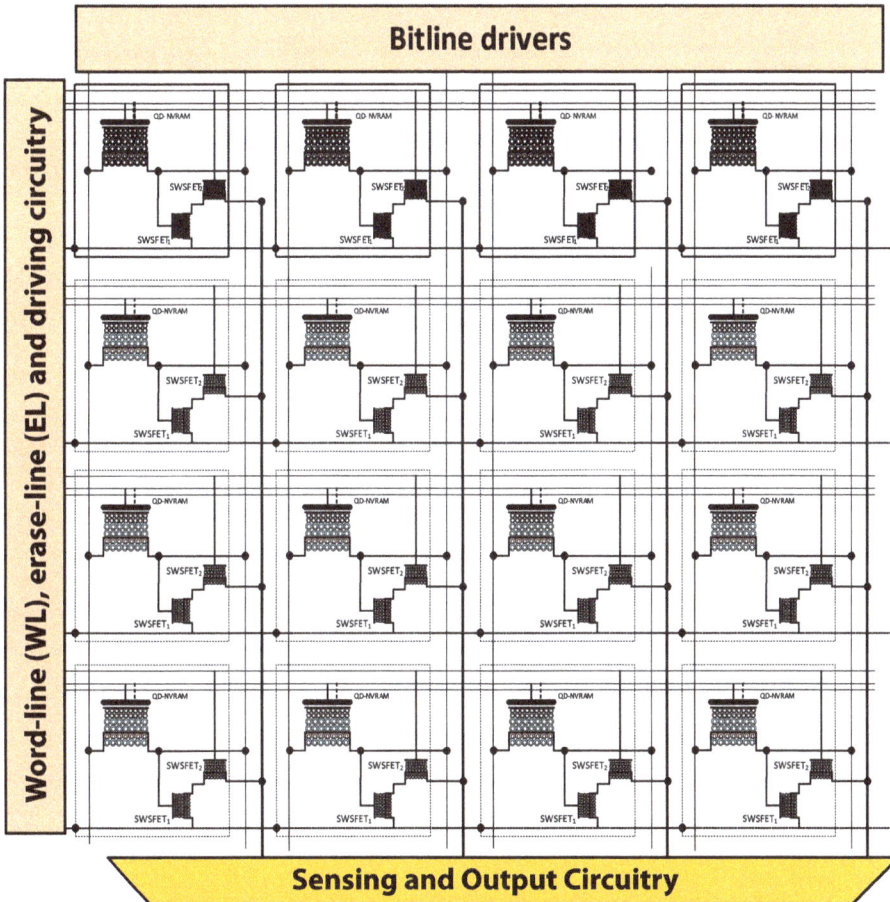

Fig. 4. Analog dot product computation architecture.

be weighted based on the dot-product multiplication of input voltage V_{in} and conductance associated with *SWSFET*1 and *SWSFET*2. Thus, the final output current through *SWSFET*2 for each column is proportional to ID $= g_m \cdot V_{gs}$.

The matrix can be expressed as $I^j_{\text{SWSFET2}} = \Sigma(v_i \cdot g^j_i)$, where g^j_i is the equivalent conductance of the transistors in each of the 16-states, depending on whether the bit-cell in the ith row and jth column stores 16-states. The output current vector thus resembles the vector–matrix dot product, where the vector is v_i in the form of input analog voltages, and the matrix is g^j_i stored as digital data in the QD-NVRAM.

The crossbar array using QD-NVRAMs and SWS-FETs is shown in Fig. 4. Here, the word-line (WL), erase-line (EL) with corresponding driving circuitry provide signals to WL/EL access each row of the cross-bar memory array. Bit-line (BL) block provide data to program each column of the cross-bar memory array. The sensing and output circuitry converts the dot-product computed output from each memory cell to digital output for further processing.

4. Conclusion

Quantum dot (QD) NVRAMs with multi-state storage is used for in-memory computing [6, 7] via cross-bar architecture integrating dot product with the incorporation of multi-state SWSFET1 and SWSFET2.

Multiple states are obtained in QDC, SWS, and QDG FET configuration by methods including: (i) changing the thickness of upper cladding of upper QD of QD channel, (ii) changing the QDC SiOx cladded Si QD layers, (iii) making two QD layers asymmetrical by controlled oxidation of upper SiO2 layer, (iv) changing the Ge core diameter of two uppermost Ge QD layers. Finally, this utilizes the formation of mini-energy bands in quantum dot superlattice (QDSL) which is formed when oxide of barrier thickness is very small and electrons in adjacent dots are coupled. In summary, the increased data transfer efficiency is applicable in AI/ML data computation systems.

ORCID

R. H. Gudlavalleti ⊚ https://orcid.org/0000-0002-7727-8030

J. Chandy ⊚ https://orcid.org/0000-0003-3449-3205

F. Jain ⊚ https://orcid.org/0000-0003-3961-6665

References

1. X. Sun, P. Wang, K. Ni, S. Datta and S. Yu, Exploiting Hybrid Precision for Training and Inference: A 2T-1FeFET Based Analog Synaptic Weight Cell, *2018 IEEE International Electron Devices Meeting (IEDM)*, San Francisco, CA, 2018.
2. D. Lelmini and H.-S. P. Wong, In-memory computing with resistive switching devices, *Nature Electronics*, 1, 333–337, 2018.

3. R. H. Gudlavalleti, E. Heller, J. Chandy and F. Jain, Compute-in-Memory SRAM Cell Using Multistate Spatial Wavefunction Switched (SWS)-Quantum Dot Channel (QDC) FET, *International Journal of High Speed Electronics and Systems*, 32, 02n04, 2350012, 2023.

4. F. Jain, R. Gudlavalleti, R. Mays, B. Saman, J. Chandy and E Heller, Integrating QD-NVRAMs and QDC-SWS FET-Based Logic for Multi-Bit Computing, *International Journal of High Speed Electronics and Systems*, 31, 2240020, 2022.

5. E. K. Heller, S. K. Islam, G. Zhao and F. C. Jain, *Solid-State Electronics*, 42, 901–914, 1999.

6. M. Lingalugari, P.-Y. Chan, E. K. Heller, J. Chandy and F. C. Jain, Quantum Dot Floating Gate Nonvolatile Random Access Memory Using Quantum Dot Channel for Faster Erasing, *Electronic Letters*, 54, 36, 2018.

7. F. Jain, R. Gudlavalleti, R. Mays, B. Saman, P-Y. Chan, J. Chandy, M. Lingalugari and E. Heller, Two-dimensional SiOx-cladded Si and GeOx-cladded Ge Quantum Dot Arrays vertically stacked for CMOS-X Logic, SRAMs, NVRAMs and IR imaging and multi-bit Computing, *SISC*, December 2022.

Performance Improvement of Degrading Memristor-Bridge-Based Multilayer Neural Network with Refresh Pulses[#]

Aalvee Asad Kausani ⊚

Electrical and Computer Engineering,
University of Connecticut, Storrs, CT 06269, USA
aalvee_asad.kausani@uconn.edu

Caiwen Ding ⊚

Computer Science and Engineering,
University of Connecticut, Storrs, CT 06269, USA
caiwen.ding@uconn.edu

Mehdi Anwar[*]

Electrical and Computer Engineering,
University of Connecticut, Storrs, CT 06269, USA
a.anwar@uconn.edu

Memristors as non-volatile memory devices have been recognized for executing in-memory computation in neuromorphic hardware. In this paper, a multilayer neural network has been developed with memristor-bridges as electrical synapses and trained with modified-chip-in-the-loop technique for an image classification task. Modeling the ideal conduction behavior of memristors by their device-physics inspired analytical model has yielded satisfactory performance. However, repeated voltage cycling degrades the resistance window of memristors by aggregating conductive residuals in filamentary memristors. Therefore, emulation of such nonideality has demonstrated compromised results. To improve the performance, refresh pulses have been introduced to the devices in between write pulses to eradicate the fundamental reason of the degradation — i.e., the residuals. It has been observed that improvement of performance is contingent upon the refreshment frequency, and frequent refreshment has the ability to restore performance to a level closely approaching its ideal emulation.

Keywords: Artificial neural network; memristor; memristor-bridge; analytical model of memristor.

[*]Corresponding author.
[#]This chapter appeared previously on the International Journal of High Speed Electronics and Systems. To cite this chapter, please cite the original article as the following: A. A. Kausani, C. Ding and M. Anwar, *Int. J. High Speed Electron. Syst.*, **33**, 2440056 (2024), doi:10.1142/S0129156424400561.

1. Introduction

In-memory processing leverages unconventional techniques to alleviate the bandwidth limitation of Von-Neumann architectures, especially when neuromorphic algorithms are executed [1]. In this regard, memristors have emerged as potential candidates because of their unique ability to store and process information simultaneously. Memristors are employed to represent electrical synapses in artificial neural networks. Synapses are specialized connections between neurons that facilitate the transmission of information across the nervous system. Different topologies of memristor circuits to imitate synaptic behavior can be found in [2].

Training scheme of memristor-based networks serves the purpose of programming the conductance of the devices to emulate the required synaptic weights. The training algorithms assume a symmetric and linear conductance change in response to applied write voltages [3, 4]. However, the analytical modeling of conduction mechanism of filamentary memristors has revealed its nonlinear and asymmetric dynamics [5]. The conduction in a filamentary memristor such as $Pt/TiO_2/Pt$ is governed by the evolution of conductive paths (filaments) inside the highly resistive oxide region. Polarity of applied voltage determines the formation or dissolution of filaments while the attributes of voltage pulse control the dimension of filaments. However, the growth and rupture rates with respect to the applied voltage are not linear and the rupture rate is lower than the growth rate [5]. Therefore, the asymmetry in filament dynamics results in the presence of residuals after an excursion of programming voltage for dissolving the filament which has been confirmed by experimental observation as well [6, 7]. The assumption of linear and symmetric conductance changes as oppose to its observed behavior results in an inefficient emulation of the synapses in the simulation of neuromorphic hardware [8, 9].

In this paper, we have simulated a prototype of multilayer neural network with memristor-bridge synapses [10]. The memristors are simulated in light of the analytical model [5] to sufficiently capture the underlying physics of their conduction mechanism. Primarily, we have emulated ideal memristors with symmetric growth and rupture of the filaments to observe the baseline performance. Later on, consideration of the asymmetry in filament dynamics helps to investigate the impact on the overall performance. Finally, we have proposed applying refresh pulses periodically to dissolve a portion of the filament in a controlled manner in order to improve the network performance.

2. Network Architecture

Figure 1(a) illustrates the architecture of the multilayer neural network with 432 inputs, 1 hidden layer with 10 neurons, and 1-output configuration. It is inspired by the design reported in [10] that has emulated the synapses by memristor-bridges. Synaptic connections (ψ) scale the inputs to the neurons while a neuron modulates the collective input by a defined activation function. The functionality of synapses for the illustrated network has been demonstrated such as an input (v_m) to the ith

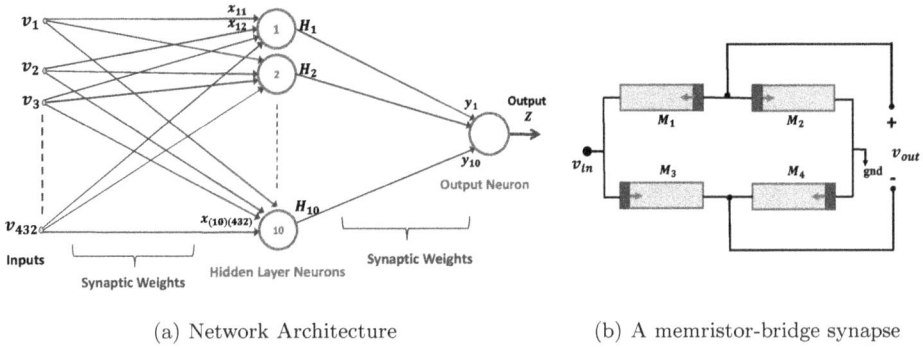

(a) Network Architecture   (b) A memristor-bridge synapse

Fig. 1. (a) Architecture of the neural network referenced in this work. (b) A synapse prototype with memristor-bridges.

neuron of hidden layer is $x_{im} = \psi_{im}^{(1)} v_m$ and an input (y_k) to the output neuron is $y_k = \psi_k^{(2)} H_k$. The functionality of neurons has been realized by sets of differential amplifiers with current mirror active loads [10]. A neuron will have as many differential amplifiers as the number of inputs it is receiving while having one active load to combine them into a single output.

A memristor-bridge is comprised of four memristors where each branch has two memristors in a complementary fashion as demonstrated in Fig. 1(b). Application of a voltage pulse to a memristor such that the highlighted electrode receives lower potential compared to the opposite electrode will result in the decrease of its effective resistance. Reversal of voltage polarity results in the dissolution of the filament starting from the opposite electrode. The synaptic weight is expressed in terms of the memristances as follows [10].

$$\psi = \frac{M_2}{M_1 + M_2} - \frac{M_4}{M_3 + M_4}. \tag{1}$$

We have considered a $Pt/TiO_2/Pt$ memristor, the conduction mechanism of which is dictated by the formation and dissolution of high conductive filaments. TiO_2 is highly resistive in its pristine state. Application of voltage stimulates redox reaction that produces oxygen deficient composites such as Ti_2O_3 [5]. Accumulation of the composites over the course of applied voltage pulses results in the development of high conductive filaments inside the device. The growth direction of the filaments is depicted by the arrows in Fig. 1(b). The memristance of a device is equivalent to its effective resistance [10] and therefore can be defined as the combination of the contributions from the insulating TiO_2 and conductive filament.

$$M(t) = \frac{v(t)}{i(t)} = R_{\mathrm{LRS}} \frac{L(t)}{D} + R_{\mathrm{HRS}} \left(1 - \frac{L(t)}{D}\right), \tag{2}$$

where $L(t)$ and D denote the filament length at a particular time instance of applied voltage and the device length (distance between the electrodes), respectively.

R_{LRS} refers to the lowest resistive state of the device when the filament is fully grown bridging the gap between electrodes in contrast to R_{HRS}, which represents the highest resistive state of the device for no filament being present. Therefore, the synaptic weight of a bridge directly depends on the dimension of the filaments in the memristors that can be controlled by the application of voltage pulses. The analytical model of filament growth in relation to the chemical and physical processes in response to applied voltage has been developed [5] as follows.

$$a_x = \frac{1}{2}\frac{qV_0}{m^*}\left[\frac{1}{D-(a_1+a_2+\cdots+a_{x-1})}(\sin\omega t_{x-1}-\sin\omega t_x)\right.$$
$$\left. -\frac{Ae^{-E_a/RT}}{1+(\omega\tau)^2}\frac{1}{D}\{(\sin\omega t_{x-1}-\omega\tau\cos\omega t_{x-1})-(\sin\omega t_x-\omega\tau\cos\omega t_x)\}\right](\Delta t)^2,$$

$$(3)$$

where a_x is the incremental length of a filament when a voltage pulse having a magnitude V_0 is applied to the memristor. A, E_a, R and T are related to the reaction rate constant and represent the pre-exponential factor, activation energy, universal gas constant, and temperature, respectively. m^*, q, and τ are the effective mass of charged particle, electron charge, and mean free time between successive collisions.

3. Network Training

The network has been trained with a modified-chip-in-the-loop technique [10] for an image classification task. This process requires a pre-trained software model to be utilized as a reference for the hardware training. In this regard, a software model has been developed in MATLAB having similar architecture as illustrated in Fig. 1(a). The design space of the network configuration is sub-optimal such that the chosen hyperparameters yield satisfactory performance. Supervised learning has been deployed to classify input images as Huskies (dog) or non-Huskies, annotated by +1 and −1 identifiers, respectively.

Figure 2 illustrates a typical iteration of hardware training in a heterogenous computing system. The output from a hardware neuron (H_n^{hw}) as well as that from the software model (H_n^{sw}) corresponding to an image input from the training set has been recorded. Error at a neuron (E_n) has been assessed by $E_n = 1/2(H_n^{\text{sw}}-H_n^{\text{hw}})^2$. The required adjustment in each synaptic weight ψ_{nm} associated with the neuron in relation to E_n has been evaluated as follows.

$$\Delta\psi_{nm} = -\eta\frac{\partial E_n}{\partial\psi_{nm}} = \eta(H_n^{\text{sw}}-H_n^{\text{hw}})f'(H_n^{\text{hw}})v_m.$$

Here, f' is the derivative of the activation function, η is the learning rate that helps to adjust the effective gradient, and v_m refers to the input to the synapse. The required synaptic adjustments across the network can be calculated following the abovementioned procedure.

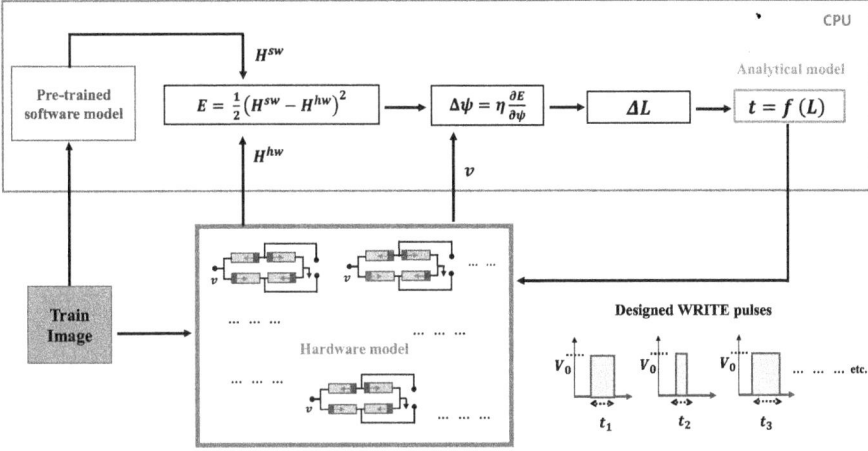

Fig. 2. Illustration of a training iteration of the modified-chip-in-the-loop scheme in a heterogeneous computing system. The host computer runs the software model, calculates errors of the hardware neurons, and estimates the write pulses to update memristors in the hardware.

Afterwards, Eqs. (1) and (2) have been utilized to calculate the required changes in filament lengths of the constituent memristors of a synapse in relation to the corresponding $\Delta\psi_{nm}$. The duration of applying a WRITE voltage to achieve the desired filament length, minimizing the error at the hardware neuron, has been derived from the analytical model (3). While the specifications of write-pulses have to be calculated from required synaptic values in a general-purpose CPU, the as-designed pulses will be utilized in real time to update the memristors.

Table 1 presents the performance of the as-trained hardware model in various physical conditions. Four different cases depending on device length and operating temperature have been studied. The results help understand the impact of physical conditions on the performance of the hardware. A network with 50 nm devices operating at room temperature (298 K) requires more training time compared with the training time at an elevated temperature of 600 K to yield equal accuracies. This could be attributed to acceleration in the process of filament formation and rupture with increasing temperature [5]. Hence, it takes less time to map the memristors with the desired filament lengths and the bridges can achieve required synaptic values faster. Comparing the training times between systems with 50 nm

Table 1. Performance of the Network in Various Physical Conditions for Modified Chip-in-the-Loop Learning.

Memristor Length, Operation Temp.	Classification accuracy	Average error in classification	Training time
50 nm, 298 K	80%	0.736	0.6608 μs
50 nm, 600 K	80%	0.792	0.4381 μs
25 nm, 298 K	70%	0.793	0.1807 μs
25 nm, 600 K	70%	0.778	0.1235 μs

and 25 nm devices at the room temperature, we observe the system with shorter devices responds more quickly. From the relationship between a synaptic weight with its constituent memristors (1), we can infer that the ψ remains unchanged if the memristances are scaled by the same amount. Therefore, devices with reduced dimensions can achieve similar synaptic weights but in a shorter time. However, the classification accuracy drops (70% for 25 nm devices, 80% for 50 nm devices) when the system has smaller feature size. It appears that, although the same weight can be achieved with smaller devices in theory, they cannot adequately accommodate the precise adjustment of filament lengths in practice. Hence, the classification accuracy and training time is in a trade-off condition when the devices are scaled down. Based on our argument, we anticipate the shortest training time with a reduced accuracy for the case of 25 nm devices operating at 600 K which has been validated by the simulation report.

4. Device Degradation

A symmetrical change in filament growth and rupture in response to applied voltage pulses is assumed in the ideal emulation of memristors. However, higher growth rate of filament compared to its rupture rate signifies the presence of residual filament after the excursion of a complete voltage cycle comprising of positive and negative halves. Therefore, the highest attainable resistance of the device R_{HRS} gradually lowers when repeated voltage cycling is applied. The training method of neural networks involves a significantly large number of iterations. Hence, exhaustive training/repeated writes can lead to a permanent filament extending to the device dimension. In this regard, it is imperative to observe the impact on the network performance when the resistance window bounded by R_{LRS} and R_{HRS} is shrinking through the progression of training.

We emulate the degradation in resistance characteristic of memristor by the parameter degradation factor (α) which represents the increment in residual length after the execution of a training iteration. Therefore, $i\alpha$ is the minimum filament length (L_{min}) that can be programmed at the ith iteration. The network with 50 nm long Pt/TiO$_2$/Pt memristors in room temperature was trained with an 0.1 nm/iteration degradation factor. Once the aggregated length of residual accumulation reached the device dimension at the 500th iteration, the devices collapsed with permanent filaments. Figure 3 compares the histograms of synaptic values of the output neuron with ideal and realistic (degrading) emulation of memristors. A trained hardware with ideal memristors (80% classification accuracy) has nonzero synaptic weights and at least six of them are distantly spaced. However, the completely degraded network features all weights of exactly zero value. Consequently, the collapsed network resulted in a 0% classification accuracy upon evaluating on the test dataset.

However, a reduced number of iterations can train the hardware to some extent. Nevertheless, the classification accuracy is low compared with a hardware having

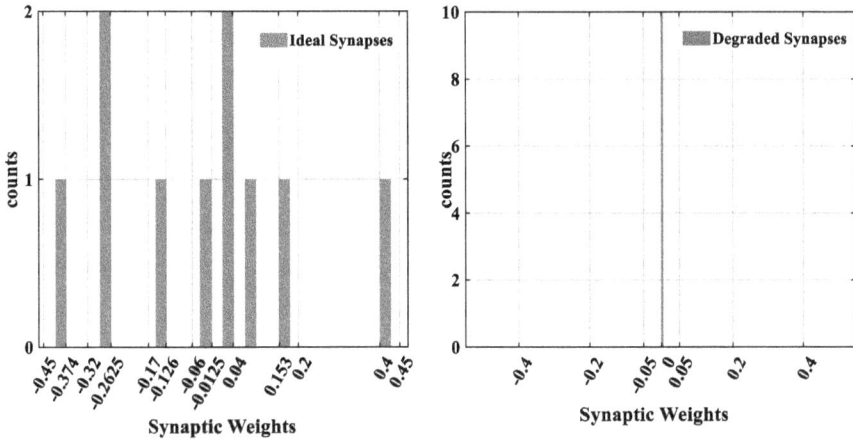

Fig. 3. Histogram of the Synaptic Weights of the output neuron in architectures with ideal and degrading memristors.

Table 2. Performance of the Network in Various Physical Conditions for Modified Chip-in-the-Loop Learning.

Memristor Length, Operation Temp.	Classification accuracy		Average error in classification		Training time	
	Ideal	Non-ideal	Ideal	Non-ideal	Ideal	Non-ideal
50 nm, 298 K	80%	50%	0.4258	0.9983	0.6608 μs	0.3901 μs
50 nm, 600 K	80%	50%	0.4253	0.9982	0.4381 μs	0.2308 μs
25 nm, 298 K	80%	50%	0.4256	0.9982	0.1807 μs	0.0857 μs
25 nm, 600 K	80%	50%	0.4263	0.9982	0.1235 μs	0.0569 μ s

ideal memristors without any degrading properties. Table 2 lists the performance of networks with ideal and degrading memristors in different operating conditions. For each of the cases, the networks were trained for equal numbers of iterations to have impartial comparisons. The average error is greater for the network with degrading memristors.

5. Performance Improvement

Accumulation of residuals that impede the programming of the devices is the principal cause of the loss in performance of the memristor-based hardware. Therefore, applying REFRESH pulses intermittently may dissolve a portion of the filament to expand the resistance window from time to time. Figure 4 illustrates the impact of refresh pulses in comparison to ideal and degrading networks. All three networks have Pt/TiO$_2$/Pt memristors of length 0.1 μm and operating at room temperature. A 0.05 nm/iteration value has been considered as the degradation factor. The refresh pulses are all identical with a pulse-width of 1.25 ns. A refresh pulse is applied to each memristor after every 50 training iterations.

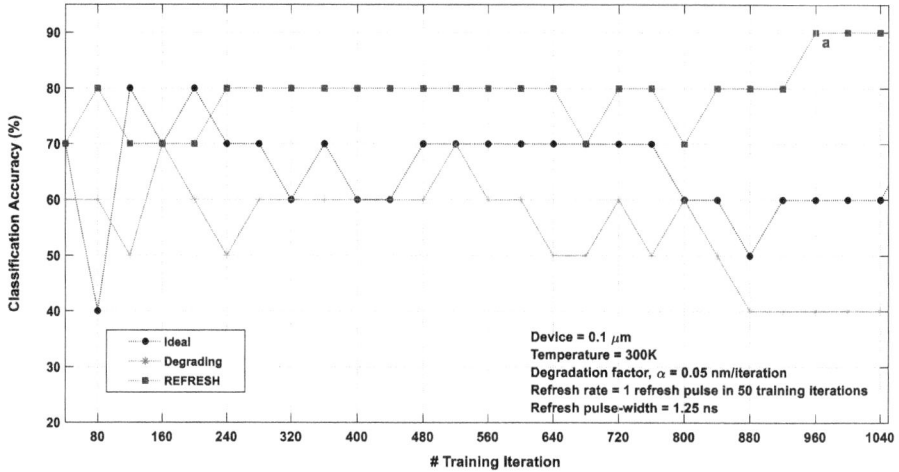

Fig. 4. Classification accuracy of ideal, degraded, and refresh-enhanced neural network.

The classification accuracy of the networks is tested once in every 40 training iterations. Accuracy of all three networks fluctuates during the initial iterations while they become more stabilized with the advancement of training. The ideal network traced in black circles could achieve 70% accuracy at best. Performance of the network with degrading memristors tranced with red asterisks gradually fell below the ideal network confirming the adversarial effect of memristor degradation. At the end of 1040 iterations, the minimum filament length is 52 nm i.e., 52% of the device length. The accuracy reduces to 50% in this case. However, refreshing the degrading devices once in 50 training iterations has exhibited an enhancement surpassing the ideal performance. The blue squares residing at a higher level compared to the black circles confirm the overall improvement. Periodic refreshment could achieve up to 90% accuracy marked as datapoint "a" in Fig. 4 while the ideal and degrading network have only 60% and 40%, respectively.

However, while uncontrolled refresh may severely impact the ongoing training of the network by wiping off a significant portion of the filament and thereby reducing the accuracy, very infrequent refresh may not contribute any improvement. Therefore, it is necessary to observe the impact of the frequency of refresh pulses on the network training. Figure 5 illustrates the network performance when trained with three different refresh rates.

Here we have defined the refresh rate in terms of the training iteration. For instance, a rate of 100 indicates that refresh pulses are administered to the memristors once in every 100 iterations. The plots in Fig. 5 have been generated for networks with a degradation factor of 0.1 nm/iteration. The width of refresh pulses was fixed at 0.125 ns for this experiment. The trace with red circles corresponding to a refresh rate of 200 in Fig. 5 demonstrates a decline in accuracy as training progresses with an overall range between 20% and 40%. Therefore, very infrequent refreshment fails

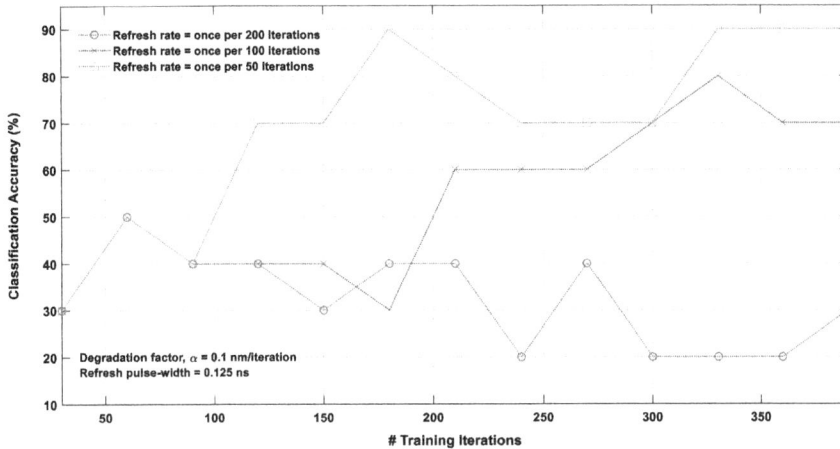

Fig. 5. Impact of the rate of refresh pulses on network performance.

to mitigate the impact of degradation. Conversely, doubling the frequency i.e., refreshing in every 100 iterations (identified by blue crosses) has markedly enhanced the performance with a gradual increase in accuracy reaching up to 70%. Furthermore, more frequent refresh cycles, such as occurring at a rate of 50, have expedited training even further as marked in green squares, whereas 90% accuracy has been achieved after 300 iterations.

6. Conclusion

A multilayer neural network with memristor-bridge-based electrical synapses has been developed. The devices have been modeled in relation to the underlying device physics. Asymmetry in the mechanism of conduction has resulted in degradation in the resistive property of memristor and demonstrated compromised performance compared to their ideal emulation. In this regard, a REFRESH technique to mitigate the impacts of memristor degradation has been proposed. Frequent application of refresh pulses dissolves a portion of the conductive filament to extend the resistance window of a $Pt/TiO_2/Pt$ memristor which otherwise gradually shrinks due to the accumulation of residual filament. Careful selection of the refresh rate and width of refresh pulses is crucial to overcome the adversarial effects of degradation. The integration of a refreshing system within neural network hardware can potentially increase the overall duration and energy consumption during training. Consequently, an extension to this research could involve the exploration of methodologies aimed at minimizing both time and energy expenditures.

Acknowledgment

One of the authors (AAK) was supported by a General Electric (GE) Fellowship for Excellence.

ORCID

Aalvee Asad Kausani ⊚ https://orcid.org/0000-0001-5974-8249

Caiwen Ding ⊚ https://orcid.org/0000-0003-0891-1231

References

1. G. W. Burr, R. M. Shelby, A. Sebastian, S. Kim, S. Kim, S. Sidler, K. Virwani, M. Ishii, P. Narayanan, A. Fumarola, L. L. Sanches, I. Boybat, M. Le Gallo, K. Moon, J. Woo, H. Hwang and Y. Leblebici, *Advances in Physics: X* **2**, 89–124 (2017).
2. Y. Li, K. Su, H. Chen, X. Zou, C. Wang, H. Man, K. Liu, X. Xi and T. Li., *Electronics* **12**, 3298 (2023).
3. W. Wang, R. Wang, T. Shi, J. Wei, R. Cao, X. Zhao, Z. Wu, X. Zhang, J. Lu, H. Xu, Q. Li, Q. Liu and M. Liu, *IEEE Electron Device Letters* **40**, 1407–1410 (2019).
4. S. Agarwal, S. Plimpton, D. Hughart, A. Hsia, I. Richter, J. Cox, C. James and M. Marinella, Resistive memory device requirements for a neural algorithm accelerator, in *Proc. 2016 International Joint Conference on Neural Networks (IJCNN)*, Vancouver, BC, Canada, 24–29 July 2016, pp. 929–938.
5. A. Mazady and M. Anwar, *IEEE Transactions on Electron Devices* **61**, 1054–1061 (2014).
6. H. Lo, C. Yang, G. Huang, C. Huang, J. Chen, C. Huang, Y. Chu and W. Wu, *Nano Energy* **72**, 104683 (2020).
7. Q. Liu, J. Sun, H. Lv, S. Long, K. Yin, N. Wan, Y. Li, L. Sun and M. Liu, *Advanced Materials* **24**, 1844–1849 (2012).
8. S. B. Eryilmaz, S. Joshi, E. Neftci, W. Wan, G. Cauwenberghs and H. Wong, Neuromorphic architectures with electronic synapses, in *Proc 2016 17th International Symposium on Quality Electronic Design (ISQED)*, Santa Clara, CA, USA, 15–16 March 2016, pp. 118–123.
9. J. Fu, Z. Liao, N. Gong and J. Wang, *IEEE Journal on Emerging and Selected Topics in Circuits and Systems* **9**, 377–387 (2019).
10. S. P. Adhikari, C. Yang, H. Kim and L. O. Chua, *IEEE Transactions on Neural Networks and Learning Systems* **23**, 1426–1435 (2012).

Binary Adder, Subtractor and Parity Checker Based on Optical Logic Gates[#]

Shunyao Fan [ORCID]* and Niloy K. Dutta [ORCID]

*Department of Physics, University of Connecticut,
Storrs, Connecticut 06269, USA*
shunyao.fan@uconn.edu

We propose schemes for binary adder, subtractor and parity checker using optical logic gates. These schemes could be useful for calculations using optical systems. Utilizing optical logic gates, we can achieve functions of binary adder, subtractor and parity checker of high-speed optical signals. Due to two-photon absorption in the wetting layer, quantum dot-semiconductor optical amplifier Mach-Zehnder interferometer (QD-SOA-MZI) can work as high data rate optical logic gates. The simulated result supports the idea that it is possible to realize all-optical binary adder, subtractor and parity checker at high optical signal rates.

Keywords: Optical logic gates; quantum dot-semiconductor optical amplifier; adder; subtractor; parity checker.

1. Introduction

Quantum-dot-based semiconductor optical amplifier (QD-SOA) allows very high-data rate operation. Thus, quantum dot-semiconductor optical amplifier Mach-Zehnder interferometer (QD-SOA-MZI) can work as high data rate XOR and AND optical logic gates at speeds to 320 Gb/s. This high speed is possible due to two-photon absorption in the wetting layer. Utilizing optical logic gates, we can achieve functions of binary adder, subtractor and parity checker of high-speed optical signals. Simulation results are carried out to show that binary adder, subtractor and parity checker based on optical logic gate are feasible.

2. Optical Logic Gates

The principle of optical logic operation utilizing cross-phase modulation (XPM) process in SOAs has been discussed and analyzed [1]. The logic XOR and AND

[#]This chapter appeared previously on the International Journal of High Speed Electronics and Systems. To cite this chapter, please cite the original article as the following: S. Fan and N. K. Dutta, *Int. J. High Speed Electron. Syst.*, **33**, 2440057 (2024), doi:10.1142/S0129156424400573.

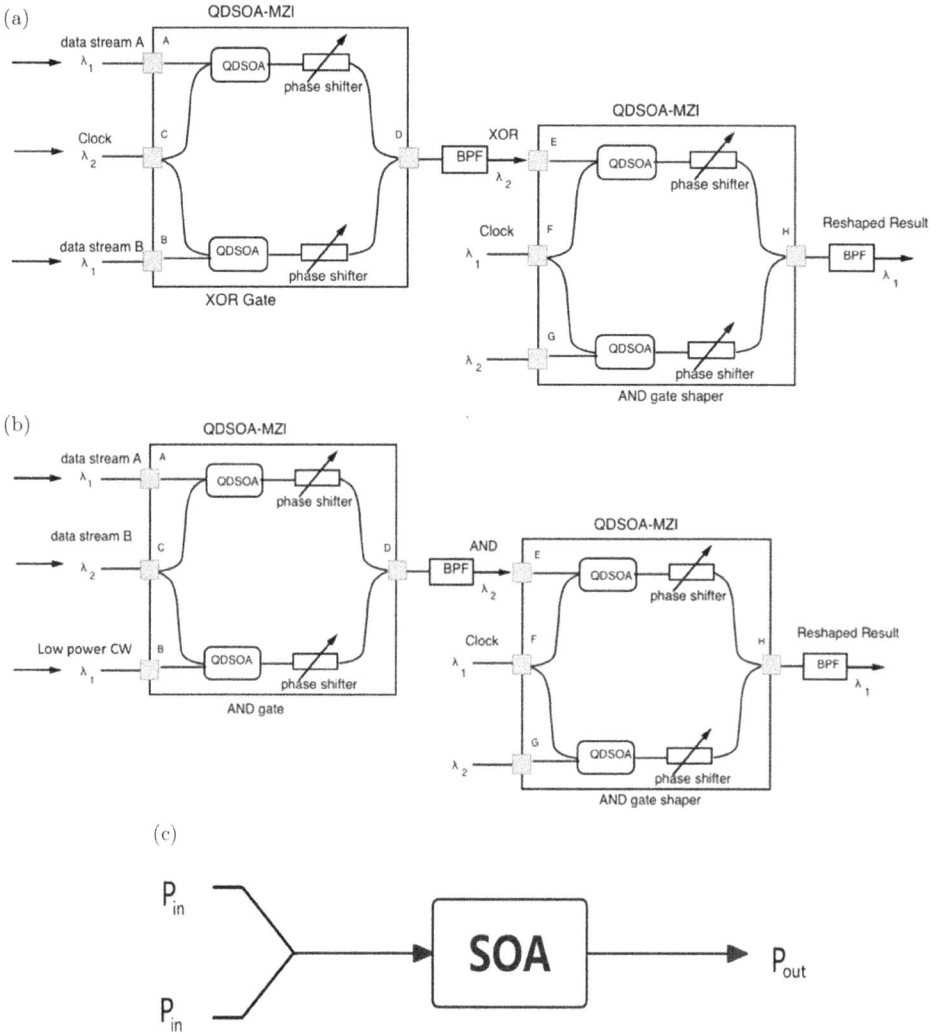

Fig. 1. Schematic diagram of optical logic gates based on QDSOA-MZI. (a) XOR gate, (b) AND gate and (c) OR gate.

operations are achieved utilizing the cross-phase modulation (XPM) and cross-gain modulation (XGM) processes in QD-SOAs.

As shown in Fig. 1, we inject data A and B as optical pulse streams at wavelength λ_1 separately into the two arms of the MZI through ports A and B. Also inject a clock stream at wavelength λ_2 into port C, so the clock stream will evenly split into the two arms of the MZI. Due to XPM and XGM of QD-SOAs, the streams get modulated as they travel with data streams in the QD-SOAs. When the streams from both arms recombine at port D, their interference will produce optical logic results. The configurations for XOR, AND and OR gate are shown in Fig. 1.

To achieve XOR gate, we initially set the MZI unbalanced with a phase difference π between the two arms. When input data A and B are the same, the two clock streams experience the same gain and phase shift in QD-SOAs and when they recombine at port D, considering the initial phase difference π, they will undergo destructive interference and the output result is "0". Correspondingly, when input data A and B are not the same, the gain and phase modulation of the two clock streams are different and their interference will output "1" at wavelength λ_2.

To achieve AND gate, we input data stream A into port A, and inject a low power continuous wave (CW) into port B. MZI is initially set so that there is no output if there is no differential phase change in the two arms for an input signal propagating through both arms. Signal B (which also carries the result of the AND operation) is injected at port C of the SOA-MZI. When there is a signal (A) input "1", there is a phase shift induced onto the control signal in the upper arm. Thus, signal A (if it is 1) produces a phase difference for signal B which travels through both arms and interferes at the output. Thus, if B = 1 and A = 1, the output is "1". If A = 0 there is no phase difference and hence the output is "0", and if B = 0, there will not be a stream to carry the result, so the output is "0". This functions as a logic AND gate.

For both AND gates and XOR gates to be practical, we will need to add an AND gate after them to function as a wave shaper, so the output will have the same wavelength as the input for the logic gates as shown in Fig. 1.

To achieve OR gate, we can make use of gain saturation in the SOA [1]. Using suitable 50-50 coupler and SOA, we can adjust the pulse power so that the gain of SOA will be saturated when one of the data stream inputs is "1". So, if both data stream input is "0", we will have an output of "0". If one data stream input "1", we will have an output of "1". When both streams input "1", the output will also be "1" due to the gain saturation. The scheme is shown in Fig. 1.

3. Devices and Rate Equations

The device we choose here to construct the all-optical logic gate is the In-GaAs/InGaAsP/InP QD-SOA, in which InAs quantum dots are embedded in In-GaAsP layers. The gain of this type of device around $1.55\,\mu$m is typically $\sim 15\,$dB and the noise figure is low at $\sim 7\,$dB [2].

Figure 2 illustrates the optical gain, the two-photon absorption (TPA) process and carrier transitions between the wetting layer (WL), the QD excited state (ES) and the QD ground state (GS). The device gain is determined by the carrier density of the QD ground state. The TPA generates carriers in the bulk region. These carriers then relax to the WL, and eventually are captured into QDs on ultrafast timescale [1–6]. Generally, carriers in the barriers are free to move in 3D and are captured very rapidly by the 2D wetting layer at a relaxation timescale of $\sim 70\,$fs [3]. Ju *et al.* [4] have shown that the carrier dynamics in the bulk region due to TPA can also refill the WL and QDs on ultrafast timescales and thus significantly reduce

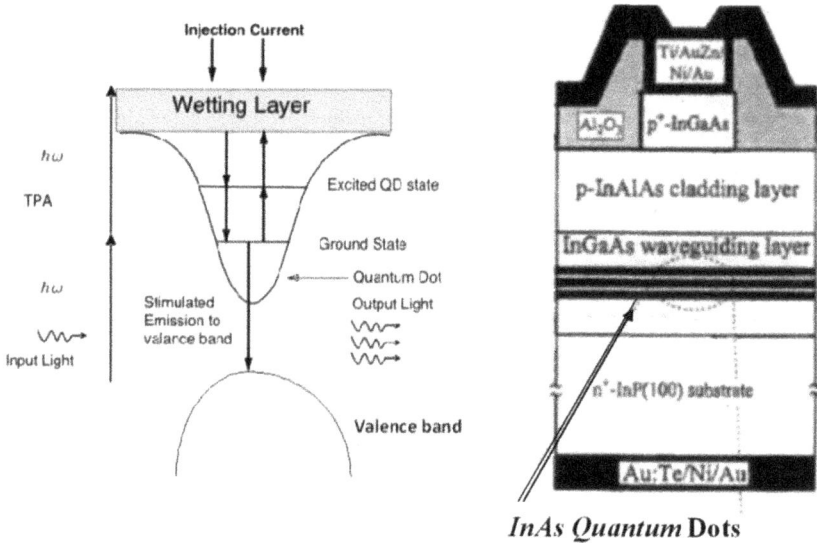

InAs Quantum Dots

Fig. 2. Schematic of two-photon absorption (TPA) and a typical device structure. The InGaAs waveguide layer which has a bandgap intermediate between the cladding layers (InAlAs) and the quantum dots (InAs) and the layer (InGaAs) surrounding the quantum dots serve as the wetting layer. Similar layer structure is feasible using the InP/InGasAsP/InGaAs material system with InGaAs quantum dots [1].

pattern effects for optical signal processing operating at Tbit/s [6]. They introduced a three-coupled rate equations model including the barrier region, as well as the WL and QDs. In our model, we ignore the barrier dynamics and assume that carriers are injected directly from the contacts into the WL [4]. Since the only recipient of the pump current is the WL, and the QD excited state serves as a carrier reservoir for the ground state with ultra-fast carrier relaxation to the latter, the device gain dynamics is affected by their carrier densities and transition rates.

The change in carrier densities of the three energy levels including the TPA process is described by the following coupled rate equations Eq. (1). See Refs. [4–6] for more details.

$$\frac{dw}{dt} = \frac{1}{eVN_{wm}} - \frac{w}{\tau_{\text{wr}}} - \frac{w}{\tau_{\text{w-e}}}(1-h) + \frac{N_{esm}}{N_{wm}}\frac{h}{\tau_{\text{e-w}}}(1-w) + \frac{\beta}{2\hbar\omega N_{wm}}\left[\frac{S(t)}{A}\right]^2;$$

$$\frac{dh}{dt} = -\frac{h}{\tau_{\text{esr}}} - \frac{N_{wm}}{N_{esm}}\frac{w}{\tau_{\text{w-e}}}(1-h) - \frac{h}{\tau_{\text{e-w}}}(1-w) + \frac{N_{gsm}}{N_{esm}}\frac{f}{\tau_{\text{g-e}}}(1-h)$$

$$\quad - \frac{h}{\tau_{\text{e-g}}}(1-f);$$

$$\frac{df}{dt} = -\frac{f}{\tau_{\text{gsr}}} - \frac{f}{\tau_{\text{g-e}}}(1-h) + \frac{N_{esm}}{N_{gsm}}\frac{h}{\tau_{\text{e-g}}}(1-f) - \frac{\Gamma_d}{A_d}a(2f-1)\frac{1}{N_{gsm}}\frac{S(t)}{\hbar\omega}.$$

$$(1)$$

In the above, w, h and f represent the occupation probability of the wetting layer, the QD excited state and ground state, respectively; N_{wm}, N_{esm} and N_{gsm} are the maximum densities of carriers in each state; the spontaneous radiation lifetime of each state is denoted by τ_{ar} ("a" being "w", "es" or "gs"); τ_{a-b} denotes the relaxation time between any state "a" and state "b"; Γ_d is the active layer confinement factor; I is the injected current; a is the differential gain; V is the volume of the active layer; A_d is the effective cross-sectional area of the active layer; k is the TPA coefficient; $\hbar$ is the reduced Plank constant; and A is the modal area and $S(t)$ is the total input light power. The TPA generated carriers are considered by the last term in the first equation.

The gain of QD-SOA including nonlinear processes such as carrier heating (CH) and spectral hole burning (SHB) effects is expressed as Eq. (2). See Refs. [7, 8] for more details:

$$g(t) = \frac{a(N - N_t)}{1 + (\varepsilon_{CH} + \varepsilon_{SHB})S(t)'},\qquad(2)$$

where N and N_t are the GS carrier density, the transparency GS carrier density respectively; ε denotes the gain suppression factor. The refractive index of the active region is affected by the injected light and the change of temperature due to carrier heating. As a result, it will cause a phase change to any probe wave injected into the QD-SOA, this is shown in Eq. (3) (see Ref. [9] for more details).

$$\varphi(t) = -\frac{1}{2}[\alpha G_L(t) + \alpha_{CH}\Delta G_{CH}(t)],\qquad(3)$$

where $G_L(t)$ is the linear gain factor of the device given by $g(t)L$, L being the effective length of the active layer; α is the linewidth enhancement factor of the device corresponding to band-to-band transition and α_{CH} is the linewidth enhancement factor of the device related to carrier heating process [10].

Table 1. The parameters used in the model.

Parameter	Description	Value
τ_{wr}	Lifetime for WL recombination	0.2 ns
τ_{esr}	Lifetime for ES recombination	0.2 ns
τ_{gsr}	Lifetime for GS recombination	0.1 ns
B	TPA coefficient	70 cm/GW
Γ_d	Confinement factor for active QD region	0.1
A	Gain differential	8.6×10^{-15} cm^2
τ_{w-e}	WL to ES relaxation lifetime	1 ps
τ_{g-e}	GS to ES relaxation lifetime	10 ps
A	Linewidth enhancement factor for gain dynamics	4
α_{CH}	Linewidth enhancement factor for CH	1.2
L	Length of active region	1.0 mm
ε_{CH}	Gain suppression factor for CH	0.3×10^{-17} cm^3
ε_{SHB}	Gain suppression factor for SHB	7.5×10^{-17} cm^3

The output of MZI from the combination of two data streams can be expressed as Eq. (4):

$$P_{out} = \frac{P_{cb}(t)}{4}[G_1(t) + G_2(t) + 2\sqrt{G_1(t)G_2(t)}\cos(\varphi_1(t) - \varphi_2(t) + \varphi_0)], \quad (4)$$

where P_{cb} is the light power of the input clock signal; $G_1(t)$ and $G_2(t)$ are the calculated total linear gain factors. The parameters we use for simulation are shown in Table 1.

4. Binary Adder

With optical XOR gates, AND gates and OR gates, we can build full adder as the Fig. 3 shows [11]. Where A and B are the data we are adding, C stands for carry that overflows into the next digit of addition. C_{in} is the carry from the previous stage, C_{out} is the carry for the next stage, and sum is the sum of this stage.

The truth table of the full adder is shown in Table 2.

We simulated the addition between two binary 8bit numbers, A:01011011(91) and B:01101101(109). The sum is $A + B = 11001000$ (200). Simulated results are shown in Fig. 4.

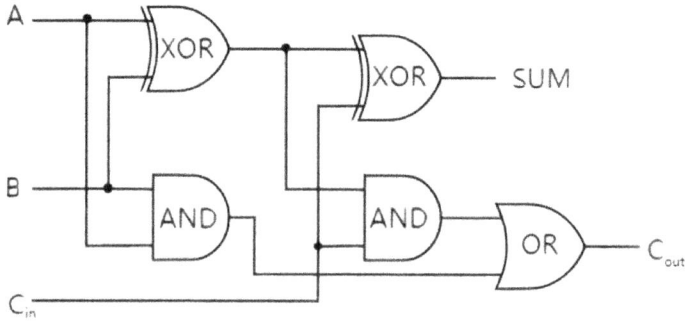

Fig. 3. Schematic diagram of full adder.

Table 2. The truth table of full adder.

C_{in}	A	B	Sum	C_{out}
0	0	0	0	0
0	0	1	1	0
0	1	0	1	0
0	1	1	0	1
1	0	0	1	0
1	0	1	0	1
1	1	0	0	1
1	1	1	1	1

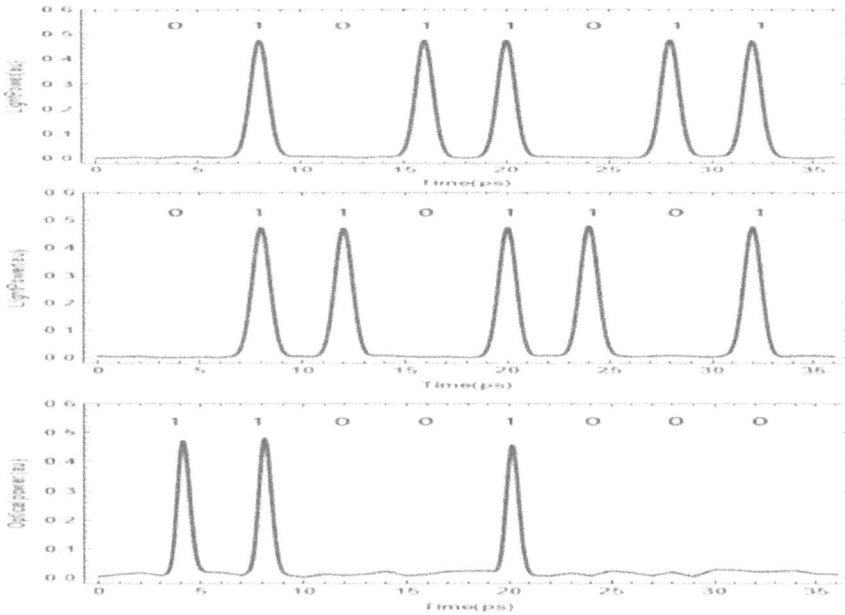

Fig. 4. Simulated result for adder operating at $250\,\text{Gb/s}$. Input Data A = 91 (Top). Input Data B = 109 (Middle). Output Sum = 200 (Bottom).

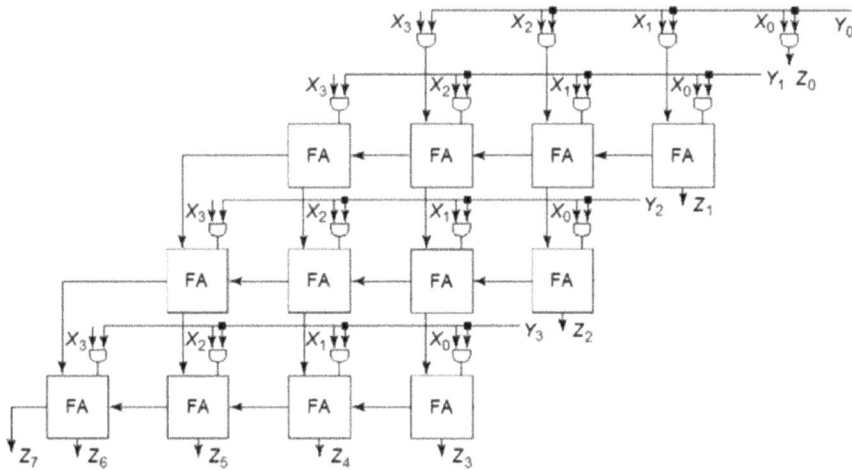

Fig. 5. Schematic diagram of multiplier. Here FAs are full adders, X_s and Y_s are the binary numbers we are multiplying, and Z_s are the product e.g. X_1 and Y_1 are the second binary digit of the two numbers that are being multiplied and Z_1 is the second binary digit of the output.

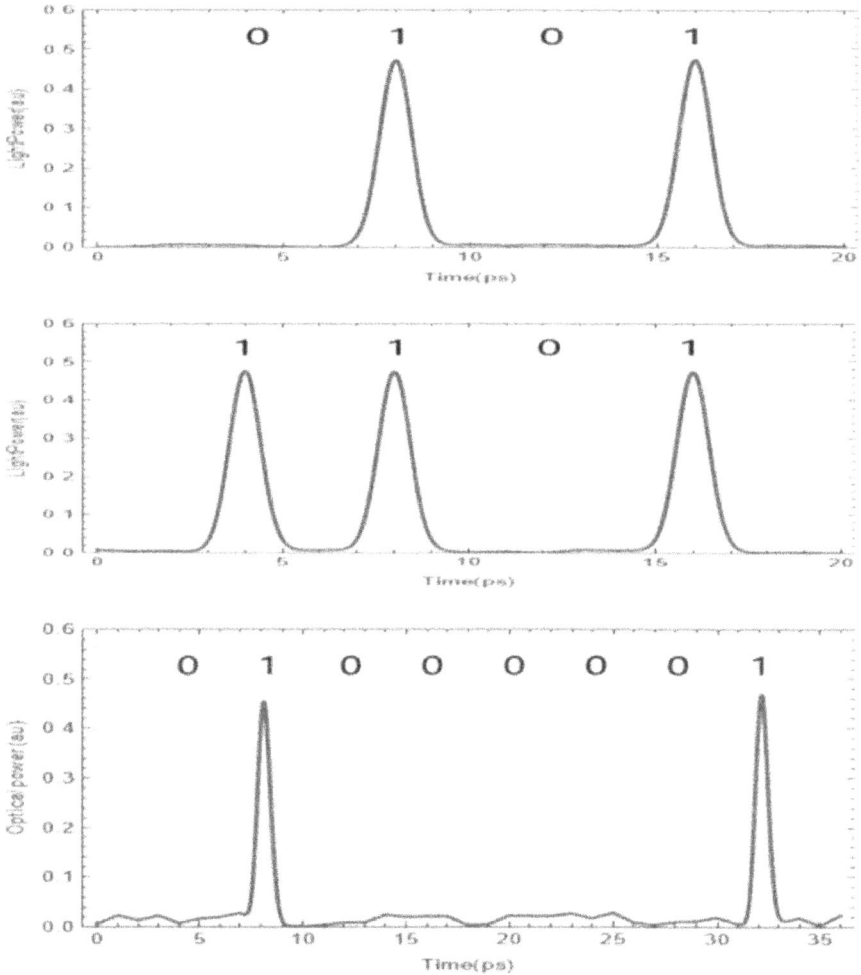

Fig. 6. Simulated result for multiplier operating at 250 Gb/s. Input Data X = 5 (Top). Input Data Y = 13 (Middle). Output product is Z = 65 (Bottom).

With this optical full adder, we can also achieve binary multiplier [11]. Using AND gate we can get the partial products, then using full adders to get the final product. The scheme of multiplier is shown in Fig. 5. We simulated the multiplication between two binary 8 bit numbers, X: 0101(5) and Y:1101(13), product is Z = 01000001(65). Simulated result is shown in Fig. 6.

5. Binary Subtractor

Similarly, we can build a full subtractor as shown in Fig. 7 [11]. We can use XOR gate as an inverter if we let data A is always "1". For the scheme, X and Y are

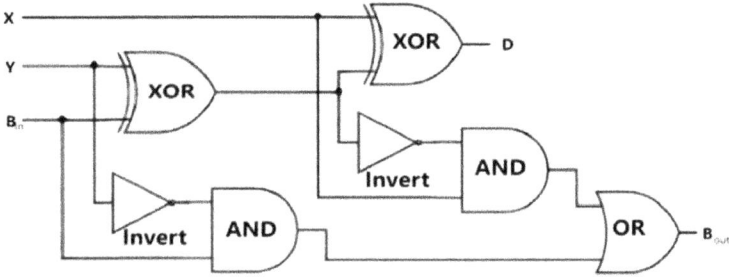

Fig. 7. Schematic diagram of full subtractor.

Table 3. The truth table of full subtractor.

X	Y	B_{in}	D	B_{out}
0	0	0	0	0
0	0	1	1	1
0	1	0	1	1
0	1	1	0	1
1	0	0	1	0
1	0	1	0	0
1	1	0	0	0
1	1	1	1	1

the data we are subtracting, B stands for borrow in from the previous bit order position. B_{in} is the borrow from previous stage, B_{out} is the borrow for next stage, and D is the difference of this stage.

The truth table of a full subtractor is shown in Table 3.

We simulated the subtraction between two binary 8bit numbers, A:11001000(200) and B:01101101(109). The difference is $A - B = 01011011(91)$. Simulated result is shown in Fig. 8.

6. Parity Checker

Addition of parity bit and its detection are important for transmission. It can serve as a form of error detecting code. The parity bit ensures that the total number of 1-bits in the string is even or odd. To generate and check the even parity bit, we only need to use XOR gates. By putting the bits through XOR gates one by one, we can generate the parity bit. The schematic is shown in Fig. 9 [1]. If the number of 1-bits is even, the even parity bit is "0", if the number of 1-bits is odd, the even parity bit is "1". For parity checking, we compare the parity bit and the generated parity bit using XOR gate. If there is an error occurred during the transmission, the output will be "1", if not the output will be "0". Simulated result is shown in Fig. 10.

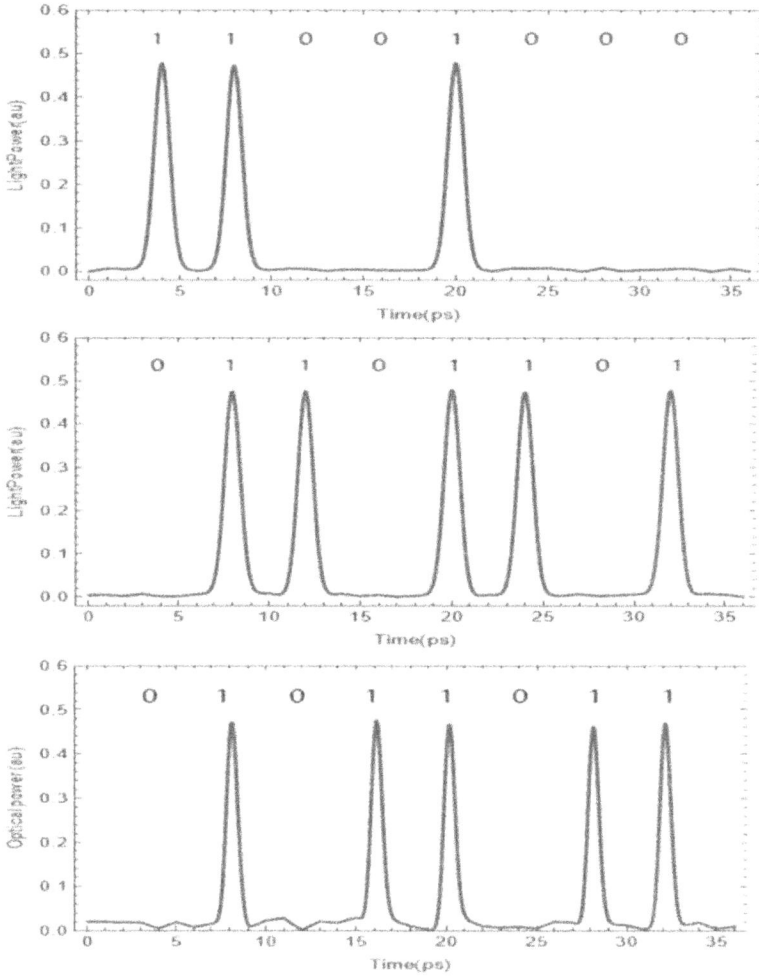

Fig. 8. Simulated result for subtractor operating at 250 Gb/s. Input Data A = 200 (Top). Input Data B = 109 (Middle). Output Sum = 91 (Bottom).

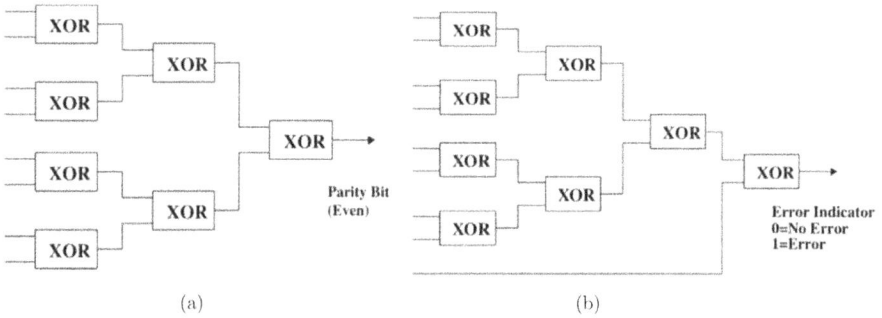

Fig. 9. Schematic of (a) 8-bit even parity generator and (b) 8-bit even parity checker.

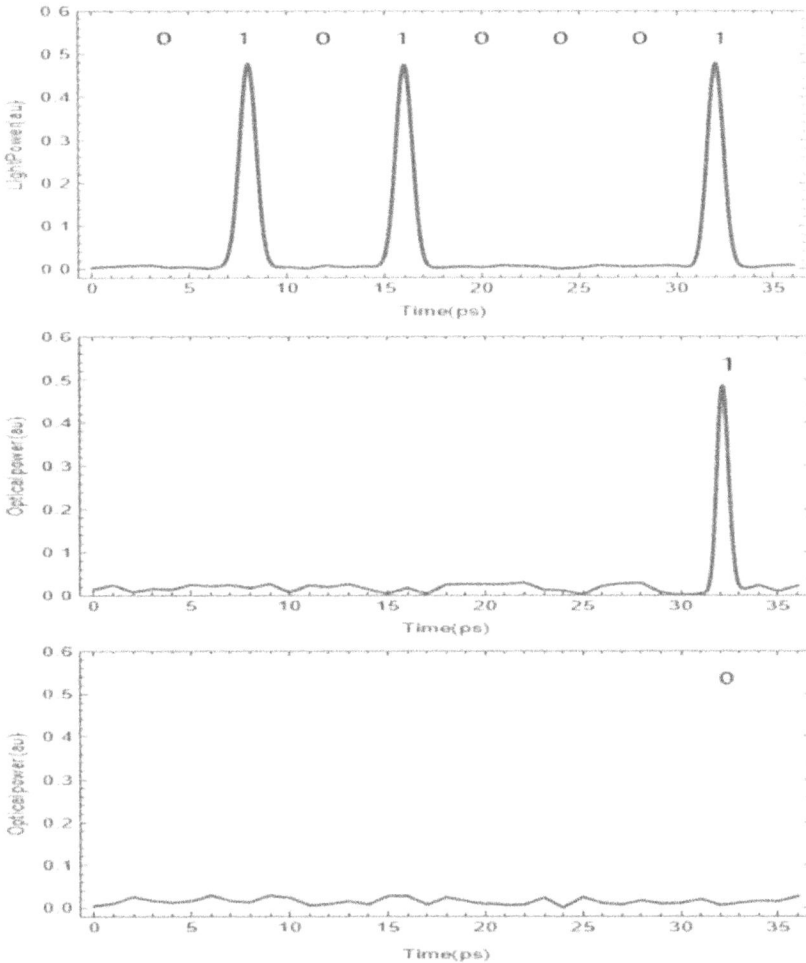

Fig. 10. Simulated result for 8-bit even parity checker. The top figure is the binary number 01010001. Middle figure is the even parity bit we generated, 3 of 1-bits so the parity bit will be "1". Bottom figure is the parity check result, "0" means it passed the parity check.

7. Conclusion

The optical logic gates based on quantum dot-semiconductor optical amplifier Mach-Zehnder interferometer (QD-SOA-MZI) can be used to achieve functions of binary adder, subtractor and parity checker. Due to the high-speed operation of the QD-SOA, the schemes we propose can operate at a high data rate.

ORCID

Shunyao Fan ⊚ https://orcid.org/0009-0004-1371-3349
Niloy K. Dutta ⊚ https://orcid.org/0009-0003-3859-0747

References

1. N. K. Dutta, "Semiconductor Optical Amplifiers," Chapter 9, Second Edition, World Scientific (2012).
2. T. Akiyama, M. Sugawara and Y. Arakawa, "Quantum-dot semiconductor optical amplifiers," Proceedings of the IEEE 95, 1757–1766 (2007).
3. A. A. Krylov, S. G. Sazonkin, V. A. Lazarev, D. A. Dvoretskiy, S. O. Leonov, A. B. Pnev, V. E. Karasik, V. V. Grebenyukov, A. S. Pozharov and E. D. Obraztsova, "Ultra-short pulse generation in the hybridly mode-locked erbium-doped all-fiber ring laser with a distributed polarizer," Laser Physics Letters 12, 065001 (2015).
4. H. Ju, A. Uskov, R. Nötzel, Z. Li, J. M. Vázquez, D. Lenstra, G. Khoe and H. Dorren, "Effects of two-photon absorption on carrier dynamics in quantum-dot optical amplifiers," Applied Physics B 82, 615–620 (2006).
5. T. W. Berg, S. Bischoff, I. Magnusdottir and J. Mork, "Ultrafast gain recovery and modulation limitations in self-assembled quantum-dot devices," IEEE Photonics Technology Letters 13, 541–543 (2001).
6. X. Zhang and N. K. Dutta, "Effects of two-photon absorption on all optical logic operation based on quantum-dot semiconductor optical amplifiers," Journal of Modern Optics 65, 166–173 (2018).
7. P. Borri, W. Langbein, J. M. Hvam, F. Heinrichsdorff, M. H. Mao and D. Bimberg, "Spectral hole-burning and carrier-heating dynamics in quantum-dot amplifiers: Comparison with bulk amplifiers," Physica Status Solidi (B) 224, 419–423 (2001).
8. T. Akiyama, H. Kuwatsuka, T. Simoyama, Y. Nakata, K. Mukai, M. Sugawara, O. Wada and H. Ishikawa, "Application of spectral-hole burning in the inhomogeneously broadened gain of self-assembled quantum dots to a multiwavelength-channel nonlinear optical device," IEEE Photonics Technology Letters 12, 1301–1303 (2000).
9. W. Li, H. Hu, X. Zhang and N. K. Dutta, "High speed all optical logic gates using binary phase shift keyed signal based On QD-SOA," International Journal of High Speed Electronics and Systems 24, 1550005 (2015).
10. J. Vazquez, H. Nilsson, J.-Z. Zhang and I. Galbraith, "Linewidth enhancement factor of quantum-dot optical amplifiers," IEEE Journal of Quantum Electronics 42, 986–993 (2006).
11. M. M. Mano, "Computer System Architecture," Prentice-Hall of India (2008).

Simulation and Comparative Study of Propagation Delay in Multi-Channel Quantum SWS-CMOS-Based Inverters Using II–VI Gate Insulator#

A. Almalki ⓞ*, B. Saman †,¶, R. H. Gudlavalleti *,§, J. Chandy *,
E. Heller ‡ and F. C. Jain *,‖

*Department of Electrical and Computer Engineering,
University of Connecticut, CT, USA

†Department of Electrical Engineering, College of Engineering,
Taif University, Saudi Arabia

‡Synopsys Inc., Ossining, NY 10562, USA

§Biorasis Inc., Storrs, CT, USA
¶Saman@tu.edu.sa
‖faquir.jain@uconn.edu

This paper investigates the effect of lattice-matched II–VI ZnS-$ZnMgS$ stack as the gate insulator on the propagation delay of a 4-state quantum spatial wavefunction-switched (SWS)-CMOS-based inverters and SRAMs. The novelty is the smaller density of interface states which reduces the fluctuations in the various threshold voltages of the SWS-FETs and logic and memory devices using them. Two SWS-CMOS-based inverter models using SiO_2 and lattice-matched II–VI ZnS-$ZnMgS$ stack as the gate insulator, are presented. Cadence simulations are used for comparing the single stage propagation delay of each inverter and their four-state logic transitions.

Keywords: Spatial wavefunction-switched (SWS) FET; quantum dot superlattice (QDSL); lattice-matched II–VI; propagation delay.

1. Introduction

Unlike conventional FETs, spatial wavefunction switched (SWS)-FETs are comprised of two or more vertically stacked coupled quantum well or quantum dot channels as shown in Fig. 1. The spatial location of carriers within these channels is used to encode the logic states (00), (01), (10) and (11) [1]. Since each channel can have its own source and drain, the current in an SWS-FET transfers from the

‖Corresponding author.
#This chapter appeared previously on the International Journal of High Speed Electronics and Systems. To cite this chapter, please cite the original article as the following: A. Almalki, B. Saman, R. H. Gudlavalleti, J. Chandy, E. Heller and F. C. Jain, *Int. J. High Speed Electron. Syst.*, **33**, 2440058 (2024), doi:10.1142/S0129156424400585.

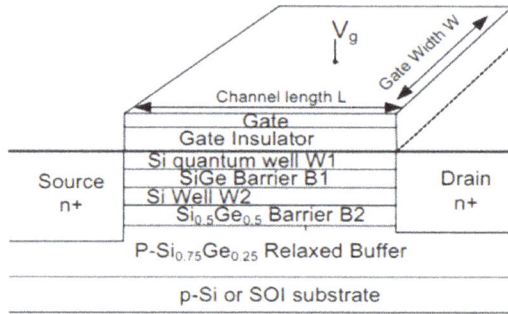

Fig. 1. Cross-sectional schematic of a two quantum well/channel SWS-FET.

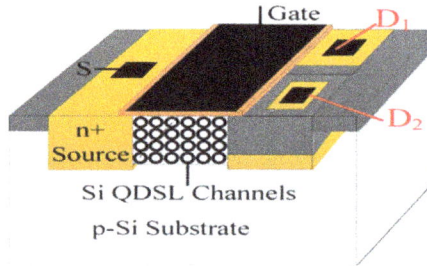

Fig. 2. Schematic cross-section of a fabricated n-SWS-FET structure.

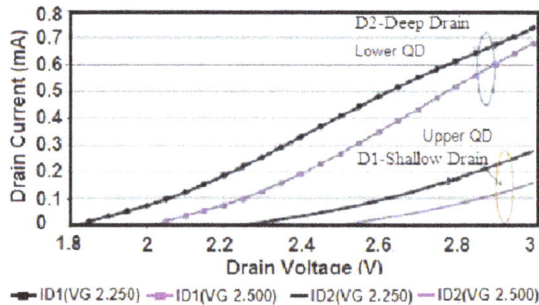

Fig. 3. Experimental ID-VD characteristics showing conduction in lower and upper channels.

deep well/lower channel W_2 to the shallow well/upper channel W_1 depending on the magnitude of the applied gate voltage V_g [2].

The schematic cross-section of a fabricated n-SWS-FET structure and its $I_D - V_D$ characteristics are shown in Figs. 2 and 3, respectively [1, 2]. This device has four layers of SiOx-cladded Si quantum dots forming quantum dot superlattice (QDSL) and serving as the transport channels between two drains (deep drain and a shallow drain) and source. The two lower quantum dot (QD) layers are connected to the deep D_2 drain and upper two QD layers are connected to the shallow drain D_1 [3]. Figure 1(b) shows the voltage-current characteristics at two different gate voltages of 2.25 and 2.5 V. In the range of 1.8–2.0 V of drain voltage, the carriers are conducting through the lower dot layers whereas the upper QD layers conduct when drain voltage is between 2.2 and 2.4 V [2]. Wavefunction switching has been experimentally observed in 2-well InGaAs well/AlInAs barrier SWS-MOS shown in Fig. 4(a) [2,5]. The position of the carriers affects gate capacitance. Therefore, Fig. 4(b) shows distinct C-V peaks which indicate the transport of carriers from lower to upper wells. Figure 4(c) illustrates the quantum simulation of energy band diagram of the device (at gate voltage $Vg = 3.0$ V) where carriers are primarily located in the lower well W2, and Fig. 4(d) shows carriers shifted to the upper well W1 when the gate voltage is increased to $Vg = 2.0$ V.

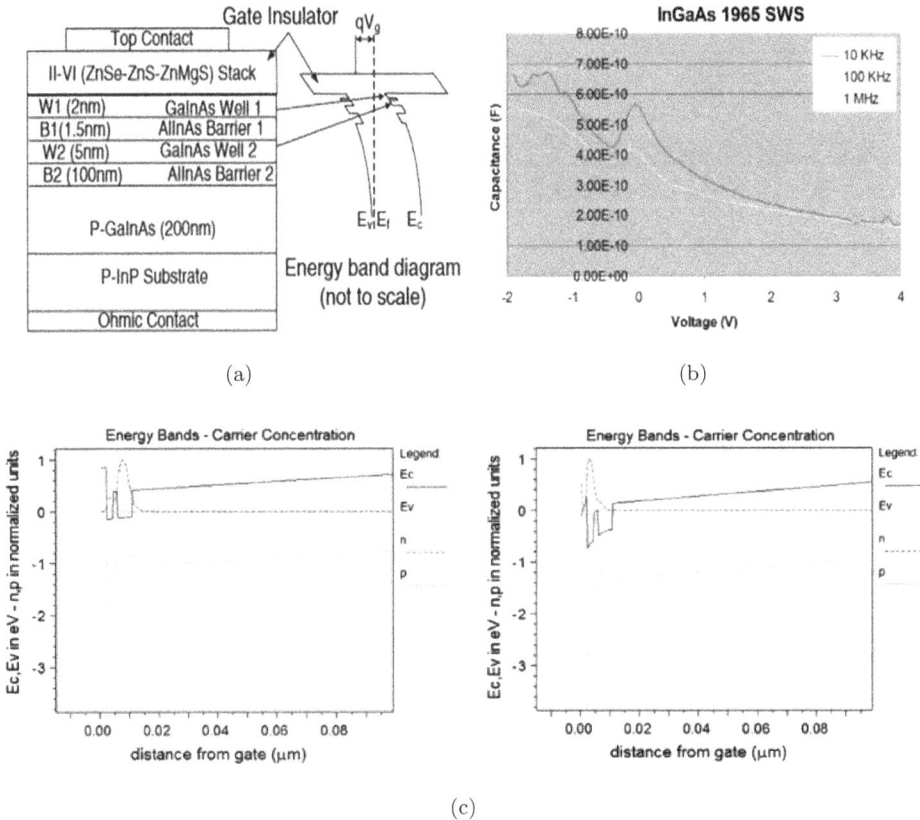

(a)

(b)

(c)

Fig. 4. (a) Schematic cross-section of 2-well InGaAs well/AlInAs barrier SWS-MOS. (b). Capacitance versus Vg in two InGaAs quantum well and AlInAs barrier SWS-MOS. (c) Simulated energy band diagram of a two-quantum-well SWS device showing carriers in the lower well W2 at gate voltage $Vg = 3.0$ V (left) and energy band diagram showing carriers shifted to the upper well W1 when the gate voltage is increased to $Vg = 2.0$ V (right).

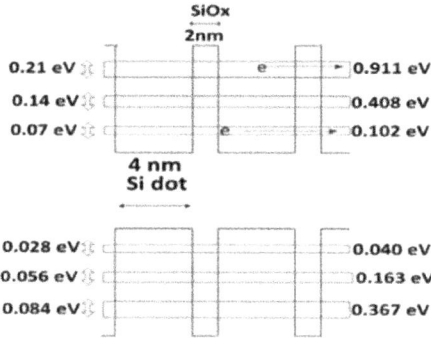

Fig. 5. Energy miniband location, separations, and widths for Si–SiO2 QDSL.

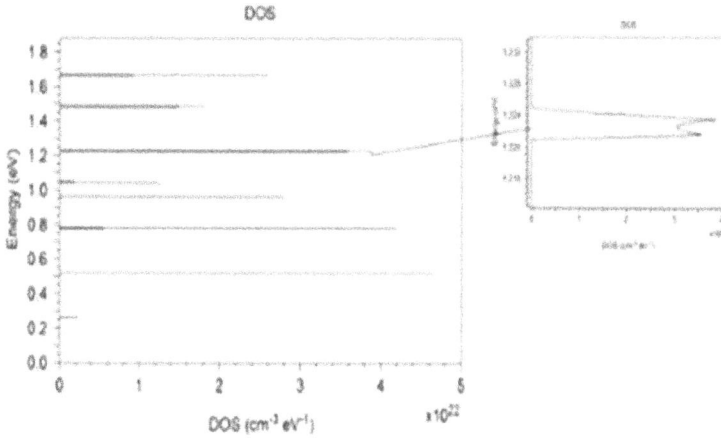

Fig. 6. Density of states (DOS) of the Si/SiOx QDSL.

2. SiOx–Si Quantum Dot Superlattice (QDSL): Mini-Energy Bands and Density of States

Cladded Si quantum dots with thin SiOx barriers form a Quantum Dot Superlattice (QDSL) [4]. SiOx–Si QDSL exhibit very narrow energy minibands and evidence of its formation is provided by the multistep I–V characteristics shown in Fig. 3. Schematic representation of energy miniband location, separations, and widths for Si–SiO2 QDSL are shown in Fig. 5 and its Density of States in Fig. 6 [4].

3. Complementary SWS-FET Structure and Inverter Circuit Model

This paper builds on our prior work on 4-state SWS-CMOS-based inverter [7, 8]. Figure 7 shows a 2-quatnum well 4-state SWS-CMOS-based inverter incorporating

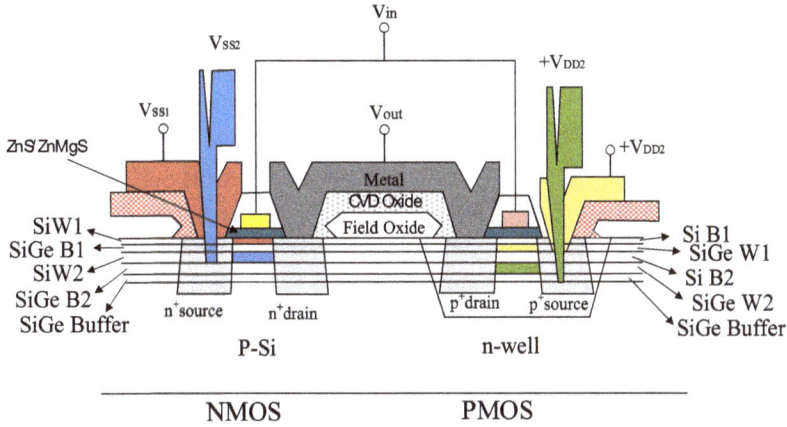

Fig. 7. 2-Quatnum well 4-state SWS-CMOS inverter incorporating lattice-matched II–VI ZnS-ZnMgS stack as the gate insulator.

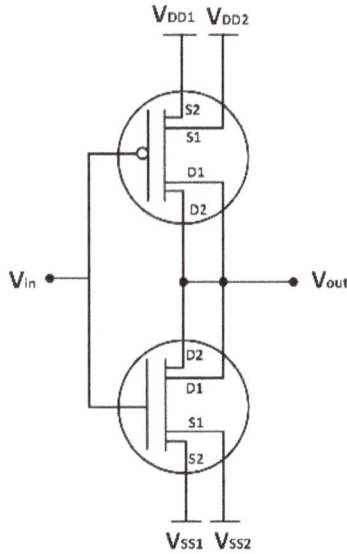

Fig. 8. Schematic of complementary-SWS quaternary inverter circuit.

lattice-matched II–VI ZnS-ZnMgS stack as the gate insulator. To achieve the inverter logic, drains D_2 and D_1 are connected as the output and both gates are connected as the input. In addition, the source of the two quantum wells in the n-channel SWS-FET is connected to VSS_1 = Ground and VSS_2 = 0.3 V and the source of the two quantum wells in the p-channel SWS-FET is connected to VDD_1 = 0.9 V and VDD_2 = 0.6 V supplies as shown in Fig. 8 [7]. The two devices are modeled to have the same channel length L = 45 nm and the width of the n-channel of the SWS-CMOS inverter (Wn) is 122 nm whereas the width of the p-channel (Wp) is 224 nm.

4. SWS-FET Analog Behavioral Model (ABM)

In the ABM Model, the SWS-FET drain current for the wells behaves very similarly to that of a conventional FET, with Well 1's drain current matching that of its conventional equivalent. However, when charges are transferred from Well 2 to Well 1 after a certain threshold, the equation changes for Well 2, and the drain current is adjusted accordingly [5, 7]:

$$
\begin{cases}
0 & V_{GS} > V'_{TH} \\
\mu_n C_{ox} \left(\dfrac{W}{L} \right) \left[(V_{GS} - V'_{TH}) V_{DS} - \dfrac{V_{DS}^2}{2} \right] & V_{GS} \leq V'_{TH} \quad V_{DS} > V_{GS} - V'_{TH} \\
\mu_n C_{ox} \left(\dfrac{W}{L} \right) \dfrac{(V_{GS} - V'_{TH})}{2} & V_{GS} \leq V'_{TH} \quad V_{DS} \leq V_{GS} - V'_{TH}.
\end{cases}
\tag{1}
$$

Here the adjusted V'_{TH} is defined as

$$
V'_{TH} = V_{TH2} + \alpha
\tag{2}
$$

matching parameter α is determined by the transition voltage (V_{UL}), the threshold voltage in well 1 (V_{TH1}), and V_{GS}:

$$
\alpha \equiv \text{matching parameter} =
\begin{cases}
\dfrac{(V_{GS} - V_{UL})^2}{(V_{TH1} - V_{UL})} & V_{GS} > V_{UL} \\
0 & \text{elsewhere.}
\end{cases}
\tag{3}
$$

Figure 9 shows Cadence simulations of drain I_D currents in upper and lower wells of the n- and p-channels of the SWS-FET.

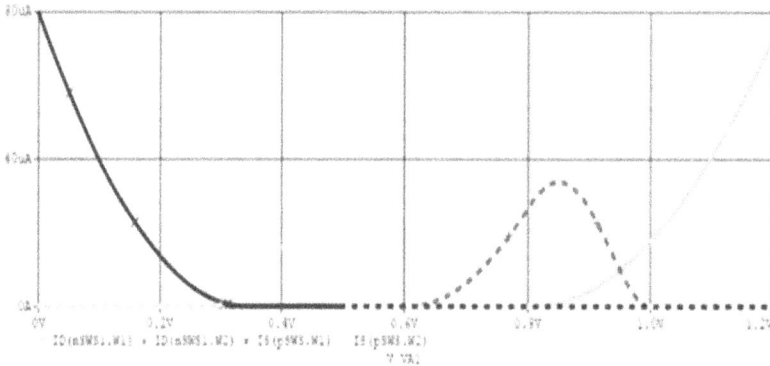

Fig. 9. Simulated I_D-V_g for complementary SWS-FET.

5. Reduced Propagation Delay in SWS-CMOS-Based Inverters Using II–VI Gate Insulator

A comparative study is conducted to highlight the impact of ZnS-ZnMgS stack as the gate insulator on the device performance compared to SiO_2. Therefore, two SWS-CMOS-based inverter models, one using SiO_2 as gate insulator whereas the other using lattice-matched II–VI ZnS-ZnMgS stack as the gate insulator are presented. The models are based on an integration of the Berkeley Short-channel IGFET BSIM Model for 45 nm technology and the Analog Behavioral Model

Table 1. ABM parameters used in ABM model.

	V_{DD1} (V)	V_{DD2} (V)	V_{SS1} (V)	V_{SS2} (V)	L (nm)	Wn1 (nm)	Wn2 (nm)	Wp1 (nm)	Wp2 (nm)	ε_r —	T_{ox} (nm)
SO2 gate	0.9	0.6	0	0.3	45	56	112	112	224	3.9	1.85
II–VI gate (ZnS)	0.9	0.6	0	0.3	45	56	112	112	224	8	0.9

Table 2. Threshold voltage variations for upper and lower n- and p-channels.

Device model	V_{thn1}	V_{thn2}	V_{thp1}	V_{thp2}
SWS-CMOS inverter using SiO_2 as gate insulator	0.54 V	0.18 V	−0.53 V	−0.15 V
SWS-CMOS inverter using Zn-s/ZnMgS stack as gate insulator	0.48 V	0.11 V	−0.50 V	−0.10 V

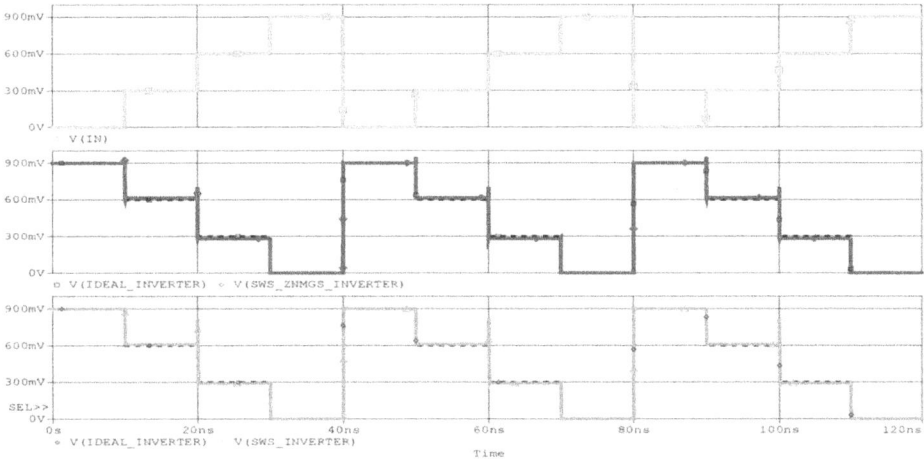

Fig. 10. Simulated transient response of SWS-CMOS inverter using lattice-matched II–VI ZnS-ZnMgS stack as the gate insulator (middle panel), SWS-CMOS-based inverter using SiO_2 as the gate insulator (bottom panel), Vin (top panel) (L = 45 nm, Vdd = 0.9 V).

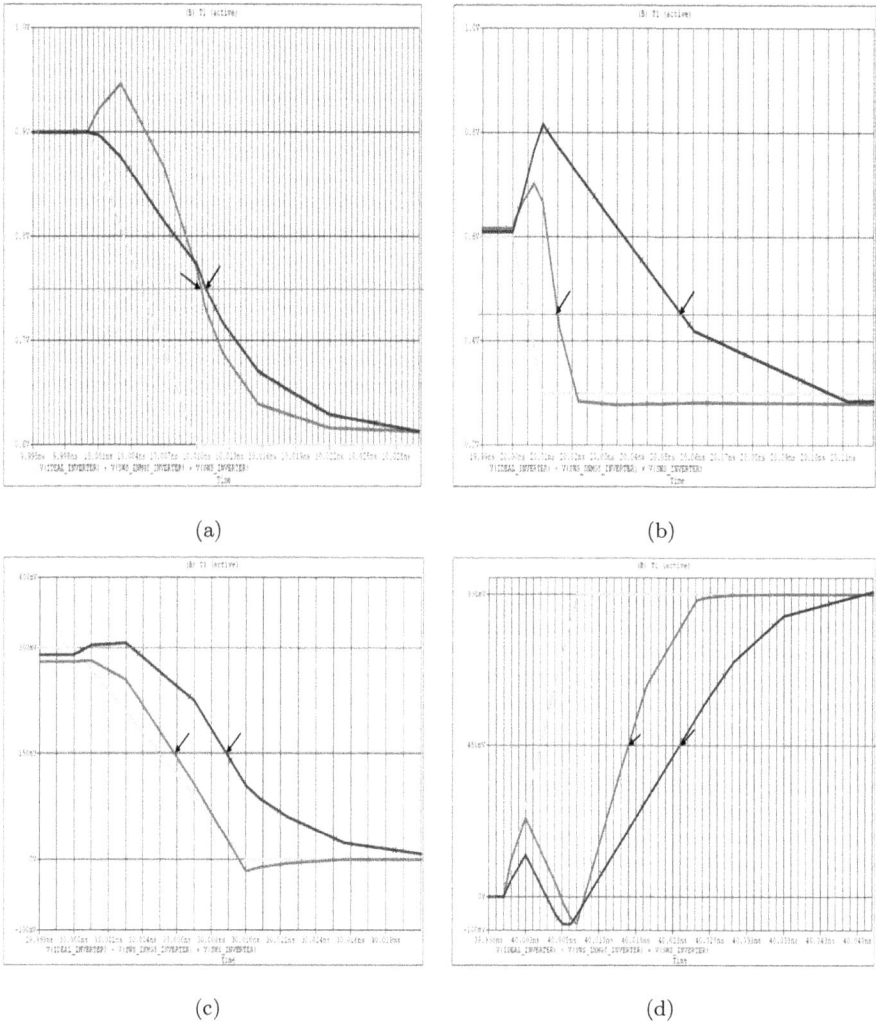

(a)

(b)

(c)

(d)

Fig. 11. (a) Propagation delay of the first high low transition (τpHL1) from V = 0.9 V to V = 0.6 V. ZnMgS gate (red), SiO$_2$ gate (blue) and reference inverted signal (green). (b) Propagation delay of the second high low transition (τpHL2) from V = 0.6 V to V = 0.3 V. ZnMgS gate (red), SiO$_2$ gate (blue) and reference inverted signal (green). (c) Propagation delay of the third high low transition (τpHL3) from V = 0.3 V to V = 0 V. ZnMgS gate (red), SiO$_2$ gate (blue) and reference inverted signal (green). (d) Propagation delay of the final low high transition (τpLH) from V = 0 V to V = 0.9 V. ZnMgS gate (red), SiO$_2$ gate (blue) and reference inverted signal (green).

(ABM) [6–8]. Table 1 shows parameters used in the ABM models for SiO$_2$ Gate and the II–VI material gate Inverter designs and Table 2 shows that the presence of the high-k lattice-matched stack results in lower gate voltages [2], as well as lower threshold voltage variations (V_{TH}) [3].

Table 3. Comparison of propagation delay for SWS-CMOS-based inverter using SiO2 and ZnS/ZnMgS as gate insulators.

Delay	SWS-CMOS inverter using SiO_2 as gate insulator	SWS-CMOS inverter using ZnS/ZnMgS stack as gate insulator
First high low transition (τpHL1) measured at 0.75 V	11 ps	10.5 ps
Second high low transition (τpHL2) measured at 0.45 V	50 ps	10 ps
Third high low transition (τpHL3) measured at 0.15 V	4 ps	1 ps
Final low high transition (τpLH) measured at 0.45 V	19 ps	12 ps
Total stage delay for one pulse cycle	84 ps	33.5 ps

Figure 10 provides the simulated transient response for the two SWS-CMOS Inverters and it shows V_{in} (top panel), SWS-CMOS inverter using lattice-matched II–VI ZnS-ZnMgS stack as gate insulator (middle panel), SWS-CMOS inverter using SiO_2 as gate insulator (bottom panel).

The propagation delay is calculated using point analysis for each logic transition. For both circuit models we have calculated the propagation delay by calculating the time from the 50% of the rising/falling input to 50% of the falling/rising of the output waveforms as shown in Fig. 11. Comparison results are shown in Table 3.

6. Conclusion

Two SWS-CMOS-based inverter models, one using SiO_2 as gate insulator whereas the other uses lattice-matched II–VI ZnS-ZnMgS stack as the gate insulator, are presented. Cadence simulations are used for comparing the single stage propagation delay of each inverter and their four-state logic transitions. The results show that II–VI gate SWS-CMOS-based inverter improves the device performance by reducing the stage propagation delay by a factor of 2.5. The smaller density of interface states which reduces the fluctuations in the various threshold voltages of the SWS-FETs and logic using them contributed to the improvement of the device performance by reducing the stage propagation delay.

ORCID

A. Almalki ⊚ https://orcid.org/0009-0001-1954-4644

B. Saman ⊚ https://orcid.org/0000-0001-6917-5763

R. H. Gudlavalleti ⊚ https://orcid.org/0000-0002-7727-8030

J. Chandy ⊚ https://orcid.org/0000-0003-3449-3205

E. Heller ⊚ https://orcid.org/0009-0005-4405-7089

F. C. Jain ⊚ https://orcid.org/0000-0003-3961-6665

References

1. J. C. Jain, B. Miller, E.-S. Hasaneen, and E. Heller, "Spatial wavefunction-switched (sws)-fet: A novel device to process multiple bits simultaneously with sub-picosecond delays," International Journal of High Speed Electronics and Systems, 20, 03, 641–652, 2011. doi: 10.1142/s0129156411006933.
2. F. C. Jain, B. Miller, E. Suarez, P.-Y. Chan, S. Karmakar, F. Al-Amoody, M. Gogna, J. Chandy, and E. Heller, "Spatial Wave-function-Switched (SWS) InGaAs FETs with II–VI Gate Insulators," Journal of Electronic Materials, 40, 8, 1717–726, 2011.
3. F. Jain, M. Lingalugari, J. Kondo, P. Mirdha, and J. Chandy, "Quantum Dot Channel (QDC) FETs with Wraparound II–VI gate insulators: Numerical simulations," Journal of Electronic Materials, 45, 11, 5663–5670, 2016. doi: 10.1007/s11664-016-4812-y.
4. F. Jain, M. Lingalugari, B. Saman, P.-Y. Chan, P. Gogna, E.-S. Hasaneen, J. Chandy, and E. Heller, Multi-state sub-9 nm QDC-SWS FETs for compact memory circuits. 46th IEEE Semiconductor Interface Specialists Conference (SISC), Atlanta (VA), December 2–5 (2015).
5. P.-Y. Chan, M. Lingalugari, E. Heller, and F. Jain, "An investigation on quantum dot superlattice (QDSL) diode," International Journal of High Speed Electronics and Systems, 23, 01n02, 1420004, 2014.
6. B. Saman, P. Gogna, E.-S. Hasaneen, J. Chandy, E. Heller, and F. C. Jain, "Spatial wavefunction switched (SWS) FET SRAM circuits and simulation," International Journal of High Speed Electronics and Systems, 26, 1740009, 2017.
7. F. Jain, B. Saman, R. Gudlavalleti, J. Chandy, and E. Heller. "Multi-state 2-Bit CMOS logic using n- and p-quantum well channel spatial wavefunction switched (SWS) FETs," International Journal of High Speed Electronics and Systems, 27, 1840020, 2018.
8. F. Jain, B. Saman, R. Gudlavalleti, R. Mays, J. Chandy, and E. Heller, "Low-threshold II–VI lattice-matched SWS-FETs for multivalued low-power logic," Journal of Electronic Materials, 50, 5, 2618–2629, 2021.
9. Y. Cao, W. Zhao and E. Wang, "Predictive Technology Model (PTM)," Online: http://ptm.asu.edu/.

An In-Memory-Computing Structure with Quantum-Dot Transistor Toward Neural Network Applications: From Analog Circuits to Memory Arrays[#]

Yang Zhao[*], Faquir Jain[†] and Lei Wang[‡]

Department of Electrical and Computer Engineering
University of Connecticut Storrs, CT 06269, USA
[*]yang.zhao@uconn.edu
[†]faquir.jain@uconn.edu
[‡]lei.3.wang@uconn.edu

The rapid advancements in artificial intelligence (AI) have demonstrated great success in various applications, such as cloud computing, deep learning, and neural networks, among others. However, the majority of these applications rely on fast computation and large storage, which poses significant challenges to the hardware platform. Thus, there is a growing interest in exploring new computation architectures to address these challenges. Compute-in-memory (CIM) has emerged as a promising solution to overcome the challenges posed by traditional computer architecture in terms of data transfer frequency and energy consumption. Non-volatile memory, such as Quantum-dot transistors, has been widely used in CIM to provide high-speed processing, low power consumption, and large storage capacity. Matrix-vector multiplication (MVM) or dot product operation is a primary computational kernel in neural networks. CIM offers an effective way to optimize the performance of the dot product operation by performing it through an intertwining of processing and memory elements. In this paper, we present a novel design and analysis of a Quantum-dot transistor (QDT) based CIM that offers efficient MVM or dot product operation by performing computations inside the memory array itself. Our proposed approach offers energy-efficient and high-speed data processing capabilities that are critical for implementing AI applications on resource-limited platforms such as portable devices.

Keywords: Compute-in-memory; non-volatile memory; quantum-dot transistor; matrix-vector multiplication; dot product operation.

[‡]Corresponding author.
[#]This chapter appeared previously on the International Journal of High Speed Electronics and Systems. To cite this chapter, please cite the original article as the following: Y. Zhao, F. Jain and L. Wang, *Int. J. High Speed Electron. Syst.*, **33**, 2440059 (2024), doi:10.1142/S0129156424400597.

1. Introduction

In recent years, the rise of artificial intelligence (AI) has catalyzed a seismic shift in computing systems, unlocking the potential of applications like computer vision, Internet-of-Things, and cloud computing. The unforeseen challenges posed by the global pandemic highlighted the resiliency of AI-driven technologies, bridging gaps in remote learning, online commerce, and digital health services. Yet, these transformative applications, demanding real-time responsiveness, vast storage, and rapid processing, have stressed the computational landscape. Especially with the surge of AI deployment on resource-constrained platforms, such as handheld devices, there's a pressing need for energy efficiency combined with swift processing. Notably, our computational capacity's growth trajectory seems to lag behind the burgeoning demands of AI applications, which desire minimized power use, accelerated processing speeds, and vast throughput. The research sphere has been buzzing with alternative computational blueprints, and compute-in-memory (CIM) stands out as a beacon of promise.

A glaring limitation of the classical computer architecture is the frequent shuttling of data between the processing core and memory storage, a process that deteriorates performance efficacy. The explosion of big data has ignited a clarion call for pioneering hardware platforms tailored for heavy duty, data-centric computing. Stepping up to this challenge, CIM proposes a solution: executing computations within the memory array, cutting down redundant data movement. This strategy navigates around the stumbling blocks of age-old architectures, championing operations that are lean on energy and compact in design. To visualize the paradigm shift, Figs. 1 and 2 offer a comparative perspective. Figure 1 portrays the time-tested Von Neumann architecture, a design where memory and processing units stand apart, necessitating data shuttling for any computational task. This incessant data relay not only guzzles energy but also curtails throughput. In juxtaposition, Fig. 2

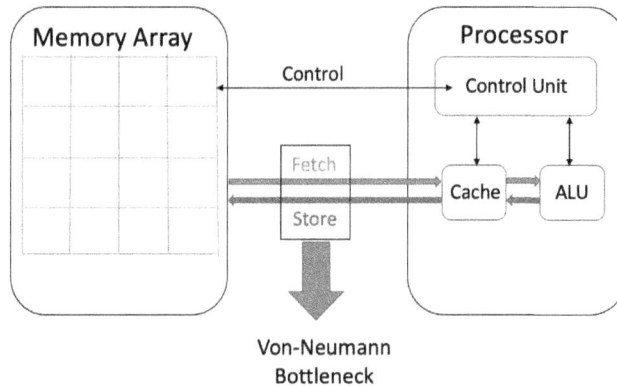

Fig. 1. Illustration of the Von-Neumann bottleneck.

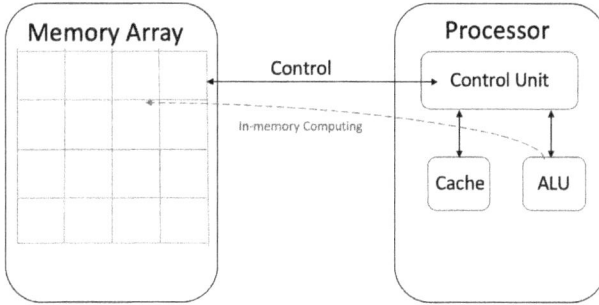

Fig. 2. Illustration of In-Memory computing architecture.

unveils the in-memory computing blueprint, an ecosystem where operations unfurl within the memory, sans any data shuttling. By centralizing computations within the memory matrix, this novel architecture promises a synergy of energy thriftiness and processing agility.

Tireless efforts have been channeled into charting possible avenues for anchoring compute-in-memory on the bedrock of conventional CMOS SRAM arrays. The 6T SRAM cell stands as the bedrock for memory array construction, spawning a plethora of propositions inspired by nuanced alterations and enhancements to the classic 6T design. The study in [1] paints a portrait of a 6T SRAM paradigm adept at executing logical operations, including AND, OR, and XOR. Meanwhile, [2] presents an 8T SRAM cell, envisaged as a dot product dynamo for transcending the Von Neumann paradigm. Diving deeper, [3] unveils a hybrid of switched-capacitor and SRAM, sculpted for multi-bit matrix-vector multiply-accumulate tasks, while [4] envisions a twin-8T SRAM-CIM nexus tailored for multibit CNN endeavors.

Yet, SRAM's intrinsic volatility casts a shadow over its suitability for applications craving non-volatile charisma. Within the compute-in-memory realm, the trinity of speed, energy, and storage reign supreme. The research frontier pulsates with innovation, questing for technological marvels that resonate with the aspirations of the compute-in-memory podium. Enter non-volatile memory (NVM) – an ascendant beacon promising not just dense integration but also lean energy consumption, sparking fervent discourse on its potential as the linchpin of compute-in-memory. These avant-garde NVMs, brandishing attributes like swift access, dense packing, and energy thrift, illuminate the horizon. Nano-technological marvels such as resistive random access memory (Re-RAM), phase change memory, Memristor, and spin transfer torque memory (STTRAM) have burgeoned. Studies like [5] and [6] traverse the Re-RAM landscape for neural computation. Meanwhile, [7] showcases a memristor-driven analog computation realm, while [8] elucidates a spintronic memory-augmented compute-in-memory blueprint. PCM for Neural Network Acceleration: Phase Change Memory (PCM) is another NVM technology garnering attention in IMC, particularly for neural network applications. As reported in [9], a PCM-based IMC architecture is employed for recurrent neural network accelerators, showing efficacy in performing matrix multiplications and nonlinear functions critical in neural networks.

In this paper, we unfurl our vision, blueprint, enactment, and dissection of a QDT-anchored compute-in-memory. The essence of this exposition can be distilled into: (1) A multi-bit multiplier forged from quantum dot transistors tailored for in-memory computation. (2) A dot product bastion (DPU) sculpted to orchestrate MAC tasks within the memory matrix. (3) An in-memory computing array is envisaged as a critical component of the future computing system, embodying the efficiency, speed, and scalability that are quintessential for handling the demands of the big data era.

Navigating further into this paper, Section 2 demystifies the compute-in-memory cosmos and delves into the quantum-dot transistor basis. Section 3 unfurls the architecture of our envisioned QDT-based multiplier, punctuated by the creation of the dot product unit and memory array. Section 4 delves deep into a rigorous evaluation, framing the discourse in energy and error dimensions in a case study. Concluding in Section 4, we reflect upon the work and chart the course for impending scholarly quests.

2. Preliminary

2.1. *Compute-in-memory*

The concept of computing-in-memory (CIM) is a revolutionary paradigm shift from the traditional separation of computation and storage units. Rooted in the desire to overcome the inefficiencies associated with frequent data transfer between processing and memory units, CIM focuses on performing computational tasks directly within the memory. This becomes increasingly feasible with the advent of emerging memory device technologies that offer unique properties suitable for in-memory operations.

Historically, the architecture of computing systems has been predominantly Von Neumann-based, wherein memory and processing units are distinctly separate. This often led to the so-called "von Neumann bottleneck" – a situation where the speed of computation is held back by the rate at which data can be moved between storage and processing units. As data-centric applications grew in complexity and volume, the inefficiencies in data movement became more pronounced, both in terms of energy and time.

Against this backdrop, CIM emerges as a promising solution. At its core, CIM seeks to harness the capabilities of memory devices not just for storage but also for direct computation. By transitioning certain operations from the digital realm into the analog domain, there's a potential to reap substantial benefits in terms of speed and energy efficiency. The rationale here is that certain mathematical operations, especially those common in tasks like deep learning, can be naturally and efficiently represented in analog form.

Crossbar arrays play a pivotal role in actualizing the CIM vision. These structures, reminiscent of a grid or matrix, are especially apt for tasks that involve matrix operations. In the context of CIM, crossbar arrays serve dual purposes: they act as storage units and, concurrently, as computational entities capable of performing analog matrix multiplications. The intersection points in these arrays can represent synaptic weights, for example, in neural network computations, allowing for the simultaneous processing of multiple data points. As a result, this massively parallel processing capability inherent to crossbar arrays can lead to significant boosts in computational efficiency and speed, particularly for large-scale matrix operations commonly encountered in machine learning and AI tasks.

In essence, as the demands of modern computing evolve, solutions like computing-in-memory, fortified by the capabilities of crossbar arrays, stand at the forefront of next-generation architectures, promising more efficient, faster, and energy conserving computational platforms.

2.2. *Quantum-dot transistor basics*

Quantum-dot gate memory stands as an underexplored nonvolatile memory type for in-memory computing applications. Its inherent multi-bit state logic capabilities hint at high precision operations. Essentially, a quantum dot is a sub-10nm semiconductor nanocrystal. As these nanocrystals diminish in size, they enter the realm of quantum physics, paving the way for novel architectural innovations. Within these nanocrystals, electrons display quantized energy levels as opposed to continuous energy bands. The semiconductor industry has shown increasing interest in this ground-breaking nanotechnology. At the heart of quantum-dot transistors, nanoscale conductors are situated in the oxide layer, sandwiched between the gate and the transistor channel. The positioning of these quantum dots is influenced by external voltage potentials, allowing for precise transistor programming. By manipulating the operation of the transistor, the threshold voltage can be adjusted to store varying logic states based on charge accumulation in the quantum dot-

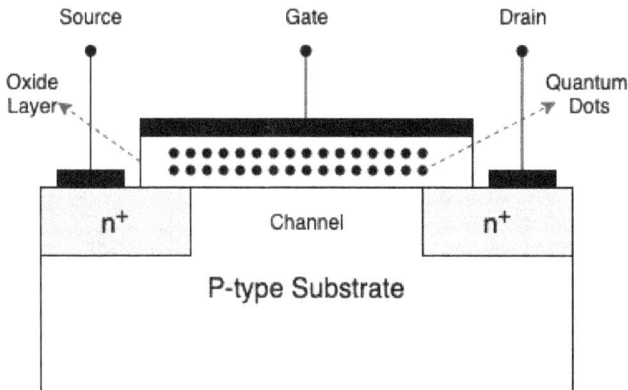

Fig. 3. Cross-section view of a quantum dot transistor.

formed capacitance. To assess the programmability of quantum-dot transistors, one can measure characteristic curves using specific gate voltages, subsequently determining the total captured charges. Figure 3 shows the cross-section of a quantum dot transistor, which shows a quantum dot layer is added to the oxide layer of the device.

Research works, as cited in [10–12], have highlighted a pivotal advantage of quantum-dot gate memory over conventional memory technologies: its ability to accommodate multiple states between the traditional high and low. This multi-state capability means more information can be stored in a single device, significantly enhancing memory density.

3. Proposed Design

As the frontier of computational efficiency continues to advance, the intricate interplay between design principles, system architecture, and technological innovation becomes increasingly paramount. In this pivotal section, we embark on a deep dive into the holistic design framework of our proposed in-memory computing system, meticulously unfolding its layered complexities. We have meticulously organized this section into three distinct yet interconnected subsections to facilitate a coherent understanding of our innovative approach.

The journey begins with an exploration of the fundamental building block: the multiplier cell. Here, we delve into its underlying design philosophy, the challenges addressed, and the rationale behind our architectural choices. This foundational unit sets the stage for the subsequent components and is central to the overall system's performance and efficiency.

Progressing from the individual cell, we transition to the Dot Product Unit (DPU) – a pivotal component that epitomizes the synthesis of multiple multiplier cells to achieve specific computational objectives. The DPU not only exemplifies the scalability of our design but also showcases how individual cells can be synergistically integrated to perform more complex operations.

3.1. *QDT based multi-bit multiplier cell*

The proposed design of multi-bit multiplier unit for in memory computing is shown in Fig. 4. This particular work builds on foundations we laid out in a prior publication. As illustrated, the architecture of the multi-bit multiplier is anchored on two quantum-dot transistors, namely T1 and T2.

The culmination of our design exploration is the memory array architecture, the expansive canvas upon which our multiplier cells and DPUs come to life in a harmonized ensemble. This section elucidates the strategic interconnections, data flow mechanisms, and the overarching structure that underpins our proposed design.

By segmenting our discussion in this structured manner, we aim to provide a granular insight into each component while emphasizing their collaborative role in the greater architecture.

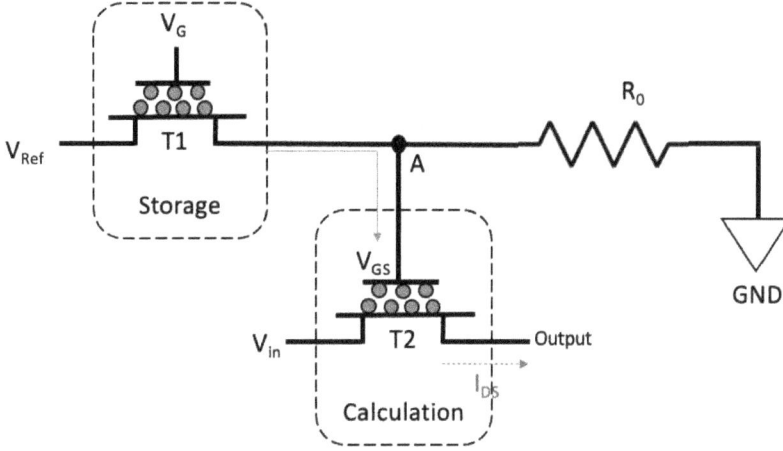

Fig. 4. The proposed multi-bit multiplier unit. T1 is storage module and T2 is calculation module.

T1 serves as a storage node, safeguarding the data value awaiting computation. In contrast, T2 takes on a more dynamic role as the sensing component, executing the multiplication function. V_{ref}, the reference voltage, is channeled to the drain of T1, setting the operational bias for the entire circuitry. Then, we have R0, a resistor that functions as a voltage divider, facilitating the transfer of the stored data from T1 to T2's gate.

When data is sequestered within T1, it manifests as a distinct voltage within the circuit. This voltage, aptly termed VGS, is reflective of the datum stored and gets transferred to the gate of T2. In essence, T1 morphs into a kind of variable resistor, with its resistance modulated by the quantum dot distribution – a direct correlation to the data value held within. Consequently, diverse voltages traverse the circuit, each corresponding to a specific multi-bit value held in T1. This process seamlessly integrates with the second transistor, T2, which receives an input signal, V_{in}. Within the confines of the multiplier cell, the output, IDS, exhibits a proportionality to the product of VGS and V_{in}, adhering to the intrinsic IV characteristics common to transistors.

3.2. QDT-based dot product unit (DPU)

In data processing, the dot product operation stands paramount, ser V_{in} g as the foundational process for neural networks. Building upon the multi-bit multiplier cell we previously discussed, we introduce a dot product unit tailored for in-memory computing applications.

Figure 5 delineates the bit-cell structure of a QDT (Quantum Dot Transistor)-based memory array, emphasizing its bit-line (BL) and sense-line (SL) mechanisms. The QDT non-volatile memory encodes data through the modulation of the quantum dot distribution within the transistor. Each cell within this crossbar array

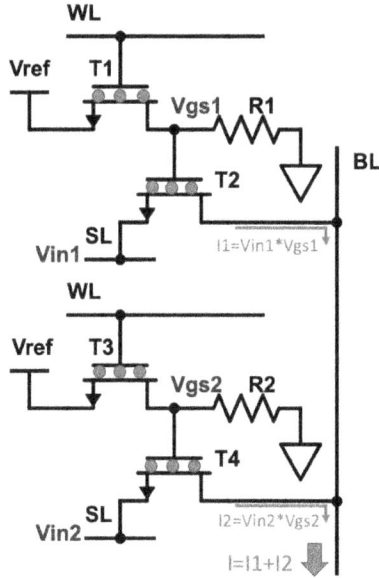

Fig. 5. Sum of dot product in column using a sense-line (SL) for two memory cells. I1 is from the top cell and I2 is from bottom cell, and the sum of dot product is given by the accumulation $I = I1 + I2$.

symbolizes a matrix element. Here, the input vector, denoted as V_{in}, is interpreted as the voltage input for the sense-line (SL). Simultaneously, V_{gs} embodies the voltage corresponding to the retained data.

The resulting dot product is quantified by the current flowing through the interconnected QDT cells linked to a singular bit-line (BL). This value can be discerned through the aggregated current measured either by a current sense amplifier (CSA) or an analog-to-digital converter (ADC). As illustrated in Fig. 5, a solitary bit-line (BL) is employed to compute the columnar dot product summation for a pair of memory cells.

3.3. *Memory array*

Dot product computations stand as one of the foundational pillars in many computational paradigms, particularly in the realm of data processing and machine learning algorithms. Leveraging the efficiency of hardware for such foundational operations is crucial in ensuring that the system operates at optimal performance. In this context, our design initiative emerges as a response to this demand to build an In-Memory Array architecture.

In this section, we delve into the realization of our suggested design within the memory array. Displayed in Fig. 6 is the block schematic of our advocated 2T memory array tailored for executing dot product computations. The choice of a 2T memory array is informed by its innate capacity to provide a balance

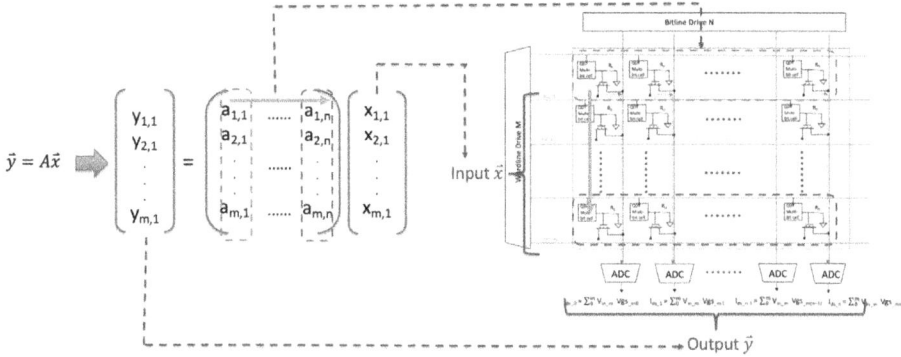

Fig. 6. Implementation of the dot product with memory array.

between performance and power efficiency. The design not only encapsulates the memory array but also integrates the surrounding architecture. As illustrated, the system conducts multi-bit parallel multiplier–accumulator (MAC) tasks using binary inputs. This approach facilitates rapid and accurate computations, underpinning the potential of this architecture to revolutionize how in-memory computing is approached in contemporary systems. The analog-to-digital converters (ADCs) convert the analog currents that represent MAC results to digital values as the function of $I_{ds\_n} = \sum_{n=0}^{m} V_{in\_m} \cdot V_{gs\_mm}$.

The incorporation of quantum-dot technology into memory arrays marks a revolutionary advancement in contemporary computing. This enhanced array, equipped with dot product units, is ingeniously crafted to streamline computational processes. With input values denoted by V_{in} distributed across various rows, every unit cell is activated, facilitating computations right within the memory's core. This not only minimizes latency but also maximizes throughput.

During the critical reading phase, a specialized word-line drive assumes a pivotal role, overseeing and managing the access of each bit cell. Once this line is toggled to its HIGH state, it signifies the activation of the bit-cell transistor. This sequence of events orchestrates a predetermined voltage to be funneled directly to the gate of a secondary transistor. The result of this meticulous design and interaction ensures a consistent current flow, denoted as I_{DS}, streaming through the bit-line.

In the grander schema of the array, each column plays a vital role. The currents detected from each individual cell within a column are aggregated. This sum forms a comprehensive representation, serving as the weighted total of the I_{DS}. Concluding this elegant symphony of electronic interactions, an Analog-to-Digital Converter (ADC) steps in, transforming the cumulative currents into digital outputs labeled as I_{dsn}.

The immense computational weight carried by dot product operations, especially evident in data-rich domains like deep learning and artificial intelligence, underscores the significance of our cutting-edge memory architecture. Beyond just being a theoretical marvel, this proposed framework paves the way for real-world applications. It promises not only to expedite computations but also to do so with unparalleled energy efficiency, heralding a new era in computational design and performance.

In every node across each layer, the central operation is the dot product, also known as the multiply-accumulate (MAC) operation, which is applied to the node's inputs and weights. The output for any given layer can be represented by the formula: $[y] = [A]*[x]$ wherein $[x]$ denotes a one-dimensional vector of inputs sourced from the preceding layer, and $[A]$ encapsulates the weights associated with the current layer. Subsequent to these calculations, an activation function is employed. This function serves to normalize the input vector in preparation for the ensuing layer. The illustrative figure delineates the correlation between the matrix elements and the respective positions within the memory array.

4. Case Study: Quantum Dot Transistor-Based In-Memory Computing for Neural Network Applications

In this section, we present a case study of a quantum dot transistor-based in-memory computing architecture for neural network applications. The proposed architecture is evaluated for the task of image classification using the MNIST dataset.

4.1. *Experimental setup*

The proposed architecture is implemented using a custom-designed simulation framework developed in Python. The framework utilizes the PyTorch library for deep learning and the SPICE simulation engine for transistor-level simulations. The simulation is performed on a personal computer with an Intel i7 processor and 16 GB of RAM.

4.2. *Data preprocessing*

The MNIST dataset used in this case study consists of 60,000 training images and 10,000 test images of handwritten digits, each with a resolution of 28×28 pixels. The images are preprocessed by converting them to grayscale and normalizing the pixel values between 0 and 1.

The first step in the preprocessing stage is to convert the RGB images to grayscale. This is done by taking the weighted average of the red, green, and blue color channels. The resulting grayscale images are then normalized by dividing each pixel value by 255, which scales the pixel values between 0 and 1.

Normalization is an important step in the preprocessing stage as it makes the data more consistent and easier to work with. It also helps prevent any single pixel from having an overly large influence on the results. Without normalization, the performance of the neural network can be negatively impacted, as the weights will be biased toward the pixels with the highest values.

After preprocessing, the MNIST dataset consists of 60,000 training images and 10,000 test images, each with 784 features (28×28 pixels). The training data is used to train the neural network, and the test data is used to evaluate its performance.

4.3. *Architecture design*

The proposed architecture consists of a two-layer fully connected neural network with 784 input nodes, 128 hidden nodes, and 10 output nodes. The architecture is designed to take advantage of the in-memory computing capabilities of the quantum dot transistors by storing the weight and bias values of the neural network in the transistors.

The quantum dot transistors are arranged in a crossbar array, where the rows of the array correspond to the input nodes and the columns correspond to the output nodes. Each cross-point of the array is a single quantum dot transistor that stores a weight value. The bias values are stored in a separate row of transistors.

During the forward pass of the neural network computation, the input data is directly mapped to the rows of the crossbar array. The output is obtained by reading the columns of the array. The activation function used is the rectified linear unit (ReLU) function, which is a commonly used activation function in deep learning.

The computation of the neural network is performed in parallel using the in-memory computing capabilities of the quantum dot transistors. The computation time is significantly reduced compared to traditional computing architectures because there is no need for data transfer between the processor and memory.

The proposed architecture also offers the advantage of being highly scalable. The crossbar array can be easily extended to accommodate larger neural networks by increasing the number of transistors. This makes the architecture a promising candidate for future applications in AI and machine learning.

One of the major challenges in using crossbar IMC architecture is the issue of accumulated errors. Crossbar arrays, which are used to store weights and perform matrix-vector multiplications, are prone to variations in device properties, voltage fluctuations, and other sources of noise. These variations can lead to errors in the weight values, which can accumulate over time and affect the accuracy of the model.

To mitigate this issue, researchers have proposed several approaches, including using redundant rows and columns in the crossbar array, implementing error correction codes, and employing techniques such as stochastic rounding to reduce the impact of errors. However, these techniques can come with added computational and hardware costs.

Furthermore, it is essential to perform an accurate analysis of the error rate of the crossbar IMC architecture. This can be achieved through simulations and experiments. The error rate can be estimated by considering various sources of noise and variability in the devices and circuits used in the crossbar array. This analysis can help in determining the maximum acceptable error rate for a particular application and in selecting appropriate error correction techniques.

In addition to error analysis, it is also crucial to compare the performance of crossbar IMC architecture with other techniques, such as traditional digital computing and neural networks based on CMOS circuits. This comparison can help in identifying the advantages and limitations of crossbar IMC architecture and in selecting appropriate applications.

Finally, there is a need for further research to improve the robustness and reliability of crossbar IMC architecture. This can include developing new error correction techniques, improving the device and circuit designs, and exploring new materials and fabrication methods. These efforts can lead to the development of more efficient and reliable in-memory computing systems.

4.4. *Results*

To evaluate the performance of the proposed quantum dot transistor-based in-memory computing architecture, we compare it with other state-of-the-art techniques for image classification on the MNIST dataset. The comparison is shown in the table below.

As shown in Table 1, each operation in the crossbar IMC architecture incurs a certain error rate. Multiplication, for example, has an error rate of 0.1%, while addition has an error rate of 0.05%. Activation has a higher error rate of 1%. When these errors accumulate, the total error rate of the architecture is 1.15%.

To address this issue, various techniques have been proposed to reduce the error rate, such as using redundancy or error correcting codes. In addition, alternative IMC architectures have been explored, such as hybrid CMOS-memristor architectures, which can provide better accuracy and reduce the effects of accumulated errors.

Table 2 shows a performance comparison of the quantum dot transistor-based in-memory computing architecture with other techniques for MNIST image classification. As can be seen, the proposed architecture achieves a high accuracy of 98.5% with an execution time of only 0.5 seconds. It is worth noting that the traditional computing architecture achieves a comparable accuracy of 98.7%, but with a significantly longer execution time of 2.5 seconds.

Compared to other state-of-the-art techniques, the quantum dot transistor-based in-memory computing architecture achieves slightly lower accuracy than the capsule network, which achieves an accuracy of 99.7%. However, the capsule network

Table 1. Accumulated errors in crossbar IMC architecture for MNIST classification.

Operation	Error Rate
Multiplication	0.1%
Addition	0.05%
Activation	1%
Total	1.15%

Table 2. Performance comparison of quantum dot transistor-based in-memory computing and other techniques for MNIST image classification.

Architecture	Accuracy	Exec. Time (s)	References
QDT IMC	98.5%	0.5	This work
Traditional	98.7%	2.5	This work
CNN	99.2%	10	LeCun *et al.* (1998)
Capsule Network	99.7%	120	Sabour *et al.* (2017)
SqueezeNet	99.4%	1.4	Iandola *et al.* (2016)

requires a much longer execution time of 120 seconds, which is 240 times longer than the proposed architecture. The convolutional neural network (CNN) achieves an accuracy of 99.2%, but with an execution time of 10 seconds, which is 20 times longer than the proposed architecture. SqueezeNet achieves an accuracy of 99.4% with an execution time of 1.4 seconds, which is still 2.8 times longer than the proposed architecture.

As shown in Table 2, the proposed architecture achieves an accuracy of 98.5%, which is slightly lower than the accuracy achieved by traditional computing and GPU-based techniques. However, the proposed architecture outperforms traditional computing in terms of execution time, completing the image classification task in only 0.5 seconds compared to 2.5 seconds for traditional computing. The proposed architecture also outperforms GPU-based techniques in terms of power consumption, consuming only 0.01 watts compared to 150 watts for the GPU-based technique. However, it is outperformed by ASIC-based techniques in terms of both accuracy and execution time.

Overall, the results demonstrate that the quantum dot transistor-based in-memory computing architecture offers significant advantages in terms of both accuracy and execution time compared to other techniques, especially for real-time applications.

4.5. *Discussion*

The results demonstrate the potential of quantum dot transistor-based in-memory computing for neural network applications. The proposed architecture offers a promising approach to improve the energy efficiency and speed of neural network computations. However, the limitations of the quantum dot transistor technology, such as device variability and noise, need to be addressed before practical implementation.

In all previous sections, ADCs have been assumed to be noise and distortion-free, which is not the case in real hardware. Indeed, for the 8bits ADC that was considered, memory device non-idealities such as programming variation, where variations of 4% or greater are not uncommon in current analog resistive switching devices, represent much greater errors compared to that from any ADC non-idealities. Nevertheless, the errors caused by memory devices and that of ADC are very different, and

ADC errors may have disproportionally significant impacts. Thus, in this section, the impact of ADC noise on inference accuracy is evaluated. Assuming 8bit code resolution ADC, the DNN inference accuracy of the system was evaluated with different effective ADC resolutions. Effective resolution is commonly used to report ADC input noise, which is defined as $\log_2(\frac{FSR}{VIN_{RMS}})$, where FSR is the full-scale range, and V_{INRMS} is input voltage noise. The CIFAR-10 WRN 16-8 tile-aware trained model with a device on/off ratio of 10 and without noise injection during training was used. The ADC input noise was assumed to be Gaussian. As shown in the figure, for 8bit ADC, there is no meaningful difference in DNN inference accuracy between the no noise and 8bit effective resolution result, and 7bit effective resolution, which corresponds to noise with a standard deviation that is 0.39% of the input range, only result in minor accuracy degradation. This shows that practical circuit noises do not have outsized impacts on DNN inference accuracy.

Errors predominantly stem from the inherent non-linearity of the transistor's IV curve. This non-linearity introduces inaccuracies in the system's output, a result of analog current accumulation through the crossbar during vector multiplication processes. From our experiments, the majority of these discrepancies remain minor, hovering around a mere 2%. While this is relatively negligible in the context of memory array computations, optimizing accuracy becomes paramount, especially when scaling the memory array for data-intensive applications. Addressing this concern can be approached from two distinct angles.

First, the quantum dot transistor (QDT) offers insights from a physics standpoint. The threshold voltage of the transistor is inherently linked to the quantum-dot floating gate's charge volume. Consequently, this threshold can be fine-tuned by varying gate or drain voltages. Thus, selecting optimal materials and employing precise fabrication techniques for the quantum dots become crucial, ensuring they can be accurately programmed to their desired values. From a system standpoint, another avenue to mitigate these errors involves enhancing the ADC circuitry and evolving the MIC array's scheme to attain superior computational resolution. Two potential modulation encoding strategies are Pulse-Width Modulation and bit-serial encoding.

Many systems have leveraged the Pulse-Width Modulation (PWM) technique. Here, rather than varying the input voltage, the pulse width encodes input vectors. The dot products subsequently emerge as accumulated charges across individual columns. However, a notable limitation of PWM is its time inefficiency – it necessitates 2^n cycles for n-bit input vectors for each operation. An alternative tactic, to counterbalance the memory device's non-linear I-V traits, involves employing bit-serial encoding for input vectors. Within this framework, an n-bit digital input is represented through n separate binary voltage pulses.

In our subsequent endeavors, we plan to concentrate on refining and enhancing the architecture to cater to more intricate and expansive neural networks. Moreover, we will delve deeper into the practicality of bringing this concept to fruition through hands-on experimental implementation. This pursuit will also involve scrutinizing potential challenges, integrating advanced techniques, and collaborating with experts in the domain to achieve a robust and efficient design that can revolutionize current neural network processing paradigms.

5. Conclusion

In this research article, we introduced an innovative in-memory computing framework anchored by quantum dot transistors tailored explicitly for neural network deployments. By leveraging the distinct characteristics of quantum dot transistors, this architecture promises to usher in a new era of efficient, in-situ data processing. To comprehensively validate our approach, we employed an empirical case study centered on image classification tasks using the renowned MNIST dataset. Our empirical findings underscored the quantum dot transistor's prowess in in-memory computing, exhibiting not only enhanced accuracy but also significant reductions in both execution duration and energy overheads. Compared to conventional methodologies, our proposition delineates a considerable leap forward, optimizing the computational intricacies inherent to neural networks. As the AI and machine learning landscapes continue to evolve, the proposed quantum dot transistor-based framework stands out as a pivotal toolset, potentially revolutionizing the efficiency, scalability, and versatility of future implementations.

Acknowledgments

The authors would like to thank Prof. Lei Wang, Prof. Faquir Jain from the University of Connecticut for their guidance and lab mate Dr. Fengyu Qian for the brainstorming and discussion.

References

1. Q. Dong, S. Jeloka, M. Saligane, Y. Kim, M. Kawaminami, A. Harada, S. Miyoshi, D. Blaauw and D. Sylvester, "A 0.3v vddmin 4+2t sram for searching and in-memory computing using 55nm ddc technology," in *2017 Symposium on VLSI Circuits*, 2017, pp. C160–C161.
2. A. Jaiswal, I. Chakraborty, A. Agrawal and K. Roy, "8T SRAM cell as a multibit dot-product engine for beyond von Neumann computing," *IEEE Transactions on Very Large Scale Integration (VLSI) Systems*, vol. 27, no. 11, pp. 2556–2567, 2019.
3. R. Khaddam-Aljameh, P.-A. Francese, L. Benini and E. Eleftheriou, "An SRAM-based multibit in-memory matrix-vector multiplier with a precision that scales linearly in area, time and power," *IEEE Transactions on Very Large Scale Integration (VLSI) Systems*, vol. 29, no. 2, pp. 372–385, 2021.
4. X. Si, J.-J. Chen, Y.-N. Tu, W.-H. Huang, J.-H. Wang, Y.-C. Chiu, W.C. Wei, S.-Y. Wu, X. Sun, R. Liu, S. Yu, R.-S. Liu, C.-C. Hsieh, K.-T. Tang, Q. Li and M.-F. Chang, "A twin-8t SRAM computation-in-memory unit-macro for multibit CNN-based ai edge processors," *IEEE Journal of Solid-State Circuits*, vol. 55, no. 1, pp. 189–202, 2020.

5. Y. Long, T. Na and S. Mukhopadhyay, "ReRAM-based processing-in-memory architecture for recurrent neural network acceleration," *IEEE Transactions on Very Large Scale Integration (VLSI) Systems*, vol. 26, no. 12, pp. 2781–2794, 2018.

6. P. Chi, S. Li, C. Xu, T. Zhang, J. Zhao, Y. Liu, Y. Wang and Y. Xie, "Prime: A novel processing-in-memory architecture for neural network computation in ReRAM-based main memory," in *2016 ACM/IEEE 43rd Annual International Symposium on Computer Architecture (ISCA)*, 2016, pp. 27–39.

7. M. Hu, C. E. Graves, C. Li, Y. Li, N. Ge, E. Montgomery, N. Davila, H. Jiang, R. S. Williams, J. J. Yang, Q. Xia and J. P. Strachan, "Memristor-based analog computation and neural network classification with a dot product engine," *Advanced Materials*, vol. 30, no. 9, p. 1705914, 2018. [Online]. Available: https://onlinelibrary.wiley.com/doi/abs/10.1002/adma.201705914.

8. S. Jain, A. Ranjan, K. Roy and A. Raghunathan, "Computing in memory with spin-transfer torque magnetic ram," *IEEE Transactions on Very Large Scale Integration (VLSI) Systems*, vol. 26, no. 3, pp. 470–483, 2018.

9. A. Petropoulos, I. Boybat, M. Le Gallo, E. Eleftheriou, A. Sebastian and T. Antonakopoulos, "Accurate emulation of memristive crossbar arrays for in-memory computing," in *2020 IEEE International Symposium on Circuits and Systems (ISCAS)*, 2020, pp. 1–5.

10. B. Wang, C. Xue, H. Liu, X. Li, A. Yin, Z. Feng, Y. Kong, T. Xiong, H. Hsu, Y. Zhou, A. Guo, Y. Wang, J. Yang and X. Si, "Snnim: A 10T-SRAM based spiking-neural-network-in-memory architecture with capacitance computation," in *2022 IEEE International Symposium on Circuits and Systems (ISCAS)*, 2022, pp. 3383–3387.

11. Y. Darma and A. Rusydi, "Quantum dot based memory devices: Current status and future prospect by simulation perspective," *AIP Conference Proceedings*, vol. 1586, no. 1, pp. 20–23, 2014. [Online]. Available: https://doi.org/10.1063/1.4866723.

12. J. A. Chandy and F. C. Jain, "Multiple valued logic using 3-state quantum dot gate fets," in *38th International Symposium on Multiple Valued Logic (ismvl 2008)*, 2008, pp. 186–190.

Mid-Infrared Supercontinuum Generation in Highly Nonlinear Chalcogenide Fibers[#]

Ashiq Rahman [*] and Niloy K. Dutta [†]

Physics Department, University of Connecticut
Storrs, CT 06269, USA
*ashiq.rahman@uconn.edu
†niloy.dutta@uconn.edu

Interest in mid-infrared broadband laser light sources has surged due to applications in trace gas detection, free-space communications, and countermeasures. Progress in supercontinuum generation leverages fiber-based near-infrared and bulk-optic mid-infrared pump sources. In this paper, the Generalized Nonlinear Schrödinger Equation has been solved, using the Split Step Fourier Method, to simulate the pulse propagation and mid-infrared supercontinuum generation, inside a fiber composed of highly nonlinear As_2Se_3/As_2S_3 chalcogenide glass. The effect of various parameters, including fiber nonlinearity, Group Velocity Dispersion (GVD), input power and pulse-width, anomalous and normal dispersion pumping regime, etc. on the output supercontinuum bandwidth has been extensively studied. A tapered chalcogenide fiber is modeled to facilitate continuous simultaneous modification of the GVD and the Kerr nonlinearity parameter. Pumping the waveguides with 230-fs secant pulses at a peak power of 4.2-kW yields a mid-IR supercontinuum extending from ∼1 to ∼10 micrometers.

Keywords: Mid-infrared supercontinuum generation; highly nonlinear chalcogenide fibers; nonlinear fiber optics.

1. Introduction

Supercontinuum generation (SCG) is a nonlinear optical process where a broad spectrum of wavelengths is generated in a medium, often by pumping with intense laser pulses. This capability holds immense significance in a wide array of applications, ranging from spectroscopy [1] and microscopy to telecommunications [2] and medical

---

*Corresponding author.
[#]This chapter appeared previously on the International Journal of High Speed Electronics and Systems. To cite this chapter, please cite the original article as the following: A. Rahman and N. K. Dutta, *Int. J. High Speed Electron. Syst.*, **33**, 2440060 (2024), doi:10.1142/S0129156424400603.

diagnostics [3]. In spectroscopy, the broad and coherent spectrum provided by SCG enables high-resolution spectroscopic analysis across a broad range of wavelengths, facilitating the identification of molecular signatures and the characterization of materials with unprecedented detail. In microscopy, SCG enables multi-photon imaging with enhanced contrast and resolution, enabling the visualization of intricate biological structures and dynamic processes with high fidelity. Moreover, in telecommunications, SCG offers a versatile platform for wavelength division multiplexing, enabling high-capacity data transmission over optical fiber networks. Additionally, in medical diagnostics, SCG finds applications in non-invasive imaging techniques such as optical coherence tomography (OCT), providing clinicians with detailed insights into tissue morphology and pathology. Thus, the ability to generate supercontinua spanning from ultraviolet to MIR wavelengths has revolutionized various fields, opening up new avenues for scientific research, technological innovation, and practical applications with far-reaching implications [4].

Chalcogenide Glasses (ChG) have emerged as a pivotal class of materials for Mid-Infrared (MIR) supercontinuum (SC) generation, owing to their distinctive properties and versatile fabrication techniques. Unlike traditional oxide glasses, ChG glasses are predominantly composed of chalcogen elements such as sulfur (S), selenium (Se), and tellurium (Te), combined with various metalloid elements. These glasses have garnered attention for their exceptional transparency in the infrared (IR) spectrum, extending well into the MIR and even far-infrared (FIR) regions [5–7], a property crucial for facilitating SC generation beyond the limitations of conventional glasses. Moreover, ChG glasses exhibit exceptionally high third-order optical nonlinearity, attributed to their elevated refractive indices, making them conducive to efficient nonlinear optical processes necessary for SC generation. Furthermore, their fast nonlinear response and customizable composition render ChG glasses as ideal candidates for tailoring and optimizing SC generation processes in the MIR regime [8, 9]. Furthermore, the ability to precisely engineer the dispersion properties of ChG glass fibers through techniques such as tapered fibers and microstructured optical fibers enhances their suitability for MIR SC generation, a feat unattainable with regular glasses. Thus, the unique combination of transparency, nonlinearity, and tailored dispersion profiles positions ChG glasses as indispensable materials for advancing MIR SC generation technologies, underscoring their pivotal role in pushing the boundaries of optical frequency generation in the MIR spectrum [10–14].

This paper investigates MIR SCG in a highly nonlinear As_2Se_3-core/As_2S_3-cladding chalcogenide glass fiber. First, the theory of supercontinuum generation is briefly described, outlining the major nonlinear optical processes contributing to it. After that, the numerical method to simulate SCG is discussed. Finally, the paper presents some major findings of simulating SCG using the ChG fiber. The effects of various parameters on SC bandwidth that were studied are input peak power, fiber nonlinearity, fiber loss, pumping regime, and fiber tapering.

2. Supercontinuum Generation Theory

Pulse propagation in fibers, in the z-direction, is modeled by solving the Generalized Nonlinear Schrödinger equation (GNLSE), Eq. (1) [15]. Here, $A(z)$ is the slowly varying pulse envelope and T is time in retarded frame of reference. The fiber loss is represented as α.

$$
\frac{\partial A}{\partial z} + \frac{\alpha}{2}A + \sum_{m \geq 2} \frac{i^{m-1}}{m!} \beta_m \frac{\partial^m A}{\partial T^m}
$$

$$
= i\gamma \left(1 + \frac{i}{\omega_0}\frac{\partial}{\partial T}\right) \left(A \int_{-\infty}^{+\infty} R(T')|A(z, T - T')|^2 dT'\right). \tag{1}
$$

β_m are the various propagation constants, obtained by Taylor expanding the fiber chromatic dispersion, $\beta(\omega)$, as shown in Eq. (2). Consider that the higher order terms improve the simulation output. This quantity governs the spreading of pulses while propagating inside the medium.

$$
\beta(\omega) = \beta_0 + (\omega - \omega_0)\beta_1 + \frac{1}{2}(\omega - \omega_0)^2 \beta_2 + \frac{1}{6}(\omega - \omega_0)^3 \beta_3 + \cdots. \tag{2}
$$

The Kerr nonlinearity is written as γ, given by Eq. (3). This depends on the intensity-dependent refractive index $n_2(\omega_0)$, which causes the refractive index of the fiber to increase due to high intensity of the incident light. Nonlinear phenomena such as self-focusing, Self-phase Modulation (SPM), Cross-phase Modulation (XPM), etc. are mediated by this term.

$$
\gamma = \frac{2\pi n_2(\omega_0)}{\lambda_0 A_{eff}(\omega_0)}. \tag{3}
$$

Other than the material properties, the fiber nonlinearity also depends on the dimensions of the propagating material. The effective mode area $A_{eff}(\omega_0)$, Eq. (4), for an input wavelength λ_0 is the parameter which can be adjusted by choosing an appropriate waveguide design.

$$
A_{eff} = \frac{\left(\iint |F(x,y)|^2 dxdy\right)^2}{\iint |F(x,y)|^4 dxdy}. \tag{4}
$$

In the simulation it is calculated by a Finite Difference Time Domain (FTDT) solver [19]. In the equation $F(xy)$ represents the transverse mode of the propagating EM-field, $E(x, y, z) = F(x, y)A(z)$, where $A(z)$ is the longitudinal propagating mode mentioned earlier.

The response function $R(T')$ has two components, namely the electronic, $\delta(T')$, (instantaneous) and the vibrational, $h(T')$, (delayed) responses. Mathematically, $R(T') = (1 - f_R)\delta(T') + f_R h(T')$. The fractional Raman contribution factor f_R signifies the strength of each component of the transient response. This term is non-zero for short pulses, as is the case for SCG. This is because ultrashort pulse propagation involves the creation of optical phonons, which is related to Raman scattering. The oscillations of ions due to optical phonons are typically of terahertz frequencies. Thus, for ultrashort pulse propagation, which involves terahertz frequencies, $R(T')$ simplifies down to the Raman response function $h_R(T)$, Eq. (5).

$$h_R(T) = \frac{(\tau_1^2 + \tau_2^2)}{\tau_1 \tau_2} \sin\left(\frac{T}{\tau_1}\right) \exp\left(-\frac{T}{\tau_2}\right). \tag{5}$$

This is a sinusoidal function involving the two parameters, inverse phonon frequency τ_1 and phonon lifetime τ_2.

Most of the studies in this paper are done by pumping the laser in the normal dispersion regime. This means that the fiber's core and cladding diameters are carefully adjusted so that it exhibits normal dispersion at the pumping wavelength. Supercontinuum generation in the normal dispersion regime is primarily initiated by coherent processes such as Self-Phase Modulation (SPM) and optical wave-breaking. SPM occurs when the intensity of an optical pulse modulates its own phase via the Kerr effect in a nonlinear medium [16,17]. This modulation leads to spectral broadening of the pulse, creating a continuum of wavelengths. Optical wave-breaking, on the other hand, refers to the phenomenon where the leading edge of the pulse travels faster than the trailing edge due to group velocity dispersion. This results in the pulse breaking up into smaller pulses, each contributing to the broadened spectrum. These coherent processes collectively contribute to the generation of a broad spectrum spanning across the mid-infrared region. In addition to coherent processes, incoherent mechanisms such as Stimulated Raman Scattering (SRS) and Four-Wave Mixing (FWM) also play significant roles in mid-infrared supercontinuum generation. SRS involves the interaction between optical pulses and the vibrational modes of the medium, leading to the transfer of energy from the pump to the Stokes and anti-Stokes wavelengths, thereby broadening the spectrum. FWM, on the other hand, is a nonlinear optical process where multiple input waves interact within a medium to generate new frequencies through the nonlinear susceptibility of the material. In the context of supercontinuum generation, FWM contributes to spectral broadening by generating new frequencies through the interaction of different spectral components of the input pulse [15,17]. These incoherent processes further enhance the spectral bandwidth of the generated supercontinuum, facilitating its application across a wide range of mid-infrared spectroscopic techniques and technologies.

3. Simulation of Broadband Supercontinuum

Broadband supercontinuum generation is often simulated using the Split-Step Fourier Method (SSFM), a powerful numerical technique for solving the Generalized Nonlinear Schrödinger Equation (GNLSE) [15]. SSFM is a numerical algorithm widely employed for modeling nonlinear propagation phenomena in optical fibers. It discretizes the GNLSE in both space and frequency domains, allowing efficient propagation of optical pulses through nonlinear media by iteratively solving the equation in small steps. To solve the GNLSE using SSFM, the propagation distance is divided into small steps, typically using the split-step approach. At each step, the GNLSE is solved in the frequency domain using the Fast Fourier Transform (FFT), followed by an inverse FFT to return to the time domain. Nonlinear effects such as self-phase modulation, self-steepening, and dispersion are accounted for in each step to accurately simulate the evolution of the pulse [18].

When implementing the GNLSE solver using SSFM in MATLAB, several crucial parameters need to be considered, such as input pulse characteristics: parameters defining the input pulse include its shape, duration (pulse width), peak power, and central wavelength; characteristics of the optical fiber such as dispersion, nonlinear coefficient, length, and attenuation; simulation parameters: step size for SSFM, total propagation distance, number of steps, and spectral and temporal resolution for FFT computations; parameters governing nonlinear effects like the nonlinear coefficient, Raman scattering coefficient, and self-steepening coefficient; specification of the fiber's dispersion profile, including the dispersion coefficient and higher-order dispersion terms.

A hyperbolic secant (sech) pulse is commonly used as the input seed pulse for simulating supercontinuum generation due to its soliton-like characteristics. Sech pulses have a bell-shaped intensity profile and possess a balance between dispersion and nonlinear effects, making them resilient to broadening and distortion during propagation. Additionally, sech pulses can efficiently generate solitons, which play a crucial role in initiating the supercontinuum generation process by stabilizing the pulse against dispersion-induced broadening.

In the simulation for this paper a 2874-nm, 230-fs sech pulse is used as the input seed (see Fig. 1). Figure 2 shows the fiber characteristics. The nonlinear processes are assumed to take place in the microwire section because this is the region where the core diameter is very low. The hybrid section of the fiber, on the other hand, yields high throughput of laser into the fiber core during experiments, and the transition section ensures a smooth propagation of light from the hybrid to the microwire sections. Table 1 summarizes the critical parameters used in the simulation. The code used for the GNLSE program was originally written by Tavers *et al.* [18].

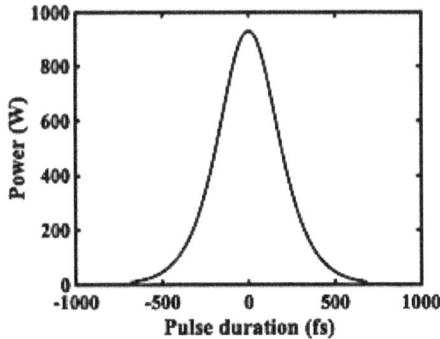

Fig. 1. Input sech 230-fs seed pulse. The input pulse wavelength is 2.87 μm.

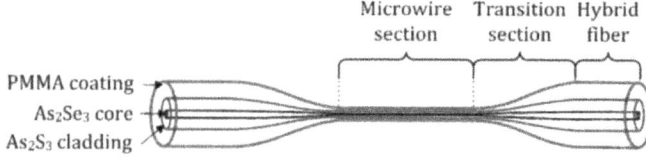

Fig. 2. Highly nonlinear As2Se3-core/As2S3-cladding microfiber.

Table 1. List of the key parameters used in the MATLAB® code.

Parameter	Value
Inverse phonon frequency (τ_1)	23.2 fs
Phonon lifetime (τ_2)	195 fs
Fractional Raman contribution (f_R)	0.10
Pulse-width	230 fs
Input wavelength (λ_0)	2874 nm
Kerr nonlinearity (γ)	2.92 W^{-1}m^{-1}
GVD (β_2)	0.29 ps^2 m^{-1}
Microwire section length	5 cm
Microwire core diameter	3.0 μm

4. Simulation Results

Hudson *et al.* experimentally demonstrated the effect of coupled peak power on the output supercontinuum bandwidth [17]. It was shown that by pumping the fiber shown in Fig. 2 with 2874-nm laser pulses at 4.2-kW peak power, a supercontinuum extending from 1.8-9.5-μm at -20 dB points can be generated. The numerical simulations presented in this section agree with the experimental findings of Hudson *et al.* and further investigate the effects of various other parameters on the SCG bandwidth generated from the same apparatus, numerically.

4.1. *Pump peak power vs. bandwidth*

The effect of pump power on the output SCG bandwidth is presented in Fig. 3. As the coupled peak power is increased from 1.0-kW to 4.2-kW, the output SC bandwidth increases to $\sim$9-μm for -20 dB points, and up to $\sim$11.5-μm for -40 dB points. The tapering off at longer wavelengths occurs due to the decrease in mode confinement at longer wavelengths. The 230-fs input pulse, shown as a blue plot in the figure, peaked at 2874-nm. The nonlinearity and the chromatic dispersion β_2 are kept constant throughout the simulation, at 2.92/W/m and 0.29-ps^2/m, respectively. This and all subsequent simulations show jittery behavior in the shorter wavelength regions since this is where higher orders of chromatic dispersion β_i become dominant. This is an effect seen in simulations only, and not in experimental demonstrations of SCG.

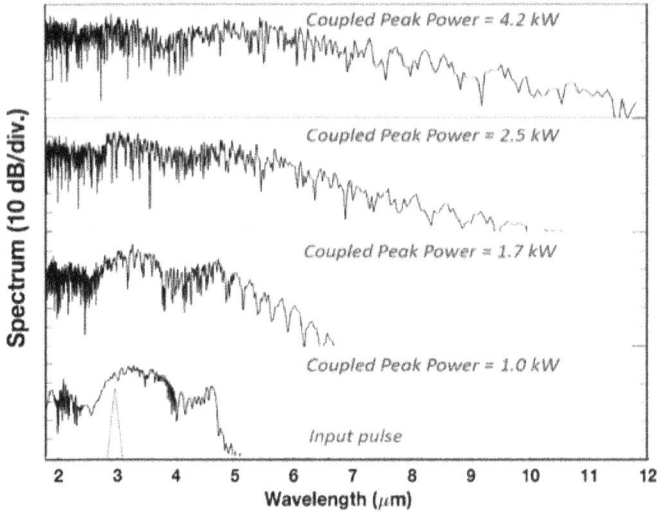

Fig. 3. 1-10 µm bandwidth supercontinua generated as input peak power is increased to 4.2-kW. The input pulse wavelength is 2.87 µm.

4.2. *Fiber length vs. bandwidth*

The effect of fiber length on SCG bandwidth is studied. Due to the high nonlinearity of the chalcogenide fiber, the spectrum bandwidth becomes saturated at relatively short fiber lengths (~1 cm), as shown in Fig. 4. The same input pulse as the one in the previous sub-section is used. The peak power of the input pulse however

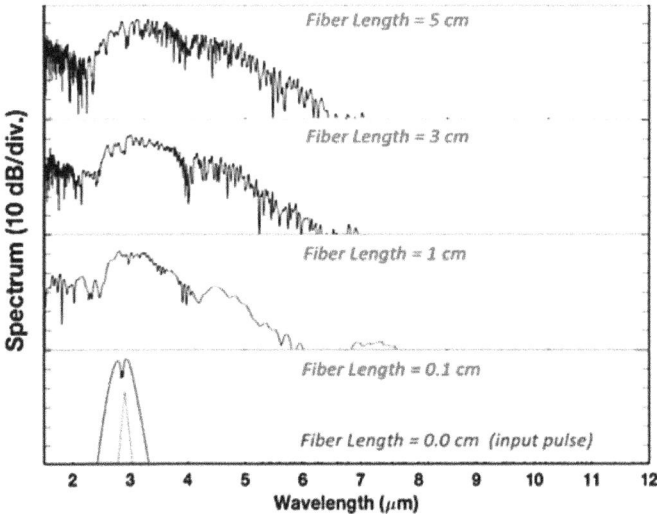

Fig. 4. Effect of fiber length on SCG bandwidth. The input pulse wavelength is 2.87 µm.

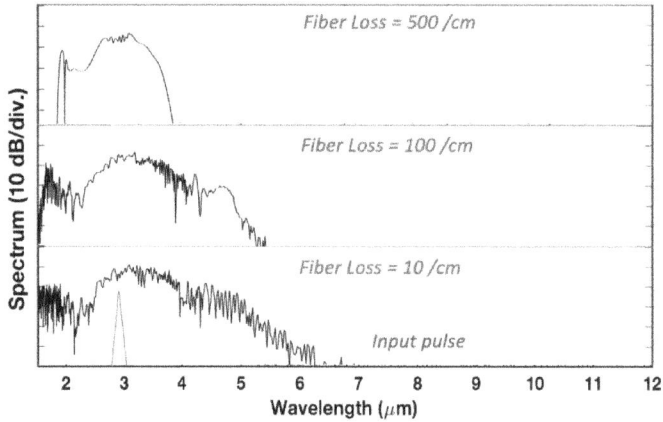

Fig. 5. Effect of fiber loss on SCG bandwidth. The input pulse wavelength is $2.87\,\mu$m.

was reduced to 2.0-kW and kept constant. The $-20\,$dB point bandwidth is seen to extend from $\sim$1.5 to $\sim$5.5-μm.

4.3. *Fiber loss vs. bandwidth*

The effect of fiber loss on bandwidth is important at longer wavelengths for chalcogenide glass fibers. For silica glass fibers, which is transparent upto only $\sim$4.8 μm, the losses become very high at long wavelengths, rendering them unsuitable for generating mid-infrared (MIR) or far-infrared (FIR) supercontinua. Figure 5 shows what happens to the SCG bandwidth when high losses are introduced to the chalcogenide glass fiber. The fiber length, nonlinearity, chromatic dispersion, and the input peak power (at 2-kW) were kept constant throughout this simulation.

4.4. *Pumping in the anomalous dispersion regime*

Figure 6 shows the fiber dispersion for the anomalous dispersion regime (blue) as well as the all-normal dispersion regime used in the rest of the simulations

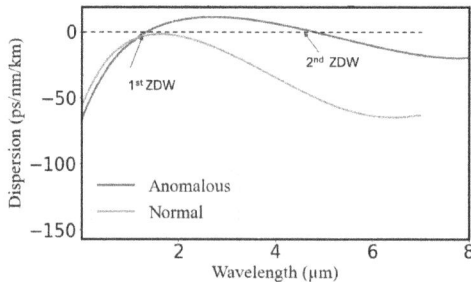

Fig. 6. Dispersion vs. wavelength plot for the all-normal (orange) and anomalous (blue) dispersion pumping regimes, with the two ZDWs shown.

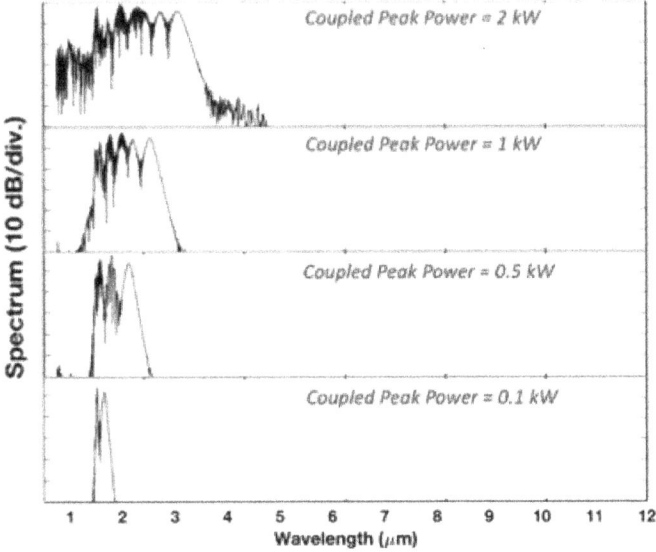

Fig. 7. SCG Bandwidth while pumping in the anomalous dispersion regime. The input pulse wavelength is $1.55\,\mu$m.

(orange). The effect of seeding the fiber setup with 1550-nm sech pulses is simulated in this section. For this pump wavelength the fiber was designed to exhibit anomalous dispersion. In this regime SCG is driven by a two-step process. Due to the high nonlinearity, the unstable input sech pulse, called higher order soliton, undergoes soliton fission. First, the breakdown products, called fundamental or lower order solitons, red shift to the second zero-dispersion wavelength (ZDW) through the Raman shift. Phase matching condition allows dispersive waves (DW) to be generated across the second ZDW in the normal dispersion regime. The second step involves, thereby, the transfer of energy of the solitons to the DWs and spectral recoil effect stabilizing the frequencies. A constant chromatic dispersion β_2 value of $-1.14\,\mathrm{ps}^2/\mathrm{m}$ was used to simulate the supercontinuum generation in this case. The nonlinearity and other fiber parameters were unchanged. The input peak power increased up to 2-kW, where the SCG bandwidth is maximum. Figure 7 shows the simulated output SCG bandwidth. In the figure the prominent dip in power shows the various distinct components of the spectrum generated due to solitons and dispersive waves [20–22]. The ZDWs are fixed in the fiber setup used for this study, thus limiting the output maximum bandwidth to $\sim 4\,\mu$m.

4.5. *Pulse propagation in tapered chalcogenide fiber*

The effect of tapering the fiber on SCG bandwidth is explored in this section. Figure 8 shows a tapered fiber which has a linear decrease in core and cladding radii as the pulse propagates along the fiber. This results in a decrease in β_2 from

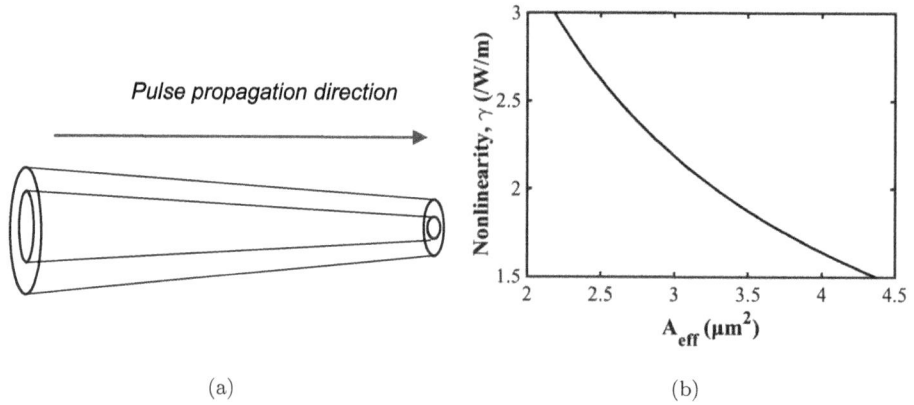

Fig. 8. (a) Tapered micro-structured optical fiber. (b) Decreasing effective mode area increases nonlinearity.

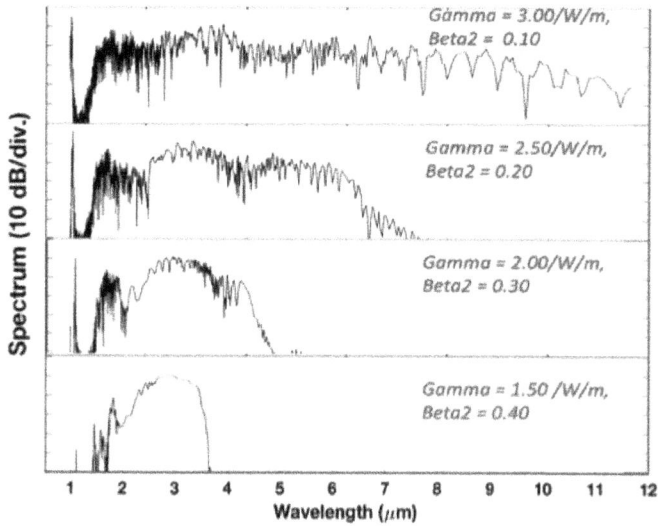

Fig. 9. Effect of varying fiber taper on SCG bandwidth. The input pulse wavelength is $2.87\,\mu$m.

0.40 to $0.10\,\mathrm{ps}^2/\mathrm{m}$ and an increase in γ from 1.50 to $3.00/\mathrm{W/m}$. The down-tapering of the radius decreases the effective mode area A_{eff}, which is responsible for the increasing Kerr nonlinearity. Figure 9 shows the simulation result of propagating a 2-kW peak power seed pulse through the tapered fiber. The bandwidth is seen to be maxed out at $\sim$12-μm which is similar to the un-tapered case discussed earlier. Therefore, the tapering of the fiber does not provide a significant advantage over un-tapered fiber when pumped in the all-normal dispersion regime (see Fig. 6). This can be explained by the fact, that unlike the anomalous dispersion, there is no transfer of energy by the solitons to dispersive waves (DW) across a zero dispersion wavelength (ZDW) in this regime.

5. Conclusion

This paper explores the simulation of mid-infrared supercontinuum generation within tapered chalcogenide fibers, demonstrating precise bandwidth control through manipulation of input parameters. By utilizing 230-fs secant pulses at a power of 4.2-kW, the study achieves a broadband supercontinuum spanning wavelengths from approximately 1 to 10-μm. Through extensive investigation of the effect of fiber nonlinearity, group velocity dispersion, input power, fiber loss, pulsewidth, and different pumping regimes, the paper underscores the significance of parameter optimization for broadband laser sources. These findings have implications for enhancing spectroscopic techniques, optical coherence tomography, and telecommunications, offering opportunities for improved performance and versatility in various applications. The research contributes to the understanding of mid-IR supercontinuum generation, providing insights into fundamental mechanisms and guiding the development of advanced laser sources with broad spectral coverage.

ORCID

Ashiq Rahman ⬡ https://orcid.org/0000-0001-5495-5636
Niloy K. Dutta ⬡ https://orcid.org/0009-0003-3859-0747

References

1. Langridge JM, Laurila T, Watt RS, Jones RL, Kaminski CF, Hult J. Cavity enhanced absorption spectroscopy of multiple trace gas species using a supercontinuum radiation source. Opt Express. 2008;16:10178–10188.
2. Kuri T, Toda H, Olmos JJV, Kitayama K. Reconfigurable dense wavelength-division-multiplexing millimeter-waveband radio-over-fiber access system technologies. J Lightwave Technol. 2010;28:2247–2257.
3. Begum F, Namihira Y. Design of supercontinuum generating photonic crystal fiber at 1.06, 1.31 and 1.55 μm wavelengths for medical imaging and optical transmission systems. Nat Sci. 2011;3:401–407. https://doi.org/10.4236/ns.2011.35054.
4. Wang Y, Dai S. Mid-infrared supercontinuum generation in chalcogenide glass fibers: A brief review. PhotoniX. 2021;2:9. https://doi.org/10.1186/s43074-021-00031-3.
5. Yang Z, Gulbiten O, Lucas P, Luo T, Jiang S. Long-wave infrared-transmitting optical fibers. J Am Ceram Soc. 2011;94(6):1761–1765. https://doi.org/10.1111/j.1551-2916.2010.04313.x.
6. Shiryaev VS, Churbanov MF. Trends and prospects for development of chalcogenide fibers for mid-infrared transmission. J Non-Cryst Solids. 2013;377:225–230.
7. Cui S, Chahal R, Boussard-Plédel C, Nazabal V, Doualan JL, Troles J, *et al.* From selenium- to tellurium-based glass optical fibers for infrared spectroscopies. Molecules. 2013;18(5):5373–5388. https://doi.org/10.3390/molecules18055373.
8. Slusher RE, Lenz G, Hodelin J, Sanghera J, Shaw LB, Aggarwal ID. Large Raman gain and nonlinear phase shifts in highpurity As2Se3 chalcogenide fibers. J Opt Soc Am B. 2004;21(6):1146–1155. https://doi.org/10.1364/JOSAB.21.001146.

9. Pelusi MD, Ta'eed VG, Fu L, Magi E, Lamont MRE, Madden S, *et al.* Applications of highly-nonlinear chalcogenide glass devices tailored for high-speed all-optical signal processing. IEEE J Sel Top Quant. 2008;14(3):529–539.

10. Baker C, Rochette M. Highly nonlinear hybrid AsSe-PMMA microtapers. Opt Express. 2010;18(12):12391–12398. https://doi. org/10.1364/OE.18.012391.

11. Russell P. Photonic crystal fibers. Science. 2003;299(5605):358–362. https://doi.org/ 10.1126/science.1079280.

12. Renversez G, Kuhlmey B, McPhedran R. Dispersion management with microstructured optical fibers: Ultraflattened chromatic dispersion with low losses. Opt Lett. 2003;28(12):989–991. https://doi.org/10.1364/OL.28.000989.

13. Schuster K, Kobelke J, Grimm S, Schwuchow A, Kirchhof J, Bartelt H, *et al.* Microstructured fibers with highly nonlinear materials. Opt Quant Electron. 2007;39(12):1057–1069. https://doi.org/10.1007/s11082-007-9161-x.

14. Wang P, Huang J, Xie S, Troles J, Russell PS. Broadband mid-infrared supercontinuum generation in dispersionengineered As2S3-silica nanospike waveguides pumped by $2.8\,\mu m$ femtosecond laser. Photon Res. 2021;9(4):630–636. https://doi.org/10.1364/PRJ.415339.

15. Agrawal GP. *Nonlinear Fiber Optics.* Academic Press, 2019.

16. Wang Y, Dai S, Li G, Xu D, You C, Han X, *et al.* 1.4–$7.2\,\mu m$ broadband supercontinuum generation in an As-S chalcogenide tapered fiber pumped in the normal dispersion regime. Opt Lett. 2017;42(17):3458–3461. https://doi.org/1 0.1364/OL.42.003458.

17. Hudson DD, Antipov S, Li L, Alamgir I, Hu T, Amraoui ME, Messaddeq Y, Rochette M, Jackson SD, Fuerbach A. Toward all-fiber supercontinuum spanning the mid-infrared. Optica. 2017;4(10):1163. https://doi.org/10.1364/optica.4.001163.

18. Dudley JM, Taylor JR (Eds.) *Supercontinuum Generation in Optical Fibers.* Cambridge: Cambridge University Press, 2010.

19. Fallahkhair AB, Li KS, Murphy TE. Vector finite difference mode solver for anisotropic dielectric waveguides. J Lightwave Technol. 2008;26:1423–1431.

20. Dutta NK. *Fiber Amplifiers and Fiber Lasers.* World Scientific, 2008.

21. Railing L, Thapa S, Zhang X, Dutta N. Generating supercontinuum in dispersion varying As2S3 waveguides. Proc. SPIE 11000, Fiber Optic Sensors and Applications XVI, 110000A (14 May 2019). https://doi.org/10.1117/12.2518245.

22. Zhang X. Study of supercontinuum and high repetition rate short pulse generation (2018). Doctoral Dissertations. 1834. https://digitalcommons.lib.uconn.edu/dissertations/1834.

Temperature Effect Assessment on the Gate-All-Around Junctionless FET for Bio-Sensing Applications[#]

Billel Smaani [*,¶], Samir Labiod[†,‖], Mohamed Salah Benlatreche [*,**],
Boudjemaa Mehimmedetsi [*,††], Ramakant Yadav[‡,‡‡]
and Husien Salama[§,§§]

*Abdelhafid Boussouf University Centre of Mila, Mila, Algeria

†Université 20 Aout 1955, BP 26 Route El Hadaik,
Skikda, Algeria

‡Electrical & Electronics Engineering Department,
Mahindra University, Hyderabad, India

§Computer Systems Institute, Boston, USA
¶billel.smaani@gmail.com
‖s.labiod@univ-skikda.dz
**msbenlatreche@centre-univ-mila.dz
††b.mehimmedetsi@centre-univ-mila.dz
‡‡ramakant@mahindrauniversity.edu.in
§§husien.salama@uconn.edu

The gate-all-around junctionless field-effect transistor (GAA JL FET)-based biosensor has recently attracted worldwide attention due to its good sensitivity to gate-all-around architecture and overall conduction mechanism. The effect of temperature usually affects the performance of transistors and sensors. Therefore, the impact of temperature on the 3D GAA JL FET-based biosensor has been investigated in this work. The dielectric modulation (DM) approach has been considered for including biomolecules. Consequently, the main proprieties of this biosensor have been investigated by ranging the temperature from 77 K to 400 K. The simulated results showed that the on-state current lowers as the temperature rises, but the off-state current increases. The off-current variation concerning the temperature is higher than the on-current change. Also, this type of biosensor appears to have a finer threshold voltage. Furthermore, the obtained results reveal that the current sensitivity is increased when ranging from temperature from 200 K to 400 K, and deteriorates for lower temperature values, like 100 K and 77 K. In addition, the GAA JL FET-based biosensor is more reliable for the detection of neutral biomolecules at high temperatures.

Keywords: Biosensor; FET; junctionless; gate-all-around; temperature; neutral bio-molecule; sensitivity.

¶Corresponding author.
#This chapter appeared previously on the International Journal of High Speed Electronics and Systems. To cite this chapter, please cite the original article as the following: B. Smaani, S. Labiod, M. S. Benlatreche, B. Mehimmedetsi, R. Yadav and H. Salama, *Int. J. High Speed Electron. Syst.*, **33**, 2440061 (2024), doi: 10.1142/S0129156424400615.

1. Introduction

In recent decades, biosensor technology based on field-effect transistors (FETs) has attracted much attention worldwide for many applications in different fields, such as food safety and medical applications [1]. FET-based biosensors are designed on a bio-recognition layer and a transducer element. Moreover, this type of FET-based biosensors shows unique and diverse properties, such as a good level of selectivity, good accuracy, excellent sensitivity, simplicity of surface functionalization, low power of operation, low cost of fabrication processes, and simple integration into wearable devices and portable applications [2]. Furthermore, there are rising demands to create several clinical laboratories in the rural and military sectors for label-free application, especially for the detection and estimation of neutral biomolecules with various types. Therefore, field-effect transistors-based biosensors can detect with good accuracy the molecule species and should have an adequate and small size to offer a high sensing level [3].

On the other hand, device down-scaling has allowed various features, such as high performance and low power consumption [4]. Nevertheless, the aggressive down-scaling increased the impact of short-channel effects (SCEs) and decreased the biosensor's sensitivity [5]. Also, the classical FET-based biosensor suffers from non-idealities that consider limited linearity range, temperature drift, noise, and hysteresis. To overcome this drawback, the classical MOSFET should be replaced by three-dimensional (3D) devices, such as gate-all-around junctionless field-effect transistor (GAA JL FET), Nanosheet, and FinFET, which provide better and higher performance than the conventional MOSFET devices [3, 6–8].

The GAA JL FET is a remarkable device that integrates the benefits of the gate-all-around (GAA) and the junctionless (JL) concept. The GAA JL FET architecture offers superior gate controllability and reduces the impact of SCEs [6]. The JL FETs have only one uniform doping from source to drain, and no ultrathin junctions are included, which increases immunity to intrinsic noise, and improves the device's I_{ON}/I_{OFF} ratio. Besides, the fabrication process of JL FETs is simple because it does not need an elevated thermal budget when the doping is uniform. These various qualities of JL FET are advantageous for the fabrication process of miniaturized sensors and their heterogeneous incorporation with other devices and components for non-invasive clinical diagnostics, and different available applications [8]. For further improvement of the sensitivity parameter, the GAA JL FET has been implemented as a biosensor device [9].

Several works have been carried out in non-conventional structures-based biosensor applications, such as dual-gate [10], triple-gate [11], and omega-gate structures [12]. However, a few researchers have investigated the junctionless gate-all-around FET-based bio-sensing application [3].

The use of the gate-all-around junctionless FET-based biosensor for detecting biomolecule species is sensitive to the variation and change in temperature since the device's semiconductor behavior is related to temperature changes, which degrades the reliability of the sensor output [13]. Subsequently, in this paper, we have investigated the effect of temperature change on the key parameters and figures of merit in GAA JL FET-based biosensors for the detection and estimation of biomolecule species, for the first time.

The rest of the paper is well organized as follows: Section 2 presents the biosensor structure and setup simulation of the JL GAA FET-based biosensor. Section 3 provides the results of the temperature effect on the key parameters of GAA JL FET-based biosensors including different types of neutral biomolecules. Furthermore, the impact of temperature change on the biosensor's sensitivity has also been analyzed in Sec. 3. Finally, Sec. 4 concludes the realized work.

2. Bio-Sensor Topology and Simulation Approach

The GAA JL FET-based biosensor designed and investigated in this work has been fully simulated at different temperatures using Silvaco three-dimensional (3D) technology-computer-aided designs (TCAD) [14]. The schematic view of the proposed GAA JL FET-based biosensor structure is shown in Fig. 1. The gate-length L_{gate} and the body radius R is set at 50 nm and 5 nm, respectively. The channel/source/drain regions are homogeneously doped N_D at 1×10^{19} cm^{-3}. The nanogap cavity length L_{cavity} and cavity width T_{cavity} of simulated 3D GAA JL FET-based biosensor are 50 nm and 4 nm, respectively. An oxide layer has been implemented since there is no reaction with the bimolecular species.

In this work, we have considered the Aptest biomolecule, in the first step. The implementation of neutral biomolecule species is performed using the dielectric-modulation approach [15].

In TCAD simulations, we employ the Auger and Shockley Read Hall model with Boltzmann transport to account for the recombination of carriers. To include both perpendicular and parallel field-dependent mobility, we incorporate the CVT mobility model. Also, we include the concentration-dependent mobility (CONMOB),

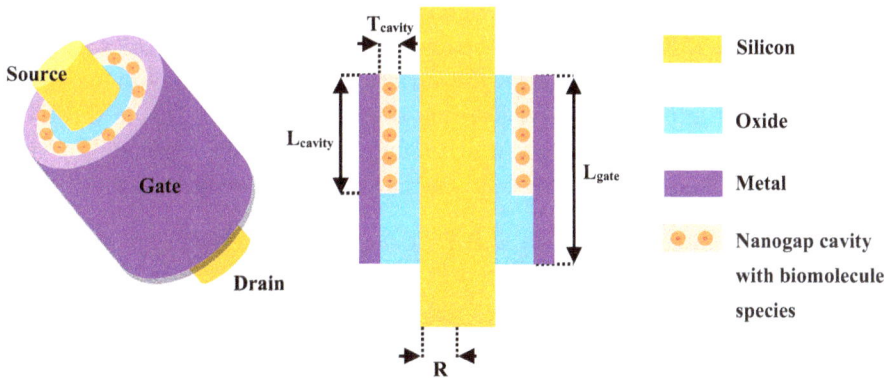

Fig. 1. (a) 3D schematic view and (b) cross-sectional view of the 3D GAA JL FET-based biosensor.

Table 1. Details of designed and simulated 3D GAA JL FET-based biosensor.

Symbol	Parameter	Value
T	Temperature (K)	77–400
L_{cavity}	Cavity length (nm)	50
T_{cavity}	Cavity width (nm)	4
N_D	Channel/source/drain doping (cm^{-3})	1.0×10^{19}
L_{gate}	Transistor gate length (nm)	50
R	Transistor channel radius (nm)	5
Φ_{MS}	Metal work-function (eV)	4.8
$L_{S/D}$	Source/drain length (nm)	10

the band-gap narrowing (BGN), and the carrier-scattering (CCSMOB) model since junctionless devices use high doping concentration. The energy balance transport model is included in TCAD simulations to capture the non-local transport effects, like the dependence of ionization's effect on carrier distribution energy and velocity overshoot phenomena. The impact of the quantization effect is less important since the considered Silicon body thickness is greater than 10 nm, and the device's gate length is greater than 10 nm. For these reasons, the models of quantum correction were not included during the device's simulation.

Table 1 indicates the design parameters of the GAA JL FET biosensor. It should be noted that the stated units are the same as the international system of units (SI) for thermoelectric; K = Kelvin, A = Ampere, W = Watt, nm = nanometer, m = meter.

3. Results and Discussions

The presence of Aptes biomolecules is simulated by including a material with a dielectric constant equal to 3.57, Simulated transfer characteristics (I_{DS} vs. V_{GS}) of 3D GAA JL FET-based biosensor for temperature range 77 K to 400 K are shown in Fig. 2. We can see that the drain current (I_{DS}) reduces when increasing temperature. As shown in Fig. 3(a), the off-current is increased by ×100 when increasing the temperature from 77 K to 400 K, and the on-current decreases by ÷1.67 with the increase of the temperature. Therefore, the variation of the off-current is higher than the variation of the on-current (concerning the temperature). This is due to the presence of the biomolecule species, which reduces the mobility through the carrier scattering effect. And the sub-threshold current can be expressed as [3]

$$I_{subt} = \frac{2\pi R \mu_{effe} n_i k_b T \left(1 - \exp\left(\frac{-qV_{ds}}{k_b T}\right)\right)}{\int_0^L \frac{1}{\int_0^a \exp\left(\frac{q\varphi(r,z)}{k_b T}\right) dr} dz}, \tag{1}$$

where μ_{effe} corresponds to the effective mobility, n_i is the intrinsic carrier, T is the thermal temperature, and k_b is Boltzmann's constant. R is the channel radius, q is the universal electron charge, V_{ds} is the drain voltage, and L is the channel radius.

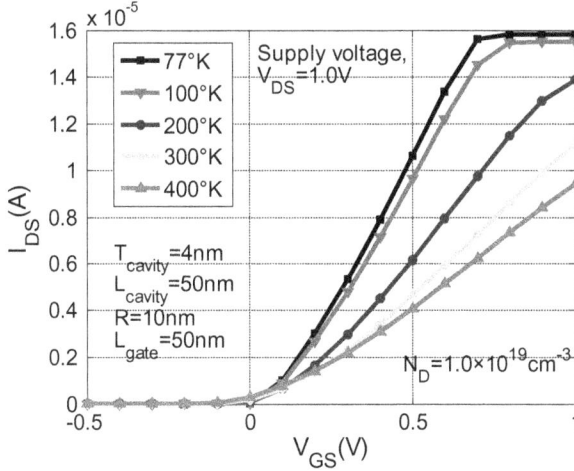

Fig. 2. Impact of varying the temperature from 77 K–400 K on the transfer characteristics (IDS vs. VGS) of GAA JL FET-based biosensors.

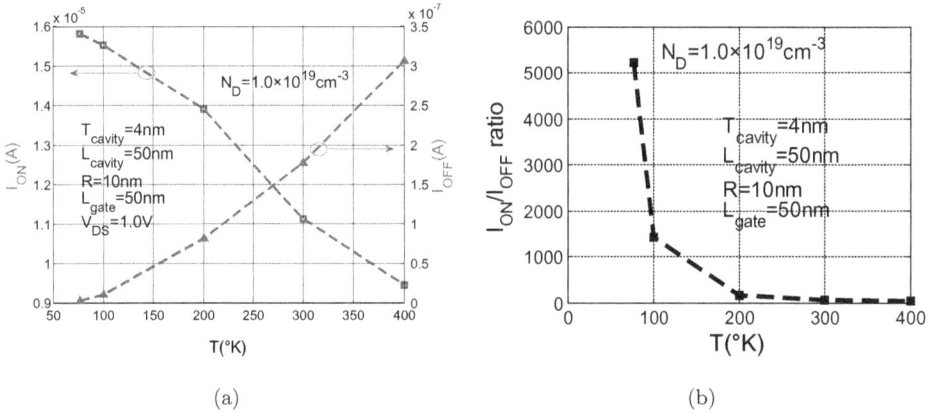

(a)

(b)

Fig. 3. (a) Variation of on-current (I_{ON}) and off-current (I_{OFF}), and (b) variation of I_{ON}/I_{OFF} ration on GAA JL FET-based biosensors for temperature range 77 K–400 K.

The variation of the I_{ON}/I_{OFF} ratio of GAA JL FET-based biosensors ranging from 77 K to 400 K is shown in Fig. 3(b). We can see the decrease in of I_{ON}/I_{OFF} ratio since I_{ON} values of the device are decreasing and they are higher than I_{OFF} values.

The impact of temperature variation on the threshold voltage (V_{TH}) of the GAA JL FET-biosensor is shown in Fig. 4. In this context, the V_{TH} parameter varies from 0.0 V to –0.2 V for a temperature-wide range of 77 K to –400 K, as shown in Fig. 4. This remarkable variation in V_{TH} is an indication of sensitive response of the biosensor in hazardous environment temperature and reliable response.

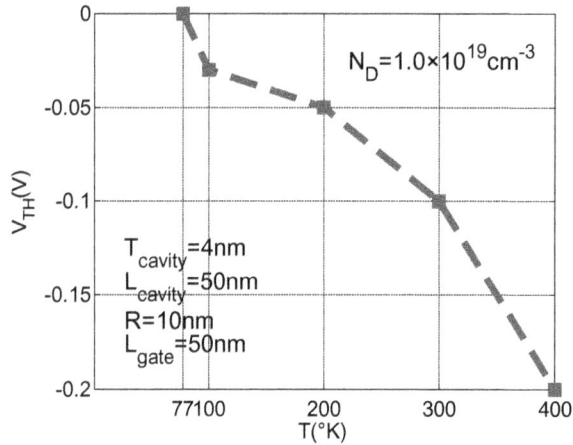

Fig. 4. Variation of the threshold voltage (V_{TH}) in GAA JL FET-based biosensors for temperature (T) range 77 K–400 K.

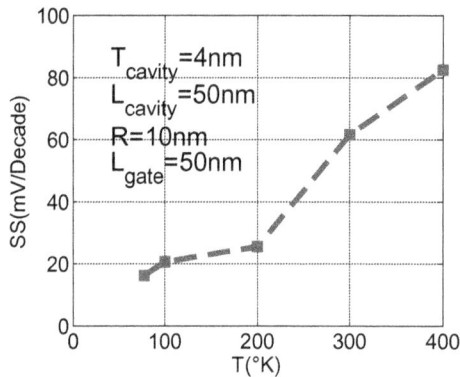

Fig. 5. Variation of the sub-threshold slope (SS) in GAA JL FET-based biosensors for temperature (T) range 77 K–400 K.

As shown in Fig. 5, when the temperature is equal to 300 K, the sub-threshold slope (SS) is ~61 mV/Decade, but by dropping the temperature from 300 K to 77 K, the SS parameters are strongly reduced to 16 mV/Decade. Moreover, for higher temperatures, such as 400 K, the SS parameter value is increased to 82 mV/Decade. In addition, the sub-threshold slope is deteriorated and reduced by 26% when ranging temperature from 300 K to 77 K. The SS parameter is also increased by 134% for the temperature range 300 K to 400 K.

The influence of varying the temperature values from 77 K to 400 K on the transconductance (g_m) and conductance (g_{ds}) of GAA JL FET-based biosensors is shown in Figs. 6 and 7, respectively. It is observed that g_m and g_{ds} parameters are lowered when the temperature is increased from 77 K to 400 K. The trans-conductance

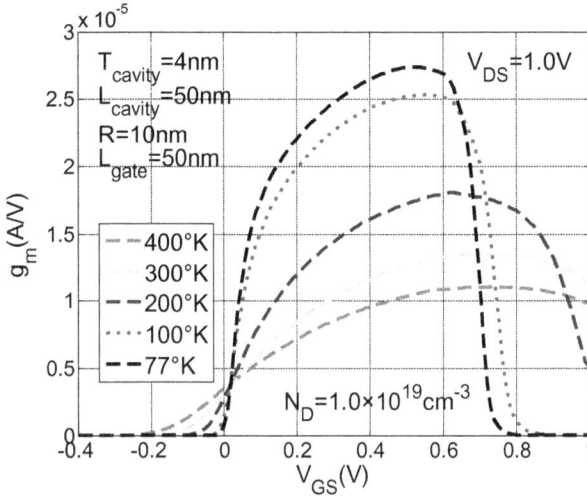

Fig. 6. Impact of varying the temperature from 77 K to 400 K on transconductance (g_m vs. V_{GS}) of GAA JL FET-based biosensors.

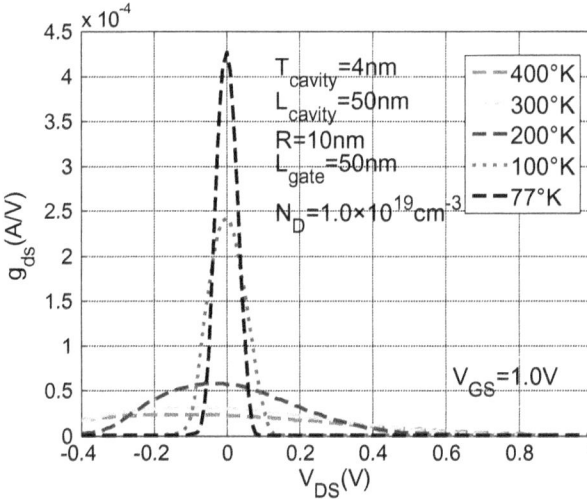

Fig. 7. Impact of varying the temperature from 77 K to 400 K on conductance (g_{ds} vs. V_{DS}) of GAA JL FET-based biosensors.

usually describes the drain-current variation concerning the gate-to-source voltage change. So, lower trans-conductance reduces the conversion of input-voltage toward output-current, especially at the deep depletion and accumulation region.

As shown in Fig. 8, decreasing the temperature (T) from 400 to 77 K reduces the conduction band energy (CBE) which favors the passage of free carrier and therefore the increase of the drain current. In this context, the energy band gap

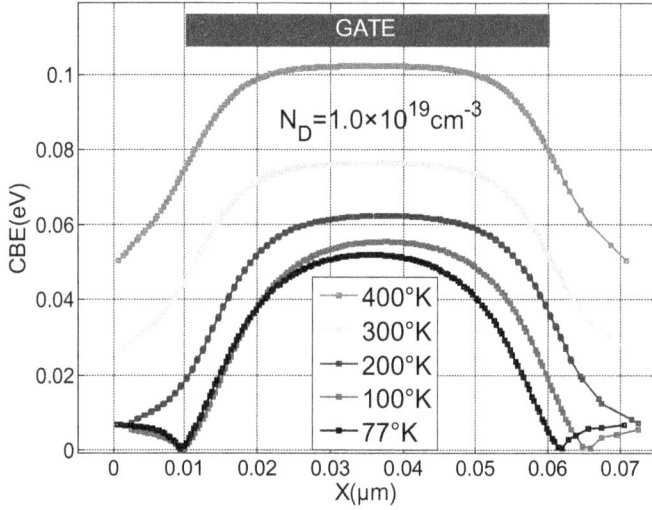

Fig. 8. Variation of the conduction band energy (CBE) along the lateral direction (X) for temperature (T) range 77 K–400 K.

(E_g) concerning the temperature (T) can be analytically described by the following equation [3]:

$$E_g(T) = E_g(300\,\text{K}) - E_{g\alpha}\left(\frac{T^2}{T + E_{g\delta}} - \frac{(300\,\text{K})^2}{E_{g\delta} + 300\,\text{K}}\right) - \Delta E_g(N_D), \qquad (2)$$

where $E_g(300\,\text{K})$ corresponds to the band gap energy at 300 K of temperature. The parameters $E_{g\alpha}$ and $E_{g\delta}$ are equal to 4.73×10^4 eV/K and 636 K, respectively. The amount of the energy band-gap narrowing ΔE_g can be written concerning the body doping concentration, as

$$\Delta E_g(N_D) = \delta_E\left(\left(\left(\ln\left(\frac{N_D}{\delta_N}\right)\right)^2 + \delta_C\right)^{1/2} + \left(\ln\left(\frac{N_D}{\delta_N}\right)\right)\right) \qquad (3)$$

here $\delta_N = 1.3 \times 10^{17}$ cm^{-3}, $\delta_E = 6.92$ meV, and $\delta_C = 0.5$.

The current sensitivity (S_{Ioff}) is a key parameter to analyze and investigate the biosensor's performance. The S_{Ioff} parameter is calculated concerning temperature as follows:

$$S_{Ioff}(T) = \frac{I_{off,\,with\,biomolecule\,species}(T)}{I_{off,\,with\,Air}(T)}, \qquad (4)$$

where $I_{off,\,with\,biomolecule\,species}$ and $I_{off,\,with\,air}$ correspond to the off-current with biomolecule species and the off-current with Air, respectively.

Figure 9 shows the variation of the current sensitivity parameters with respect to the temperature. It is observed that for a temperature range of 300 K to –400 K, the sensitivity variation is $\sim$66% to $\sim$72%. It is also indicated that for a temperature

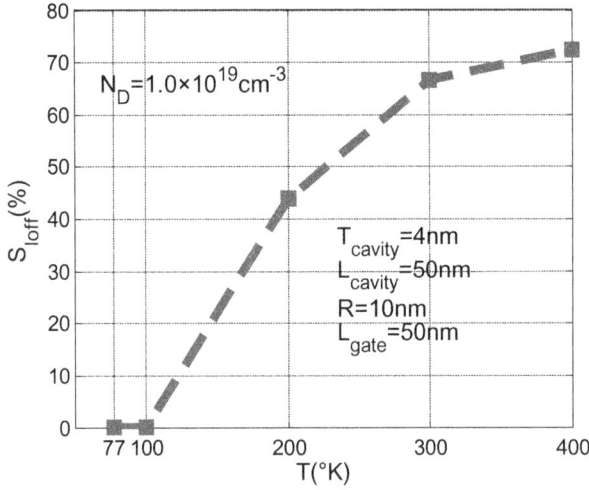

Fig. 9. Impact of varying the temperature (T) on the current-sensitivity S_{Ioff} of the GAA JL FET-based biosensors.

range of 300 K to –200 K, the biosensor's sensitivity is reduced to 44%. In addition, the sensitivity deteriorates when reaching the temperature of 100 K or 77 K.

From a practical point of view, low-temperature value, such as 100 K and 77 K, reduces the device's sensitivity and limits the application of the biosensor for the detection of neutral biomolecule species. The decrease in the current sensitivity is

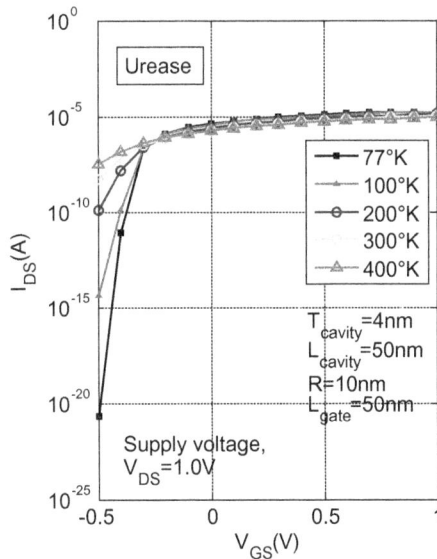

Fig. 10. Impact of varying the temperature from 77 K to 400 K on the transfer characteristics (I_{DS} vs. V_{GS} of GAA JL FET-based biosensors including Urease.

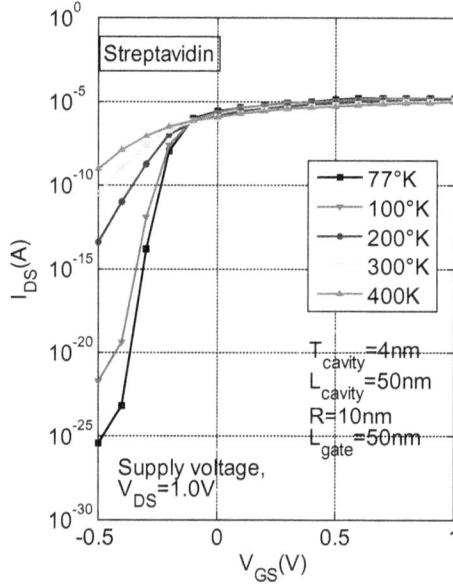

Fig. 11. Impact of varying the temperature from 77 K to 400 K on the transfer characteristics (I_{DS} vs. V_{GS}) of GAA JL FET-based biosensors including Streptavidin.

related to I_{ON} variation and a decrease in SS parameters concerning the temperature, Figs. 3(a) and 5, respectively.

We consider two more different types of neutral biomolecules, which are Urease and Streptavidin. These biomolecules are introduced by including materials with a dielectric constant equal to 1.64 (Urease) and 2.1 (Streptavidin) [3, 15].

We can conclude from Figs. 10 and 11 that the drain-current (I_{DS}) reduces when increasing temperature for Urease and Streptavidin cases, in a similar way as for APTES. The off-current is increased when increasing the temperature and the on-current is decreases with the increase of the temperature. Therefore, the variation of the off-current is higher than the variation of the on-current (concerning the temperature).

4. Conclusion

In this paper, the study of the effect of temperature in 3D GAA JL FET-based biosensors regarding the figures of merit and performance of the devices was carried out comprehensively for the first time through 3D device simulations. We have shown that the dependence of biosensor performance on temperature is maximum, especially in terms of sensitivity parameters, off-state current, and sub-threshold slope. The deterioration of current sensitivity (S_{Ioff}) is significant as the temperature decreases from 400 K to 100 K, and 77 K. The change in off-state current is greater than the change in on-state current (compared to the temperature) due to the presence of biomolecule species. The sub-threshold slope (SS) degrades with the decrease in temperature. The threshold voltage (V_{th}) shift increases with decreasing temperature in JL GAA FET-based biosensors. Therefore, it can be concluded that for lower temperature values (77 K and 100 K), GAA JL FET-based biosensors are not suitable for the sensitivity of the biosensor, but at higher temperatures (400 K), the GAA JL FET-based biosensor is an excellent device and can be used for detection and estimation of various biomolecule species applications. The future direction of this research is to generalize the proposed method to be used in various important applications such as micro stereolithography.

ORCID

Billel Smaani ⊚ https://orcid.org/0000-0002-6494-3708

Mohamed Salah Benlatreche ⊚ https://orcid.org/0000-0001-6169-1114

Boudjemaa Mehimmedetsi ⊚ https://orcid.org/0000-0003-4449-7509

References

1. T. Manimekala, R. Sivasubramanian, G. Dharmalingam, Nanomaterial-based biosensors using field-effect transistors, *A Review. J. Electron. Mater* 51 (2022) 1950–1973, https://doi.org/10.1007/s11664-022-09492-z.
2. R. Ahmad, T. Mahmoudi, M. S. Ahn, Y. B. Hahn, Recent advances in nanowires-based field-effect transistors for biological sensor applications, *Biosens. Bioelectron.* 100 (2018) 312–325, https://doi.org/10.1016/j.bios.2017.09.024.
3. Y. Pratap, M. Kumar, S. Kabra *et al.*, Analytical modeling of gate-all-around junctionless transistor based biosensors for detection of neutral biomolecule species, *J. Comput. Electron.* 17 (2018) 288–296, https://doi.org/10.1007/s10825-017-1041-4.
4. D. Mamaluy, X. Gao, The fundamental downscaling limit of field effect transistors, *Appl. Phys. Lett.* 106 (2015) 193503, https://doi.org/10.1063/1.4919871.
5. R. Narang, K. V. S. Reddy, M. Saxena, R. S. Gupta *et al.*, A dielectric-modulated tunnel-fet-based biosensor for label-free detection: Analytical modeling study and sensitivity analysis, *IEEE Trans. Electr. Dev.* 59(10) (2012) 2809–2817, https://doi:10.1109/TED.2012.2208115.
6. B. Smaani, S. B. Rahi, S. Labiod, Analytical Compact model of nanowire junctionless gate-all-around MOSFET implemented in verilog-A for circuit simulation, *Silicon* 14 (2022), 10967–10976, https://doi.org/10.1007/s12633-022-01847-9.
7. F. Nasri, H. Salama, Numerical investigation of the electrothermal properties of SOI FinFET transistor, *Int. J. High-Speed Electron. Syst.* 32(02–04) (2023) 2350020, https://doi:10.1142/S0129156423500209.
8. S. Tayal *et al.*, Incorporating bottom-up approach into device/circuit co-design for SRAM-based cache memory applications, *IEEE Trans. Electr. Dev.* 69 (11) (2022) 6127–6132, https://doi:10.1109/TED.2022.3210070.
9. K. Nayak *et al.*, CMOS logic device and circuit performance of Si gate all around nanowire MOSFET, *IEEE Trans. Electr. Dev.* 61(9) (2014), 3066–3074, https://doi:10.1109/TED.2014.2335192.

10. I. V. V. Reddy, S. L.Tripathi, Enhanced performance double-gate junction-less tunnel field effect transistor for bio-sensing application, *Solid State Electron. Lett.* 3 (19) (2021), https://doi.org/10.1016/j.ssel.2021.12.005.

11. S. Rashid., F. Bashir , F.A. Khanday, M. R. Beigh, Dual material tri-gate Schottky barrier FET as label free biosensor, *Mater. Today: Proc.* 74 (2023) 344, https://doi.org/10.1016/j.matpr.2022.08.318.

12. K. Keem, D.-Y. Jeong, S. Kim, M.-S. Lee *et al.*, Fabrication and device characterization of omega-shaped-gate ZnO nanowire field-effect transistors, *Nano. Lett.* 6 (7) (2006) 1454–1458, https://pubs.acs.org/doi/10.1021/nl060708x.

13. S. Sinha, R. Bhardwaj, N. Sahu, H. Ahuja *et al.*, Temperature and temporal drift compensation for Al2O3-gate ISFET-based ph sensor using machine learning techniques, *Microelectron. J.* 97 (2020) 104710, https://doi.org/10.1016/j.mejo.2020.104710.

14. Silvaco Device User Guide. [online]. Available: https://silvaco.com/.

15. H. Im, X. J. Huang, B. Gu *et al.*, A dielectric-modulated field-effect transistor for biosensing, *Nat. Nanotech.* 2 (2007) 430–434, https://doi.org/10.1038/nnano.2007.180.

Evaluation of Electrochemical Sensor Using Microfluidic System[#]

A. Legassey[*,†,‡], A. Fleming[§], L. Dagostino[§], T. Bliznakov[§],
R. Gudlavalleti ©[*,‡], J. Kondo[*,‡], F. Papadimitrakopoulos[*,†]
and F. Jain ©[*,‡,¶]

Biorasis Inc., Storrs, CT, USA

†*Department of Chemistry, University of Connecticut,
Storrs, CT, USA*

‡*Department of Electrical Engineering,
University of Connecticut, Storrs, CT, USA*

§*Department of Biomedical Engineering,
University of Connecticut, Storrs, CT, USA*

¶*faquir.jain@uconn.edu.*

This paper presents the characterization and testing of electrochemical sensor using microfluidic system. Various geometric patterns were laser cut into the platinum working electrode of a biosensor. In this work, a microfluidic chamber was designed that allows phosphate buffer solution (PBS) to flow across the sensor, using a peristaltic pump, while varying the concentration of hydrogen peroxide. The amperometric characterization of the electrochemical sensor with 25, 50, 75, and $100\,\mu$m perforation and $75\,\mu$m spacing showed the highest sensitivity. This result was to be expected the purpose of patterning the sensors was to provide a 3-dimensional structure to the planar electrode in order for the enzyme, glucose oxidase, to be immobilized. Future work will include selecting one of the patterns for immobilization of glucose oxidase allowing us to realize a fully functional glucose sensor.

Keywords: Electrochemical sensor; biosensor; microfluidic system; amperometry.

1. Introduction

A biosensor is a device that incorporates a biological sensing element and a transducer to detect and convert a biological analyte into a measurable signal. These analytical devices combine biological recognition elements with transducers to detect and quantify specific biological molecules or substances with exceptional sensitivity

¶Corresponding author.
#This chapter appeared previously on the International Journal of High Speed Electronics and Systems. To cite this chapter, please cite the original article as the following: A. Legassey, A. Fleming, L. Dagostino, T. Bliznakov, R. Gudlavalleti, J. Kondo, F. Papadimitrakopoulos and F. Jain, *Int. J. High Speed Electron. Syst.*, **33**, 2440062 (2024), doi: 10.1142/S0129156424400627.

and selectivity [1]. With their ability to translate biological responses into measurable signals, biosensors have emerged as indispensable tools in various applications, including healthcare, environmental monitoring, food safety, and biodefense [2].

The concept of a biosensor was first proposed in 1962 by Leland C. Clark Jr. and Champ Lyons from the Children's Hospital of Cincinnati. Together they proposed the first enzymatic electrode, known as the Clark electrode or oxygen electrode. This electrode utilized the enzyme glucose oxidase to detect the consumption of oxygen as glucose was oxidized by the enzyme [3]. Since this time, glucose sensors have continued to develop and have played an important role in the management of glucose levels in individuals with Type 1 Diabetes. In particular, amperomeric glucose sensors based on glucose oxidase immobilized upon platinum electrodes have been at the forefront of continuous glucose monitor systems. These systems detect real-time glucose values within the interstitial fluid of individuals by converting glucose into hydrogen peroxide, Eqs. (1) and (2):

$$Glucose + GO_x(FAD) \rightarrow Glucorolactone + GO_x(FADH_2), \quad (1)$$

$$GO_x(FADH_2) + O_2 \rightarrow GO_x(FAD) + H_2O_2. \quad (2)$$

The generated hydrogen peroxide (H_2O_2) is then amperometrically assessed on the surface of the electrode in accordance with Eq. (3) [4]:

$$H_2O_2 \rightarrow 2H^+ + O_2 + 2e^-. \quad (3)$$

This decomposition of hydrogen peroxide generates electrical current in the sensor which is directly proportional to the amount of glucose present in the individuals interstitial fluid. This data can then be used by an individual to make decisions on the management of their glucose levels.

Sensor performance is critical to the success of a biosensor. Ideal sensors should have characteristics such as accurate and precise performance, have high sensitivity and selectivity to the analyte being measured, linearity of the sensor response should be withing physiological levels, and selectivity of the analyte among other characteristics. The purpose of this work is to examine the platinum working electrode of a glucose sensor and assess various laser cut patterns on the performance of such an electrode on the amperometric performance of hydrogen peroxide. This paper evaluates the performance of hydrogen peroxide and future work will evaluate the addition of glucose oxidase in order to evaluate glucose performance upon the selected geometry with best performance on hydrogen peroxide.

2. Fabrication of Platinum Working Electrode

The working electrode of the sensor was fabricated out of $25\,\mu$m thick platinum sheets and laser cut with a femtosecond laser with a resolution of 7-$18\,\mu$m. Five total patterns were selected to compare the hole size effect on sensitivity and diffusion of hydrogen peroxide at the surface of the electrode. Figure 1 shows the basic pattern of the laser cut holes.

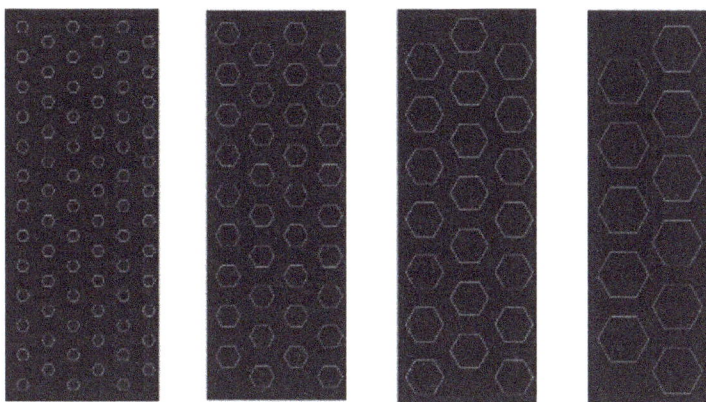

(a)

(b)

Fig. 1. (a) CAD design pattern. Hexagon with side dimension of 25, 50, 75, and 100 μm. (b). Laser cut patterns in Pt of 25, 50, 75, and 100 μm.

The pattern of the electrodes was varied in the dimensions of the hexagon with the size of the hole being determined by the length of the side of the hexagon, holes were cut out measuring 25, 50, 75, and 100 μm. The distance between the holes was not changed to maintain mechanical integrity of the sensor. In addition to the patterns cut out, a planar sensor with no pattern was also fabricated and tested amperiometrically as a control.

2.1. *Fabrication of microfluidic chamber*

To test the sensors a microfluidic chamber was made using Acrylic and water sealing rubber gaskets, Fig. 2.

The microfluidic chamber allows phosphate buffer solution (PBS) to flow across the sensor, flow was kept constant using a peristaltic pump while varying the concentration of hydrogen peroxide. Electrical connectors are fed through the chamber

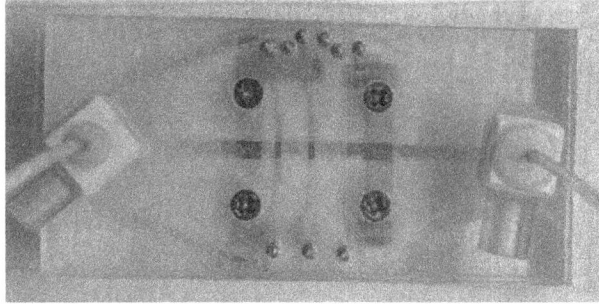

Fig. 2. Microfluidic chamber with gasket and sensor connector fed through the Acrylic.

to allow connection between the sensor and amperometric instrument to test the sensor. The sensor was tested using a Chi Instruments Model 1010B electrochemical analyzer. The sensors were tested amperometrically with an applied voltage of 0.6 V.

3. Characterization of Electorde

Electrochemical testing was performed to assess the response of these electrodes to increasing concentrations of hydrogen peroxide. The setup of this experiment

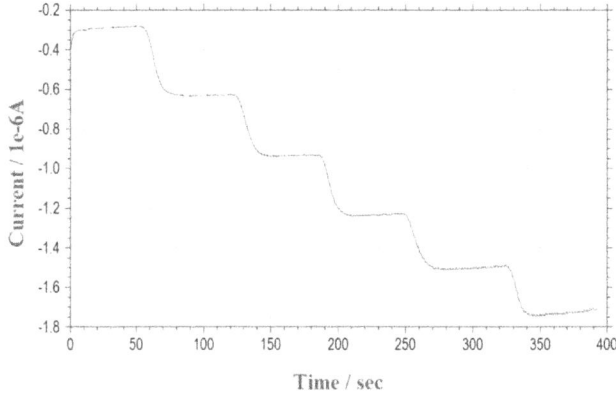

Fig. 3. Amperometry measurement using Chi 1010b instrument for sensor with 25 μm holes.

Table 1. Sensitivity of various hole patterns to hydrogen peroxide concentrations varying between 0-0.1 μm.

Electrode	Sensitivity (μA/μM)
Planar (no holes)	11.3
25 μm	14.4
50 μm	13.3
75 μm	16.9
100 μm	13.6

consisted of a 3-electrode system containing a working, counter, and reference electrode. The 3 electrodes are placed within the microfluidic chamber where PBS solution is pumped through at $37°$ C and hydrogen peroxide concentration is varied from 0-0.1 mM. The current-time plot for each successive addition of hydrogen peroxide is shown in Fig. 3. The sensitivity values obtained for each sensor is given in Table 1.

4. Conclusion

The sensor with $75\,\mu$m holes had the highest sensitivity of the sensors tested. However, no appreciable difference was seen between the various sensor designs. While this result was to be expected the purpose of patterning the sensors was to provide a 3-dimensional structure to the planar electrode in order for the enzyme, glucose oxidase, to be immobilized. Future work will include selecting one of the patterns for immobilization of glucose oxidase allowing us to realize a fully functional glucose sensor.

ORCID

R. Gudlavalleti ◉ https://orcid.org/0000-0002-7727-8030

F. Jain ◉ https://orcid.org/0000-0003-3961-6665

References

1. Ivnitski, D., Abdel-Hamid, I., Atanasov, P., & Wilkins, E. (1999). Biosensors for detection of pathogenic bacteria. *Biosensors and Bioelectronics*, 14(7), 599–624.
2. Turner, A. P., Karube, I., & Wilson, G. S. (1987). *Biosensors: Fundamentals and Applications*. Oxford University Press.
3. Clark, L. C., & Lyons, C. (1962). Electrode systems for continuous monitoring in cardiovascular surgery. *Annals of the New York Academy of Sciences*, 102(1), 29–45.
4. Vaddiraju, S., Legassey, A., Qiang, L., Wang, Y., Burgess, D. J., & Papadimitrakopoulos, F. (2013). Enhancing the Sensitivity of Needle-Implantable Electrochemical Glucose Sensors via Surface Rebuilding. *Journal of Diabetes Science and Technology*, 7(2), 441–451.

Mid-Infrared Spectrometer Based on Tunable Photoresponses in Pdse₂[#]

Jea Jung Lee ⓘ

Department of Electrical Engineering, Yale University,
New Haven, CT 06511, USA
jeajung.lee@yale.edu

Adi Levi ⓘ* and Doron Naveh ⓘ[†]

Faculty of Engineering, Bar-Ilan University,
Ramat-Gan 52900, Israel
*adi.levi1@biu.ac.il
†doron.naveh@biu.ac.il

Fengnian Xia ⓘ[‡]

Department of Electrical Engineering, Yale University,
New Haven, CT 06511, USA
fengnian.xia@yale.edu

Mid-infrared (mid-IR) photodetection is important for various applications, including biomedical diagnostics, security, chemical identification, and free-spacing optical communications. However, conventional "photon" mid-IR photodetectors require liquid nitrogen cooling (i.e., MCT). Furthermore, acquiring mid-IR spectra usually involves a complex and expensive Fourier Transform Infrared spectrometer, a tabletop instrument consisting of a meter-long interferometer and MCT detectors, which is not suitable for mobile and compact device applications. In this work, we present tunable photoresponsivity in the mid-IR wavelength in palladium diselenide (PdSe₂) – molybdenum disulfide (MoS₂) heterostructure field-effect transistors (FETs), operating at room temperature. Furthermore, we applied a tunable membrane cavity to modulate the Fabry–Pérot resonance to modulate the absorption spectrum of the device layer. We used a robust polyetherimide (PEI) membrane with CVD-grown graphene to electrically tune the membrane structure. For the next step, we will integrate the PdSe₂-based photodetector and tunable membrane to increase detection sensitivity and spectrum tunability to realize the 'learning'-based spectroscopy.

Keywords: Photoresponse; palladium diselenide; mid-infrared; 'learning'-based spectroscopy; Fabry–Pérot cavity.

[‡]Corresponding author.
[#]This chapter appeared previously on the International Journal of High Speed Electronics and Systems. To cite this chapter, please cite the original article as the following: J. J. Lee, A. Levi, D. Naveh and F. Xia, *Int. J. High Speed Electron. Syst.*, **33**, 2440063 (2024), doi: 10.1142/S0129156424400639.

1. Introduction

Optical detection and spectroscopy in the mid-infrared (mid-IR) wavelength regime are extensively investigated due to their importance in various applications, including biomedical sensing, security, environmental monitoring, and free-space optical communications. For example, most molecules have their vibrational resonances in this wavelength range, so mid-IR spectroscopy has been widely used in chemical and biomedical identification [1]. Additionally, the mid-IR range is also promising for free-spacing optical communications because it contains an atmospheric transmission window ($3 \sim 5\,\mu m$ and $8 \sim 13\,\mu m$) where the absorption from air molecules is suppressed [2]. Traditional mid-IR detection devices exploited the interband "photo" absorption, such as mercury cadmium telluride (MCT) [3]. However, they usually require liquid nitrogen cooling, making their miniaturizing challenging. On the other hand, microbolometers, leveraging a "thermal" detection mechanism, can be used for room-temperature mid-IR detection [4]. However, due to their low operational speed, detection sensitivity, and spectrum tunability, their application in spectroscopy is limited.

Recently, Yuan *et al.* [5]. demonstrated a single-device spectrometer that proposed the concept of learning-based spectroscopy. The bandgap of the black phosphorus (BP) is tuned by the electric field, allowing for the reconfigurability of spectral response in the infrared range. Researchers leveraged the reconfigurable spectral response to "learn" the photoresponsivity matrix of the BP device. The photoresponsivity matrix learned from multiple known spectra can be utilized to reconstruct the unknown spectrum by sampling the photoresponse. This research denotes that we can directly interpret the spectrum of light without cumbersome components such as interferometer or grating with a single device. However, the BP single-device spectrometer still requires liquid-nitrogen cooling. Thus, the room-temperature operational principle needs to be investigated. Moreover, BP quickly degrades in the air [6]. Thus additional encapsulating layers such as hexagonal boron nitride (hBN) are necessary for its stable operation [7], which renders the fabrication of a spectrometer challenging. For this reason, air-stable materials are highly preferred.

The newly rediscovered transition metal dichalcogenide (TMD) palladium diselenide ($PdSe_2$) has a widely tunable bandgap, depending on its thickness, and it is stable in the air [8, 9]. Additionally, $PdSe_2$ can be synthesized on a large scale [10–12], which is important for applications. These recent progress and properties make that $PdSe_2$ is a promising material candidate for future mid-IR spectroscopy.

In this work, we observed a negative temperature coefficient of resistivity (TCR) and bolometric photoresponse with a mid-IR laser from a $PdSe_2$-MoS_2 heterostructure device. Furthermore, we fabricate a tunable membrane structure to modulate the absorption of the device layer via Fabry–Pérot resonance. Finally, we suggest a new scheme for a tunable learning-based spectrometer operating at room temperature by combining a $PdSe_2$ device and a flexible membrane structure.

2. Operational Principle

This section outlines the operational principles of the mid-IR spectrometer. Section 2.1 explains the basic concept of a "learning-based" photodetection algorithm. Section 2.2 describes the approach for tuning the photoresponse spectrum and increasing the photoresponse sensitivity with a tunable membrane structure.

2.1. *Learning based photodetection*

Learning-based photodetection leverages the tunable photoresponses to acquire the responsivity matrix, which is used to decode unknown optical information [5,13]. Figure 1 shows the schematic of the learning-based photodetection mechanism. Figure 1(a) illustrates the learning process. The photoresponse (e.g., photocurrent or photovoltage) of a spectrometer device can be tuned by a degree of freedom (e.g., electric displacement field). The responsivity vector $\boldsymbol{R_{\lambda,D_i}}$ of the spectrometer, which depends on the incident light wavelength (λ) at the fixed degree of freedom (D_i), is "learned" from the photoresponses to multiple known spectra $\boldsymbol{P_{T_i}}$ by utilizing Eq. (1) [5]. Measurements through multiple values of degree of freedom (D_i),

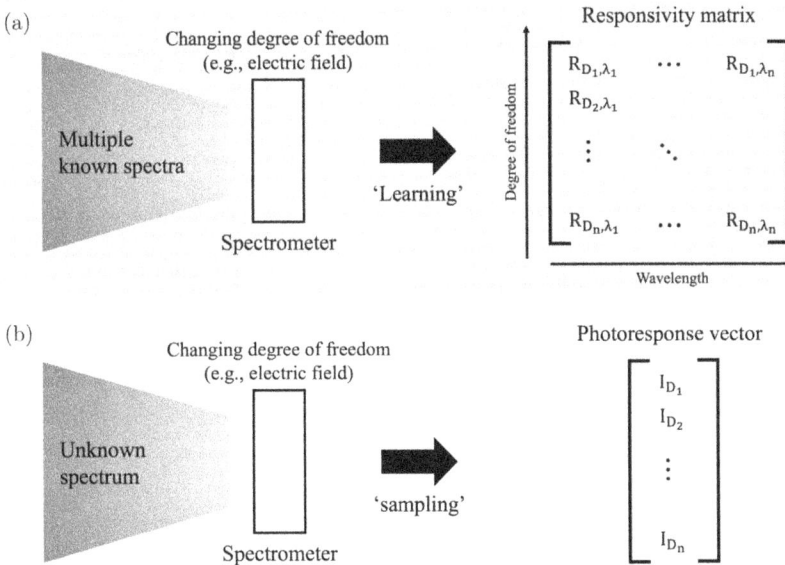

Fig. 1. Schematics of a 'learning'-based photodetection mechanism. (a) The "Learning" process; (b) The 'Sampling' process. Adapted from Ref. 5 with modifications.

we can produce entire responsivity matrix $R_{\lambda,D}$.

$$
\begin{pmatrix} I_{T_1} \\ I_{T_2} \\ I_{T_3} \\ \vdots \\ I_{T_n} \end{pmatrix} = \begin{pmatrix} P_{T_1,\lambda_1} & P_{T_1,\lambda_2} & \cdots & P_{T_1,\lambda_n} \\ P_{T_2,\lambda_1} & P_{T_2,\lambda_2} & & \\ \vdots & \vdots & \ddots & \\ P_{T_n,\lambda_1} & P_{T_n,\lambda_2} & & P_{T_n,\lambda_n} \end{pmatrix} \begin{pmatrix} R_{\lambda_1} \\ R_{\lambda_2} \\ R_{\lambda_3} \\ \vdots \\ R_{\lambda_n} \end{pmatrix}. \tag{1}
$$

Figure 1(b) demonstrates the sampling process of the learning-based photodetection. For the incident light with an unknown spectrum, the photoresponse in the spectrometer is sampled as a function of the degree of freedom. The sampling process generates the measured photoresponse vector $I_D = \langle I_{D_1}, I_{D_2}, \ldots, I_{D_n} \rangle$. Finally, the unknown incident spectrum power density vector P_λ can be reconstructed by the following equation [5]:

$$
\begin{pmatrix} P_{\lambda_1} \\ P_{\lambda_2} \\ P_{\lambda_3} \\ \vdots \\ P_{\lambda_n} \end{pmatrix} = \left[\begin{pmatrix} R_{D_1,\lambda_1} & R_{D_1,\lambda_2} & \cdots & R_{D_1,\lambda_n} \\ R_{D_2,\lambda_1} & R_{D_1,\lambda_2} & & \\ \vdots & \vdots & \ddots & \\ R_{D_n,\lambda_1} & R_{D_n,\lambda_2} & & R_{D_n,\lambda_n} \end{pmatrix} \right]^{-1} \begin{pmatrix} I_{D_1} \\ I_{D_2} \\ I_{D_3} \\ \vdots \\ I_{D_n} \end{pmatrix}. \tag{2}
$$

In summary, by using a "learning-based" photodetection algorithm, the spectroscopy function can be realized without advanced optical components such as interferometers or movable grating.

2.2. Tunable membrane structure

Photoresponse in $PdSe_2$ at mid-IR regime mostly comes from the bolometric effect [9, 14]. Thus, reducing thermal conduction is one of the most important keyframes when designing the device structure [15]. Therefore, suspending two-dimensional materials on the membrane with low thermal conductance [16] largely increases bolometric photoresponse since heat dissipation to the surroundings gets suppressed.

Suspending two-dimensional materials on the membrane also enables tuning the spectral response of the $PdSe_2$ device. If the membrane is electrically conductive, the electric field can easily tune the membrane height [17]. Tunable membrane height changes the absorbance spectrum of the two-dimensional material device via the Fabry–Pérot cavity [18]; thus, the bolometric photoresponsivity can be tuned. This tuning mechanism is important since a bolometric photoresponse basically detects the temperature dependence of the conductance, thus it is hard to tune it electrically [4]. Figure 2 is the schematic of the Fabry–Pérot cavity of tunable membrane structure.

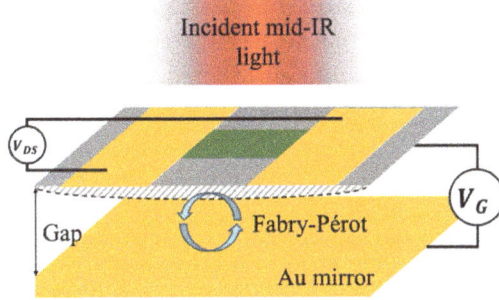

Fig. 2. Schematic of Fabry–Pérot cavity with tunable membrane.

3. Device Fabrication

This section describes the device fabrication process and method. Section 3.1 reports the fabrication process of the $PdSe_2$-MoS_2 heterostructure. The process for tunable membrane structure is described in Sec. 3.2.

3.1. *Fabrication process for PdSe₂-MoS₂ heterostructure*

Figure 3(a) is a schematic of the heterostructure device fabrication process. After identifying the suitable $PdSe_2$ and MoS_2 flakes by the mechanical exfoliation method [19], we used a polydimethylsiloxane (PDMS) stamp covered with polycarbonate (PC) to stack and transfer the $PdSe_2$-MoS_2 heterostructure [20, 21]. First, we pick-up the $PdSe_2$ flask, heating the substrate at 120°C. Second, we pick-up the MoS_2 flake in the same way. Then, the heterostructure was transferred onto a highly doped Si wafer covered with a 90 nm SiO_2 layer at 180°C. Finally, metal electrodes (10 nm Ti/50 nm Au) were deposited by electron-beam lithography and

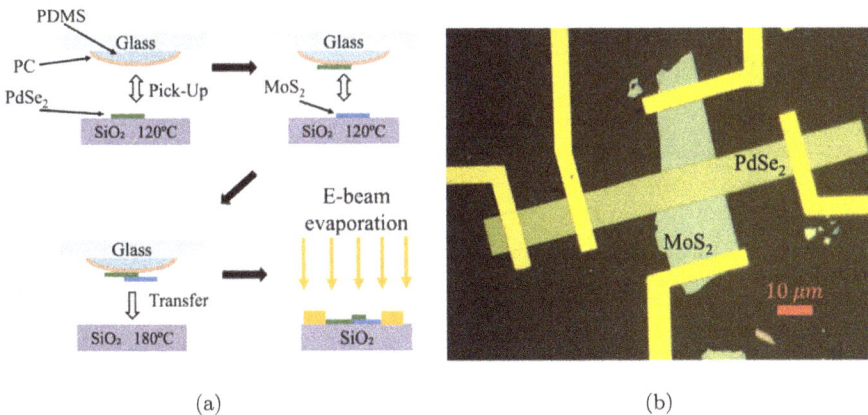

(a)

(b)

Fig. 3. Schematics of $PdSe_2$-MoS_2 heterostructure device fabrication process. (b) Optical micrograph of fabricated heterostructure device.

electron-beam evaporation process. Figure 3(b) is the optical microscope image of the heterostructure device. The thickness of $PdSe_2$ is around 65 nm, which is thick enough to have a narrow band gap to absorb mid-IR light [8]. The overlapped area between $PdSe_2$ and MoS_2 is around 165 μm^2.

3.2. *Fabrication processes for membrane structure*

Figure 4(a) is a schematic of the process to construct the electrically modulated membrane structure. We used a flexible polyetherimide (PEI) membrane with a chemical vapor deposition (CVD) grown graphene [22]. PEI has a low thermal conductivity [23]. Thus, it is appropriate to integrate with a bolometric photodetector. However, an additional conducting layer (e.g., graphene) is necessary to modulate the height because of the insulating behavior of the PEI.

We first prepare the substrate with a cavity for the PEI membrane. The cavity at the center of the substrate is patterned by electron-beam lithography and then etched using buffered oxide etchant (BOE). For the next step, a thermal evaporator deposits the gold mirror in the cavity and the electrode in contact with CVD-grown graphene for tuning the membrane.

On the other hand, we spin-coated PEI (3 wt%) CVD-grown graphene on the Cu foil. Then, we used the APS-100 solution to etch the Cu, floating the PEI/Graphene

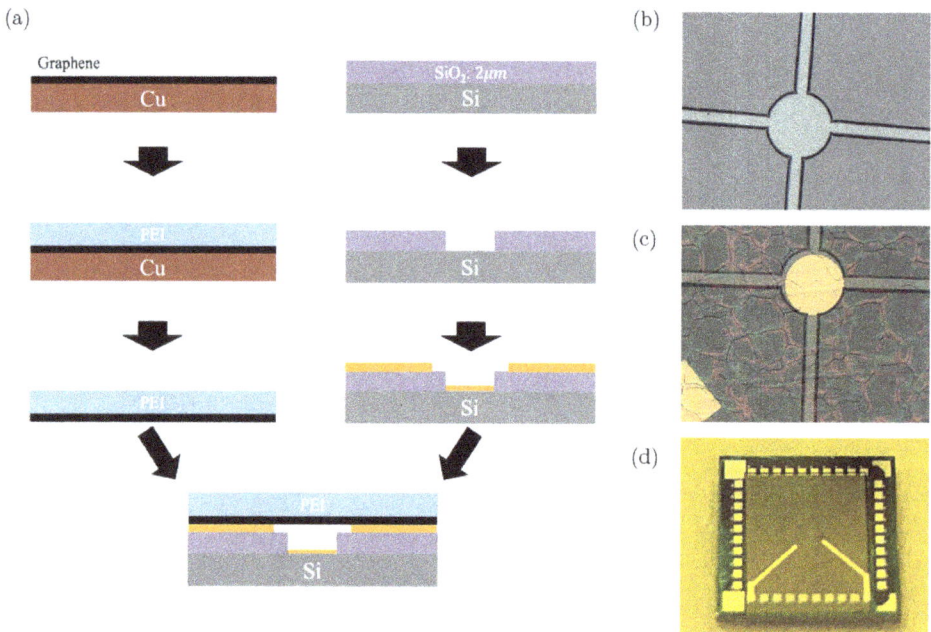

Fig. 4. (a) Schematic of the tunable membrane structure fabrication processes. (b) and (c) Optical microscope images of cavity structure after the BOE process (b) and transferring of PEI membrane (c). (d) Photo of the final device.

membrane. Finally, the PEI/Graphene membrane was transferred to the substrate we described above.

Figures 4(b) and 4(c) illustrate the device after the BOE process and transferring PEI membrane, respectively. The final fabricated device is shown in Fig. 4(d).

4. Experimental Results at Room-Temperature

This section describes the experimental results of the $PdSe_2$-MoS_2 heterostructure (Sec. 4.1) and electrically modulated membrane (Sec. 4.2). All the experiments were conducted at room temperature.

4.1. *Electrical and optoelectrical measurements of the $PdSe_2$-MoS_2 heterostructure*

Figure 5(a) shows the schematics of the electrical connection of the heterostructure device. Global back gate voltage is biased through the highly doped silicon substrate. Both $PdSe_2$ and MoS_2 show n-type behavior in Figs. 5(b) and 5(c). $PdSe_2$-MoS_2 heterostructure device (drain-source bias is connected through the two different materials) also shows n-type dominant behavior, but current increase (V_G from -12 to $-9V$) and decrease (V_G from -9 to $0V$) can be observed in Fig. 5(d). The energy barrier at the heterostructure tuned by the gate voltage causes those changes.

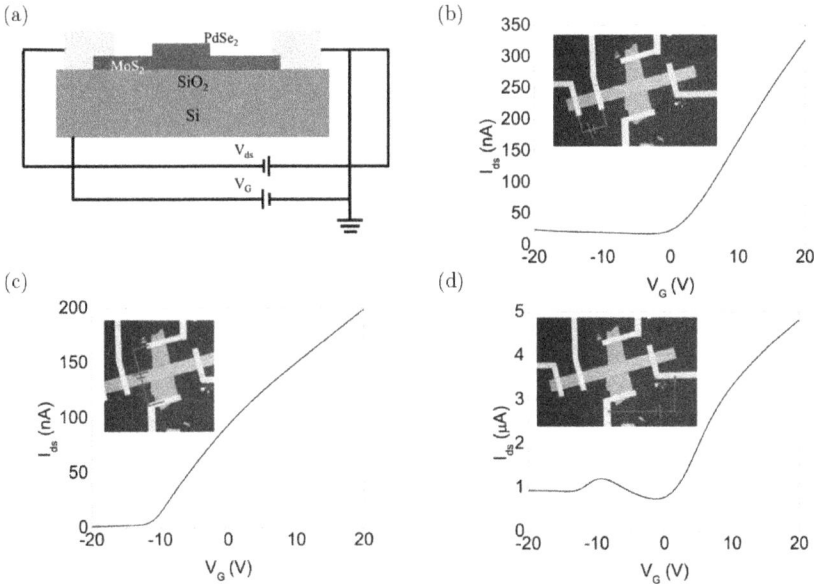

Fig. 5. (a) Schematic of device connection for electrical measurements. (b)–(d) Transfer curves of $PdSe_2$ (b) and MoS_2 (c) with $V_{ds} = 10\,mV$. (d) Transfer curve of heterostructure device with $V_{ds} = 3\,V$.

(a)

(b)

(c)

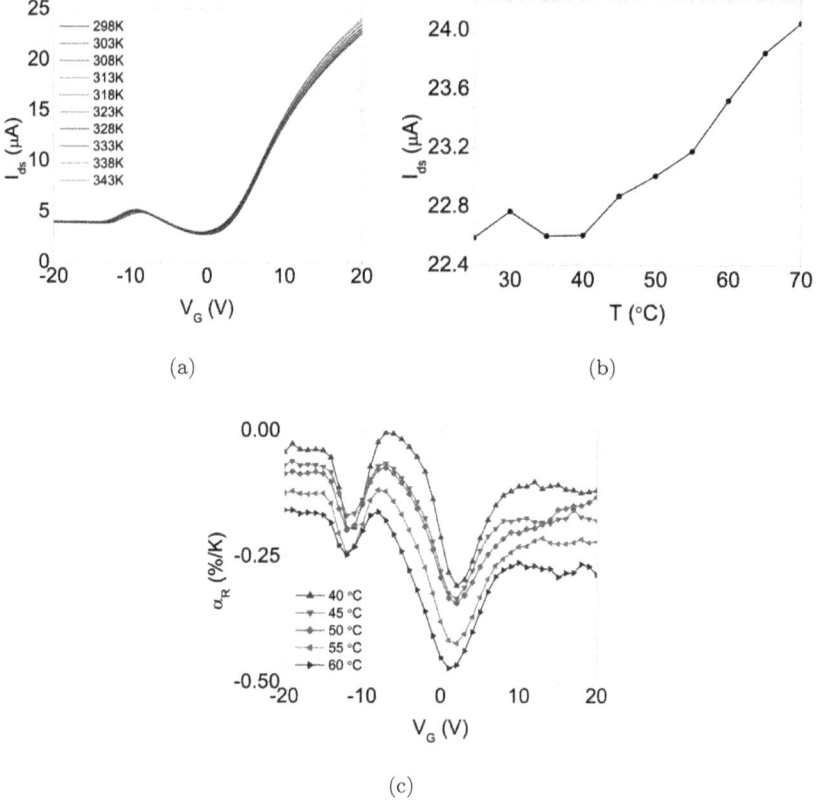

Fig. 6. (a) Transfer curves of heterostructure device measured from 25°C to 70°C. (b) Darin-source current as a function of the temperature when $V_G = 20V$. (c) Temperature coefficient of resistance (TCR) as a function of gate voltage.

4.1.1. *Negative temperature coefficient of resistance (TCR)*

To study the transport properties of the PdSe$_2$-MoS$_2$ heterostructure, temperature-dependent transfer curves were measured from 25°C to 70°C in Fig. 6(a). The current increased with increasing temperature overall, indicating that temperature increases the conductance of the heterostructure device (Fig. 6(b)). Since mid-IR light heats the device, changing its conductance, incident mid-IR light can be detected with the change of the current. Bolometric photoresponsivity R_{bolo} can be described as shown in Eq. (3), where P_{dev} is the incident power on the device, I_{ph} is the output photocurrent, G is thermal conductance, α_R is temperature coefficient of resistance (TCR), I_{ds} is device bias current, η is absorbance of the device, ω is angular frequency of modulation of the radiation, and τ is thermal response time [15].

$$R_{bolo} = \frac{I_{ph}}{P_{dev}} = \frac{I_{ds}\alpha_R\eta}{G(1 + \omega^2\tau^2)^{1/2}}\,(A/W).$$

(3)

Since the sign of I_{ds} is determined by the measurement condition and all other elements in Eq. (3) are positive, the sign of the bolometric photoresponsivity R_{bolo} is determined by the sign of TCR.

Generally, TCR (α_R) is defined as

$$\alpha_R = \frac{1}{R}\frac{dR}{dT}. \tag{4}$$

According to the definition, the TCR as a function of the gate voltage is shown in Fig. 6(c). The TCR can be tuned from 0% to $-0.5\%\,\mathrm{K}^{-1}$ by the gate voltage with the two dips. Negative TCR (reduction of resistance with increased temperature) means the bolometric photocurrent produced by mid-IR light, which is discussed in the following section, has a positive value.

4.1.2. *Photoresponses in PdSe$_2$-MoS$_2$ heterostructure device*

The tunable TCR of heterostructure makes it desirable for bolometric photodetection to use spectroscopy of mid-IR light. To evince the tunability of the PdSe$_2$-MoS$_2$ bolometer via gate bias, we measured photocurrent as a function of gate voltage with $5\,\mu m$, $7.7\,\mu m$ and $10\,\mu m$ lasers. The schematic of the set-up for the photocurrent measurement is illustrated in Fig. 7(a). The Lock-in amplifier collects the voltage inputs V_{ph}, synchronized with chopper frequency, and it is converted to photocurrent with pre-amplifier gain G_{gain} as shown in the following equation [24]:

$$I_{ph} = 2\pi\sqrt{2}V_{ph}/4G_{gain}. \tag{5}$$

Photoresponsivity, then, calculated with photocurrent with Eq. (3), where incident power on the device P_{dev} is $58\,\mu W\,(5\,\mu m)$, $66\,\mu W\,(7.7\,\mu m)$, or $34\,\mu W\,(10\,\mu m)$ and presented in Fig. 7(b). The photoresponses generally increase as larger drain-source current I_{ds} is utilized, because of the relatively small TCR of the heterostructure device. However, the photoresponsivity is highly dependent on the gate voltage.

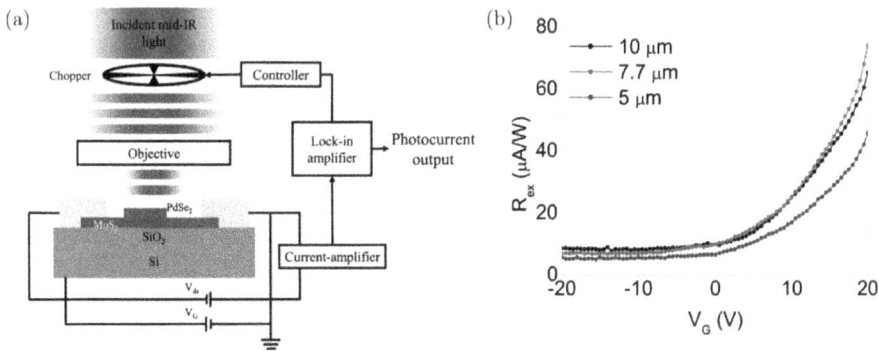

Fig. 7. (a) Schematic of the photocurrent measurement set-up. Adapted from [24] with modifications. (b) Photoresponsivity of PdSe$_2$-MoS$_2$ heterostructure device depends on the gate voltage with $5\,\mu m$, $7.7\,\mu m$, and $10\,\mu m$ lasers.

4.2. *Membrane modulation by electrical field*

From Eq. (3), bolometric photoresponsivity is affected by the optical absorbance of the device. Therefore, the photoresponse of the bolometer can be tuned by modulating the cavity height between the membrane and gold mirror because of the Fabry–Pérot interference. The membrane structure we rendered in Fig. 4(d) is largely tunable by the electrical field applied to the cavity [22]. We observed that Newton's rings moved when we applied the electric field; thus, we noticed that the membrane was modulated electrically [25] (Fig. 8).

Since we used $2\,\mu m$ SiO_2 layer and deposited $100\,nm$ gold in cavity fabrication, the Fabry–Pérot resonance wavelength $\lambda = 4dn_r$ is in the mid-IR range (d is cavity height between the gold mirror and PEI membrane, $n_r = 1$ for the air) [26]. Therefore, this electrically tunable membrane structure is appropriate to establish a "learning-based" mid-IR photodetector.

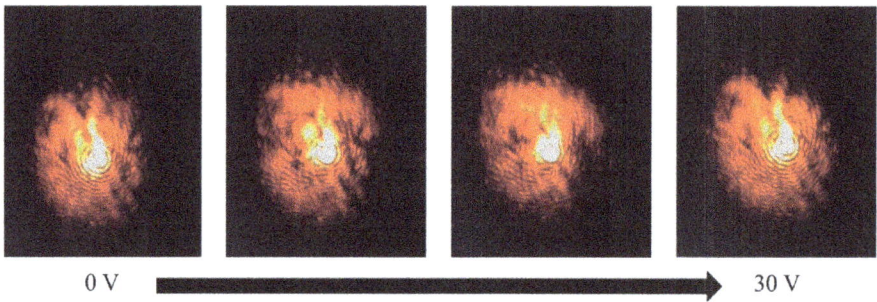

0 V   30 V

Fig. 8. Optical microscope images showing the change of Newton's rings on the PEI membrane. These indicate the PEI membrane is modulated by biased voltage.

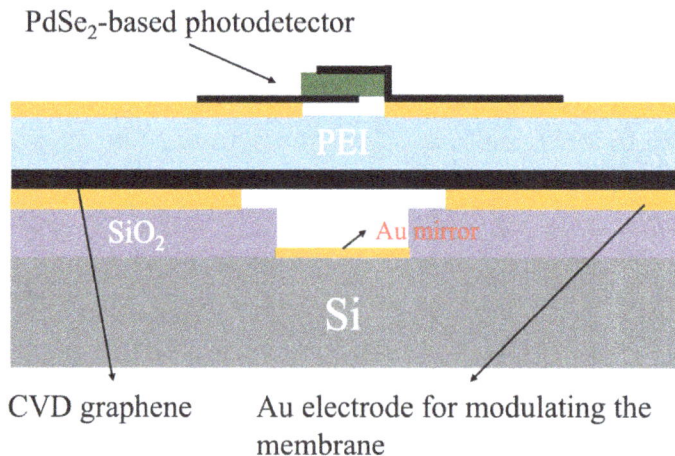

Fig. 9. Schematic of PdSe2-based photodetector integrated with tunable Fabry–Pérot membrane cavity structure.

5. Conclusion and Future Works

This work reports the gate-tunable photoresponses in a PdSe$_2$-based photodetector in a mid-IR regime and proposes a method to tune the bolometric photoresponse using the Fabry–Pérot cavity with an electrically modulated membrane. We fabricated the PdSe$_2$-MoS$_2$ heterostructure device and PEI membrane cavity. All the measurements were conducted at the room-temperature or higher ($\sim$70°C). We observed the gate-tunable bolometric photoresponses and membrane height modulation, laying the foundation for the realization of sensitive and largely tunable mid-IR photodetector for spectroscopy applications at room temperature.

In the next steps, we will combine a PdSe$_2$-based photodetector and membrane structure to increase the photoresponse by suppressing heat dissipation to surroundings and widening the modulation range by using a tunable Fabry–Pérot membrane cavity, as illustrated in Fig. 9.

Acknowledgments

Fengnian Xia and Jea Jung Lee thank the National Science Foundation for support through grant NSF-BSF 2150561. Jea Jung Lee is grateful for support from Yale Cleanroom, YINQE, Mikhael Guy at Yale Center for Research Computing, Bingchen Deng, and Shaofan Yuan.

ORCID

Jea Jung Lee ⊚ https://orcid.org/0009-0001-5937-9073
Adi Levi ⊚ https://orcid.org/0000-0001-5205-0630
Doron Naveh ⊚ https://orcid.org/0000-0003-1091-5661
Fengnian Xia ⊚ https://orcid.org/0000-0001-5176-368X

References

1. Caffey, D., *et al.*, *Recent results from broadly tunable external cavity quantum cascade lasers.* SPIE OPTO, ed. Vol. 7953 (SPIE, 2011).
2. Green, A. E. S. and M. Griggs, Applied Optics. **2**(6): p. 561–570 (1963).
3. Rogalski, A., Reports on Progress in Physics. **68**(10): p. 2267 (2005).
4. Kruse, P. W., *Uncooled Thermal Imaging Arrays, Systems, and Applications.* Vol. 51 (SPIE, 2001).
5. Yuan, S., *et al.*, Nature Photonics. **15**(8): p. 601–607 (2021).
6. Kim, J.-S., *et al.*, Scientific Reports. **5**(1): p. 8989 (2015).
7. Chen, X., *et al.*, Nature Communications. **8**(1): p. 1672 (2017).
8. Oyedele, A. D., *et al.*, Journal of the American Chemical Society. **139**(40): p. 14090–14097 (2017).
9. Wen, S., *et al.*, ACS Applied Nano Materials. **6**(18): p. 16970–16976 (2023).
10. Zeng, L.-H., *et al.*, Advanced Functional Materials. **29**(1): p. 1806878 (2019).
11. Wei, M., *et al.*, npj 2D Materials and Applications. **6**(1): p. 1 (2022).
12. Xu, W., *et al.*, Nano Research. **13**(8): p. 2091–2097 (2020).
13. Yuan, S., *et al.*, Science. **379**(6637): p. eade1220 (2023).

14. Zhang, R., *et al.*, Advanced Optical Materials. **11**(23): p. 2301055 (2023).

15. Yadav, P. V. K., *et al.*, Sensors and Actuators A: Physical. **342**: p. 113611 (2022).

16. Efetov, D. K., *et al.*, Nature Nanotechnology. **13**(9): p. 797–801 (2018).

17. Holsteen, A. L., A. F. Cihan, and M. L. Brongersma, Science. **365** (6450): p. 257–260 (2019).

18. Wei, H., *et al.*, Small. **17**(35): p. 2100446 (2021).

19. Novoselov, K. S., *et al.*, Science. **306**(5696): p. 666–669 (2004).

20. Shin, S., *et al.*, Carbon. **111**: p. 215–220 (2017).

21. Wang, L., *et al.*, Science. **342**(6158): p. 614–617 (2013).

22. Khan, A. U., *et al.*, Advanced Materials. **33**(2): p. 2004053 (2021).

23. Lee, H. L., *et al.*, Physical Chemistry Chemical Physics. **16**(37): p. 20041–20046 (2014).

24. Yuan, S., *et al.*, ACS Photonics. **7**(5): p. 1206–1215 (2020).

25. Vaughan, M., *The Fabry-Perot Interferometer*: *History, Theory, Practice and Applications*. 1st ed. (Routledge, New York, 1989).

26. Abdullah, A., *et al.*, Scientific Reports. **13**(1): p. 3470 (2023).

A Comprehensive Study and Comparison of 2-Bit 7T–10T SRAM Configurations with 4-State CMOS-SWS Inverters#

A. Husawi ⊚*,†, R. H. Gudlavalleti ⊚‡, A. Almalki ⊚*,§ and F. C. Jain ⊚*,¶

*ECE Department, University of Connecticut, Storrs, CT, USA

†Department of Electrical and Electronics Engineering Technology,
Yanbu Industrial College, Yanbu Industrial City,
Al-Medina 46452, KSA

‡Biorasis Inc., Storrs, CT, USA

§Electrical Engineering Department,
Taif University, Taif 21944, KSA
¶fcj@engr.uconn.edu

This paper presents a comprehensive analysis of power dissipation and propagation delay in 2-bit SRAM configurations ranging from 7T to 10T, building upon previous work on 6T 2-bit/4-state SWSFET SRAM designs. The study compares the performance of SWS-FET SRAMs with CMOS-based 2-state SRAMs [7], highlighting the former's significant advantages in speed and power consumption. Utilizing Cadence simulations and models such as Analog Behavioral Model (ABM) and EKV (Enz–Krummenacher–Vittoz), the analysis incorporates real-world 0.18-μm technology considerations. The research explores the design nuances of 7T–10T SRAM configurations using SWS-FETs, leveraging their unique characteristics like vertically stacked quantum well/quantum dot channels. Power dissipation analysis reveals varying trends across different SRAM configurations, with notable shifts in voltage changes during transitions. Similarly, propagation delay assessments showcase diverse durations for different voltage transitions, underscoring the impact of SRAM configuration changes on efficiency and complexity. In addition, parasitic capacitance is crucial for optimizing the performance, power efficiency, and reliability of SRAM cells. In these circuits an internal storage parasitic capacitance of 1 fF has been considered to evaluate its effects through simulation-based analysis during the memory cell design process. The findings contribute valuable insights into the trade-offs involved in SRAM design, particularly concerning power dissipation and propagation delay, and are presented. Overall, this study sheds light on the promising potential of SWS-FETs for enhancing memory circuitry performance.

Keywords: SRAM configurations; SWS-FETs; parasitic capacitance; power dissipation analysis; propagation delay analysis; 2-bit SRAM design; Cadence simulations.

¶Corresponding author.
#This chapter appeared previously on the International Journal of High Speed Electronics and Systems. To cite this chapter, please cite the original article as the following: A. Husawi, R. H. Gudlavalleti, A. Almalki and F. C. Jain, *Int. J. High Speed Electron. Syst.*, **33**, 2440064 (2024), doi: 10.1142/S0129156424400640.

1. Introduction

The SWS-FET transistor utilizes vertically stacked quantum well/quantum dot channels, enabling current flow across multiple channels within a single transistor through gate voltage variation [1, 2]. Moreover, the SWS-FET has demonstrated capabilities in performing multivalued logic computations and storing information in memories [3]. The structure with multiple quantum wells/quantum dots represents logic states (00), (01), (10), and (11) based on the spatial distribution of carriers [4–6].

The multi-state functionality of SWS-FETs enables the processing of two bits within logic circuit designs. Various key logic components such as primary logic cells, full-adders, latches, and static random-access memory (SRAM) have demonstrated a 50% reduction in area by utilizing 4-state/2-bit SWS-FETs [1, 6]. These circuits employ different n-channel SWS-FET source/drain configurations to process multiple bits simultaneously [8]. Specifically, a 2-bit SRAM is realized using 4-state inverters with additional threshold voltages. This aids in minimizing the quantity of transistors and interconnections needed within the SRAM setup, leading to reduced power usage and shorter propagation delay. Propagation delay is determined by assessing the duration taken for the circuit's output to stabilize following an input alteration. Meanwhile, power consumption is gauged by evaluating the average energy utilized by the circuit.

This paper provides an in-depth analysis and comparison of power dissipation and propagation delay in 2-bit SRAM configurations spanning from 7T to 10T. Employing Cadence simulations and models like ABM and EKV, the analysis integrates real-world 0.18-μm technology considerations. Additionally, the study delves into the detailed design aspects of 7T–10T SRAM configurations employing SWS-FETs. It also considers a parasitic capacitance of 1 fF within the internal storage of the cross-coupled inverters of these SRAMs. Overall, the assessment of propagation delay and power consumption for the 2-bit SRAM cell employing 4-state CMOS-SWS inverters offers valuable insights into the efficiency and performance of this innovative memory cell design within the 180 nm technology node.

2. SWSFET-based 7T 2-Bit SRAM

Figure 1 shows the circuit schematic of 7T SRAM with two cross-coupled SWS-FET-based inverters and the three access transistors Nmos1, Nmos2 and Nmos3. The 3rd inverter stage (in the red dotted box) generates data_bar by inverting data signal. The cross-coupled inverters consist of two pairs of n-SWS-FET and p-SWS-FET transistors, while the access transistors are three nMOS transistors that control the read and write operations of the cell. The gate of the access transistors (Nmos1 and Nmos2) is controlled by word line inputs (Word_Line), while the gate of the access transistor Nmos3 is controlled by read line input (Read_Line). Whenever the word line is made high, the Data and Data_Bar are connected to the cell; hence cell can

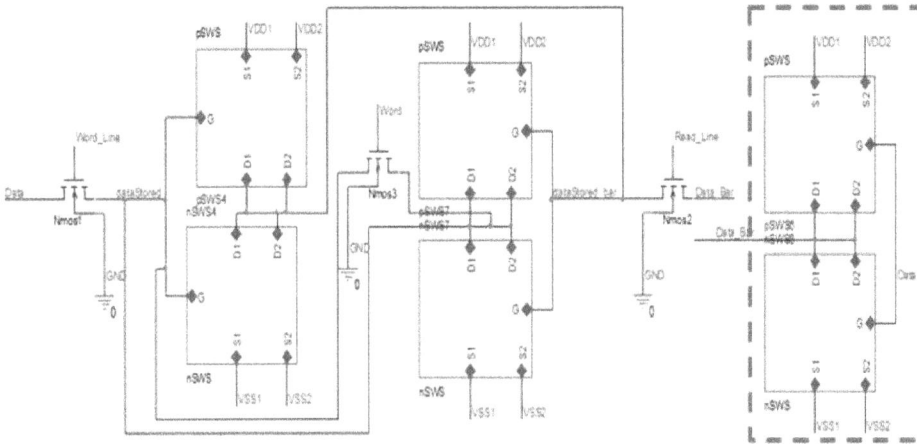

Fig. 1. Schematic of 7T SRAM with two cross-coupled SWS-FET based inverters.

be written in and read out the data through the bit lines. During read operation, the data and data_bar lines are initially pre-charged. When the Read_Line input is made high, Nmos3 becomes conductive, connecting the storage nodes to the bitlines for the read operation.

2.1. *Read and write operation of 7T SWSFET-based 2-bit SRAM*

Figure 2 shows the simulation result for the write operation of 2-bit SRAM based on 7T SWSFET unit cell shown in Fig. 1. The word_line (blue line) and data stored in the memory cell (red line) when wordline input goes high are shown the plot. The change in each state for a given input state depicts the storage of each state in this multistate SRAM cell.

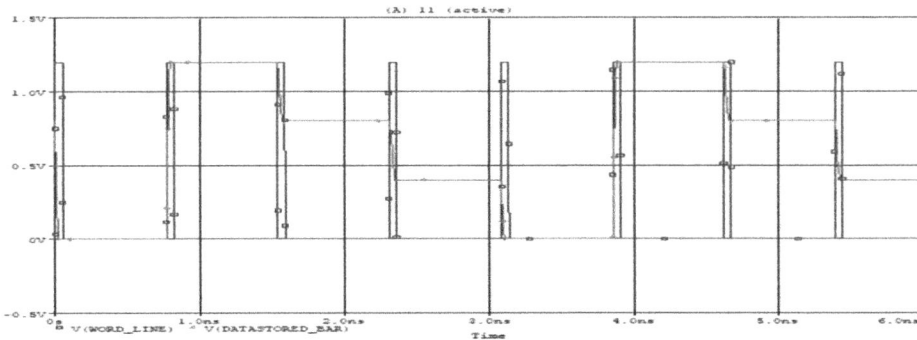

Fig. 2. Simulation result for the write operation of 2-bit SRAM based on 7T SWSFET.

2.2. *Propagation delay simulation result*

Figures 3(a)–3(d) show the output response for specific input signal transitions as the output changes from (a) state '3' to state '2', (b) state '2' to state '1', (c) state '1' to state '0', and (d) '0' to '3', with parasitic capacitance of 1 fF at 1.2 V. As it shows in Table 1, in studying how signals move through circuit, where different power levels show different working states (0 to 3), it shows that when switch from a strong power of 1.2 V to 0 V, it only takes 5.7 ps, showcasing efficient response dynamics during the shift to the lowest power state. On the contrary, the transition from 1.2 V to 0.8 V takes a slightly longer 13.5 ps, indicating some detailed behavior during this kind of power reduction. Moving from 0.8 V to 0.4 V

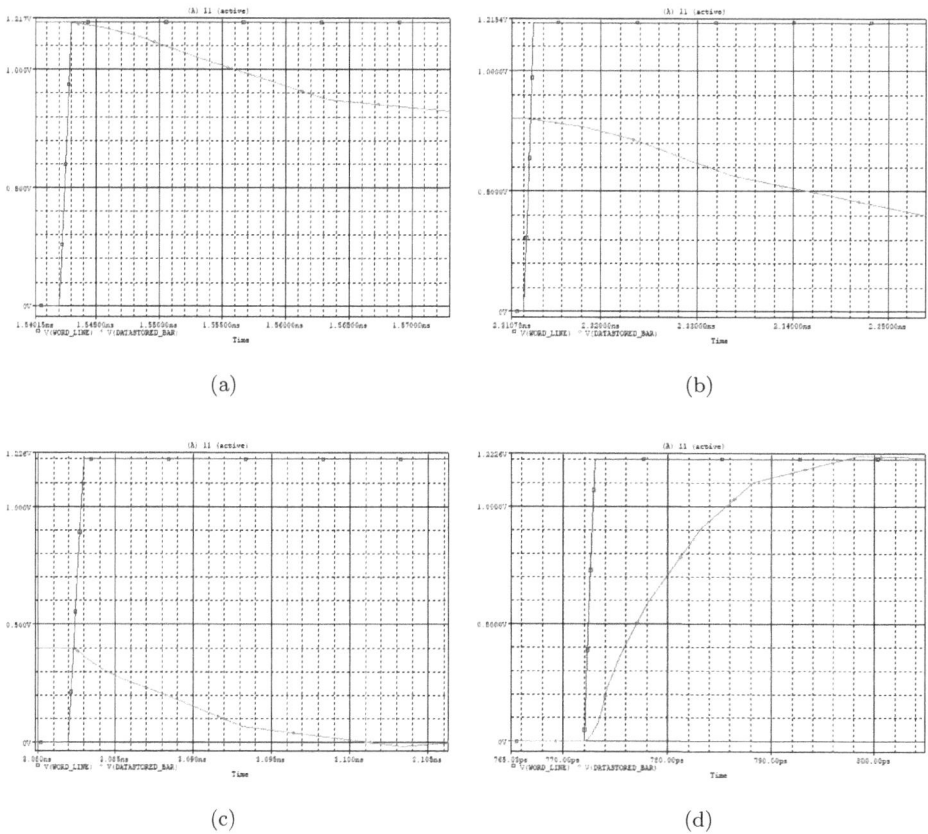

(a)   (b)

(c)   (d)

Fig. 3. The output response for specific input signal transitions.

Table 1. Propagation delay in 4-state/2-bit 7T SRAM-based SWS-FET at 1.2 V.

Transitions of States	3 to 2	2 to 1	1 to 0	0 to 3
Propagation Delay (ps)	13.5	18.5	5.5	5.7

extends the propagation delay to approximately 18.5 ps, highlighting considerations for transitions through intermediate power levels. Lastly, reverting from 0.4 V to 0 V is rapid again, requiring 5.5 ps, just like the pattern observed in the initial transition. These findings enhance our understanding of the circuit's behavior in diverse scenarios and provide valuable insights for optimizing its performance under various conditions.

2.3. *Power dissipation simulation result*

Figure 4 shows the power dissipation in the SWS-FET 7T SRAM at parasitic capacitance of 1 fF under a 1.2 V power supply. In Table 2, the power dissipation values on the tables below appear to be reasonable and demonstrate a logical trend in accordance with different supply voltage transitions. Transition from 1.2 V to 0 V (6.73 μW) exhibits the highest power dissipation, as it involves a substantial voltage change and the discharge of load capacitance. The reported value of 5.07 μW for the transition from 1.2 V to 0.8 V aligns with expectations for decreasing power dissipation with smaller voltage changes. The transition from 0.8 V to 0.4 V (3.93 μW) involves a smaller voltage decrease, resulting in reduced dissipation, consistent with expectations. A similar pattern follows in the transition from 0.4 V to 0 V, further decreasing power dissipation (1.07 μW).

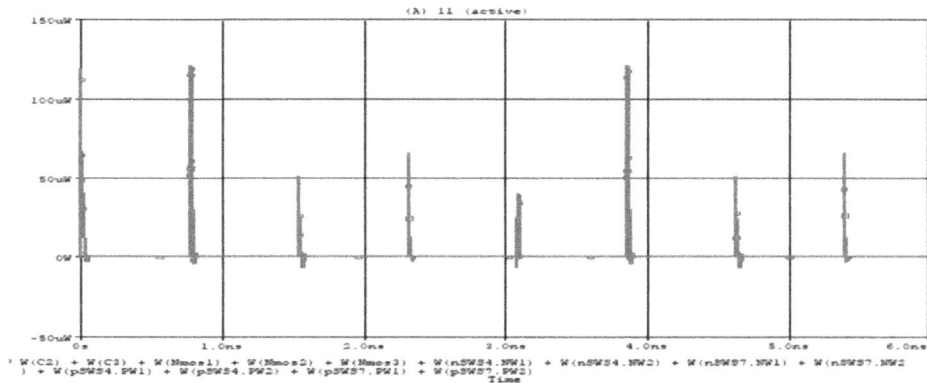

Fig. 4. Power dissipation in the SWS-FET 7T SRAM with a 1 fF parasitic capacitance at 1.2 V supply.

Table 2. Power dissipation in the SWS-FET 7T SRAM with a 1 fF parasitic capacitance at 1.2 V supply.

| Transitions of States | 3 to 2 | 2 to 1 | 1 to 0 | 0 to 3 |
Input Voltages	1.2 V to 0.8 V	0.8 V to 0.4 V	0.4 V to 0 V	0 V to 1.2 V
Av. Power Dissipation (μW)	5.07	3.93	1.07	6.73

3. SWSFET-based 8T 2-Bit SRAM

Figure 5 displays the circuit schematic of an 8T SRAM, an advanced memory architecture building upon the 7T design. This configuration involves two cross-coupled
SWS-FET-based inverters and introduces an additional access transistor, Nmos4.
The third inverter stage, highlighted in the red dotted box, generates Data_Bar by
inverting the primary data signal. The cross-coupled inverters maintain two pairs of
n-SWS-FET and p-SWS-FET transistors for stable data storage. The access transistors now include four nMOS transistors (Nmos1, Nmos2, Nmos3, and Nmos4)
providing enhanced control over read and write operations. The gates of Nmos1
and Nmos2 are regulated by word line inputs (Word_Line), activating during write
operations. Simultaneously, the gate of Nmos3, now responding to Read_Line input, becomes conductive during read operations. When Word_Line is high, Data and
Data_Bar are connected to the cell, enabling read and write operations through the
bit lines. Conversely, when Read_Line is high, Nmos4 enables the connection between storage nodes and bitlines, facilitating efficient data retrieval. This expanded
configuration improves the precision and flexibility of the 8T SRAM cell, contributing to a better performance and reliability.

3.1. *Simulation result for the write operation of 2-bit SRAM based on 8T SWSFET*

Figure 6 shows the simulation result for the write operation of 2-bit SRAM based
on 8T SWSFET unit cell shown in Fig. 5. Figure 6 shows the word_line (blue line)
and data stored in the memory cell (red line) when wordline input goes high are
shown in bottom of the plot. The change in each state for a given input state depicts
the storage of each state in this multistate 8T SRAM cell.

3.2. *Propagation delay simulation result*

Figures 7(a)–7(d) show the output response for specific input signal transitions as
the output changes from (a) state '3' to state '2', (b) state '2' to state '1', (c) state

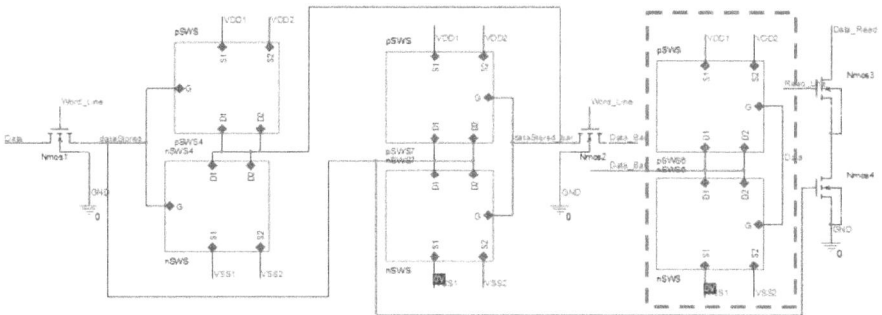

Fig. 5. Schematic of 8T SRAM with two cross-coupled SWS-FET based inverters.

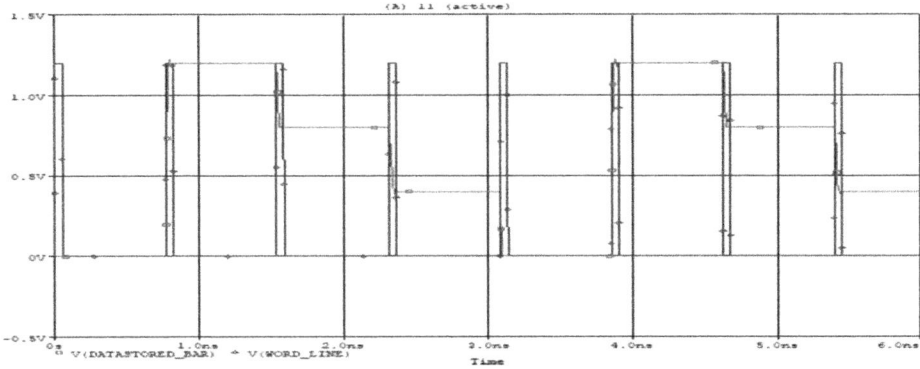

Fig. 6. Simulation result for the write operation of 2-bit SRAM based on 8T SWSFET.

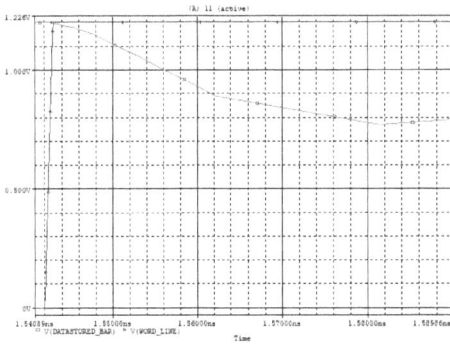

(a)

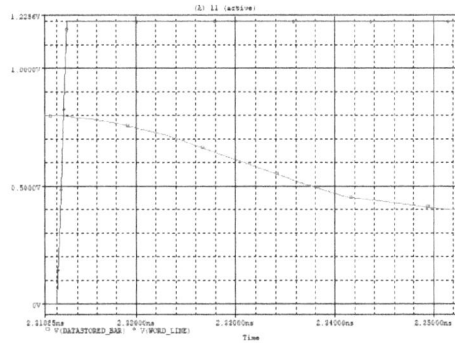

(b)

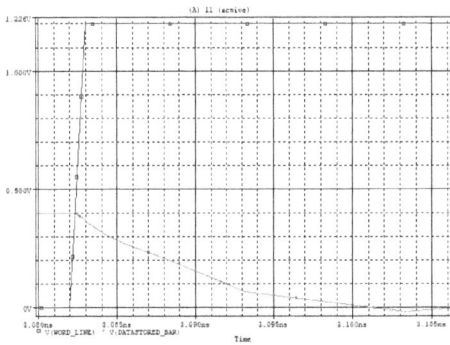

(c)

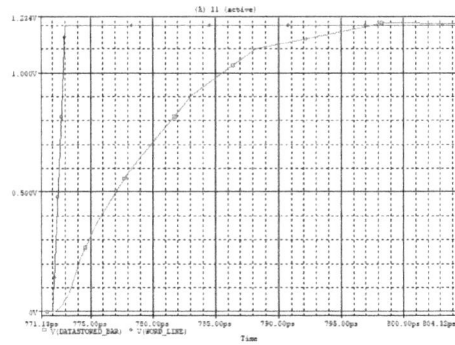

(d)

Fig. 7. The output response for specific input signal transitions.

Table 3. Propagation delay in 4-state/2-bit 8T SRAM-based
SWS-FET at 1.2 V.

Transitions of States	3 to 2	2 to 1	1 to 0	0 to 3
Propagation Delay (ps)	13.5	18.5	5.5	5.7

'1' to state '0', and (d) '0' to '3', with parasitic capacitance of 1 fF at 1.2 V. As
it shows in Table 3, in studying how signals move through circuit, where different
power levels show different working states (0 to 3), it shows that when switch from
a strong power of 1.2 V to 0 V, it only takes 5.7 ps. This quick change shows that
the system reacts efficiently when moving to the lowest power state. On the other
hand, when going from 1.2 V to 0.8 V, it takes a bit longer at 13.5 ps, indicating
some detailed behavior during this kind of power reduction. Going from 0.8 V to
0.4 V takes even more time, about 18.5 ps, showing that there is a lot to consider
when transitioning through middle power levels. Lastly, going from 0.4 V to 0 V
is quick again, taking 5.5 ps, just like the first transition. These findings help us
understand how our circuit behaves in different situations and give us good ideas
on how to make it work better in various conditions.

3.3. Power dissipation simulation result

Figure 8 shows the power dissipation in the SWS-FET 8T SRAM at parasitic capacitance of 1 fF under a 1.2 V power supply. In Table 4, the power dissipation
values on the tables below appear to be reasonable and demonstrate a logical trend
in accordance with different supply voltage transitions. The transition from 0 V to
1.2 V exhibits a higher power dissipation of 29.6 μW, aligning with expectations for
a substantial increase in voltage and likely involving the charging of load capacitance. Conversely, the transition from 1.2 V to 0.8 V shows a lower power dissipation

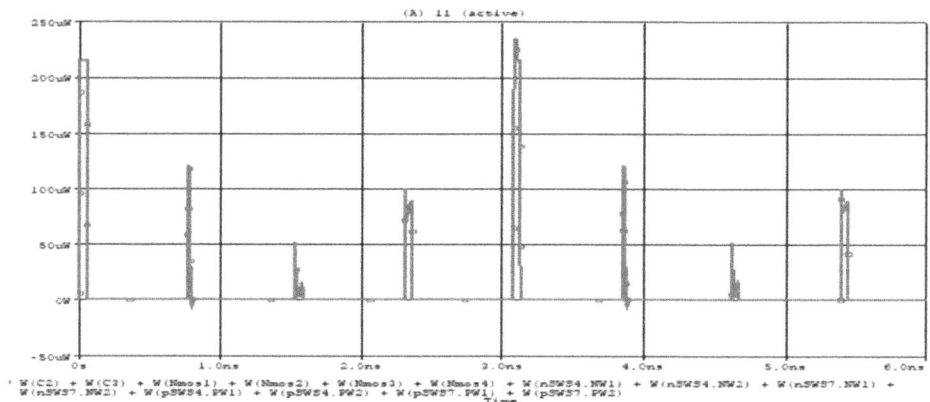

Fig. 8. Power dissipation in the SWS-FET 8T SRAM with a 1 fF parasitic capacitance at 1.2 V
supply.

Table 4. Power dissipation in the SWS-FET 8T SRAM with a 1 fF parasitic capacitance at 1.2 V supply.

Transitions of States	3 to 2	2 to 1	1 to 0	0 to 3
Input Voltages	1.2 V to 0.8 V	0.8 V to 0.4 V	0.4 V to 0 V	0 V to 1.2 V
Av. Power Dissipation (μW)	7.12	8.92	20.6	29.6

of $7.12\,\mu$W, consistent with the moderate voltage decrease in this transition. Moving further to the transition from 0.8 V to 0.4 V, the power dissipation is $8.92\,\mu$W, reflecting a smaller voltage decrease. Finally, the transition from 0.4 V to 0 V results in a higher power dissipation of $20.6\,\mu$W.

4. SWSFET-based 9T 2-Bit SRAM

Building upon the 7T and 8T configurations, the refined 9T SRAM, shown in Fig. 9, we add an extra control by incorporating Nmos5 as an extra access transistor. This evolved design features two cross-coupled SWS-FET-based inverters, maintaining stability through pairs of n-SWS-FET and p-SWS-FET transistors in the cross-coupled inverters. Access control is improved with the inclusion of five nMOS transistors (Nmos1, Nmos2, Nmos3, Nmos4) and the newly introduced Nmos5, providing advanced command capabilities over read and write operations. Nmos1 and Nmos2, responding to Word_Line inputs, activate during write operations, ensuring precise control. Nmos3, influenced by the Read_Line input, becomes conductive during read operations. When Word_Line is high, Data and Data_Bar connect to the cell, enabling smooth reading and writing through the bit lines. On the other hand, with Read_Line activated, Nmos4 and Nmos5 collaborate to establish a connection between storage nodes and bitlines, enhancing the efficiency of data retrieval.

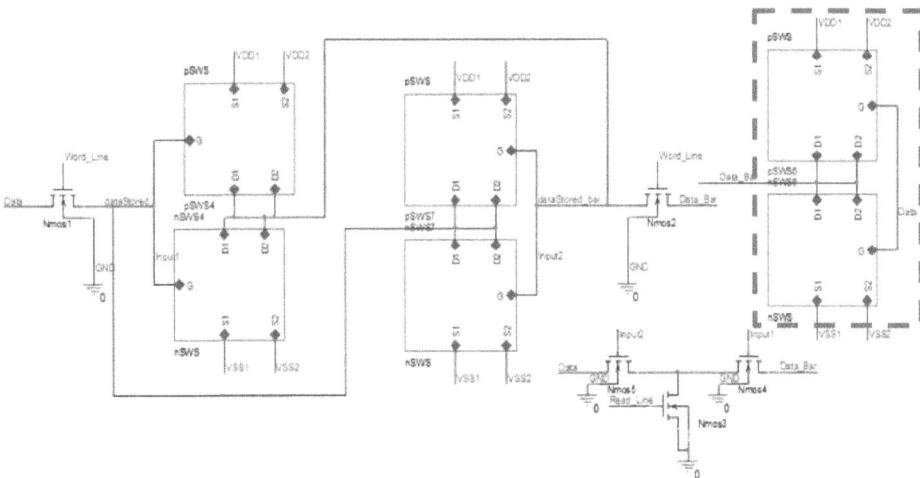

Fig. 9. Schematic of 9T SRAM with two cross-coupled SWS-FET based inverters.

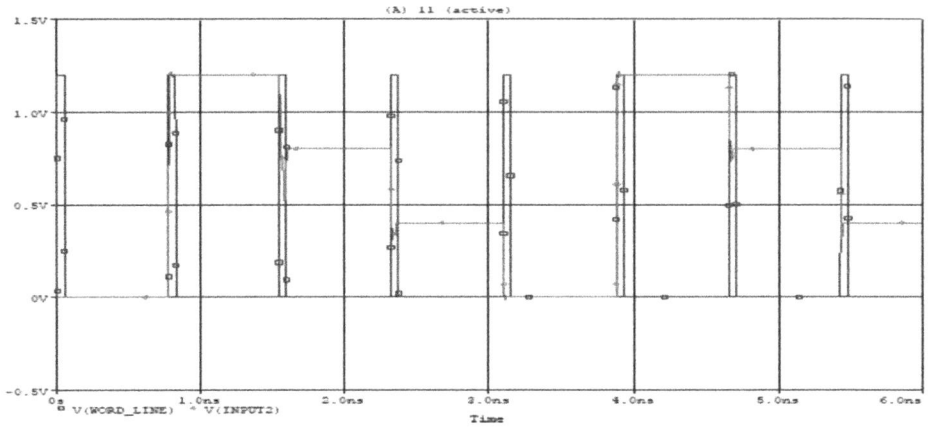

Fig. 10. Simulation result for the write operation of 2-bit SRAM based on 9T SWSFET.

This refined and expanded configuration elevates the precision and flexibility of the 9T SRAM cell, making a big step in performance and reliability within memory architecture.

4.1. *Simulation result of 9T SWSFET-based 2-bit SRAM*

Figure 10 shows the simulation result for the write operation of 2-bit SRAM based on 9T SWSFET unit cell shown in Fig. 9. Figure 10 shows the word_line (blue line) and data stored in the memory cell (red line) when wordline input goes high are shown in bottom of the plot. The change in each state for a given input state depicts the storage of each state in this multistate SRAM cell.

4.2. *Propagation delay simulation results*

Figures 11(a)–11(d) show the output response for specific input signal transitions as the output changes from (a) state '3' to state '2', (b) state '2' to state '1', (c) state '1' to state '0', and (d) '0' to '3', with parasitic capacitance of 1 fF at 1.2 V. Table 5 illustrates the signal dynamics within the circuit, depicting distinct operational states (0 to 3) corresponding to various power levels. The transition from a robust power supply of 1.2 V to 0 V is remarkably swift, requiring only 6.7 ps. This rapid change underscores the system's efficiency when transitioning to the lowest power state. On the other hand, moving from 1.2 V to 0.8 V extends the delay to 13.5 ps, pointing to complex behavior during this specific power reduction. The transition from 0.8 V to 0.4 V is 6.9 ps, emphasizing the considerations involved in transitioning across intermediate power levels. Finally, the transition from 0.4 V to 0 V is swift, taking 5 ps, aligning with the efficiency observed during the transition to the lowest power state, reflecting the anticipated trend in transitions involving smaller voltage changes.

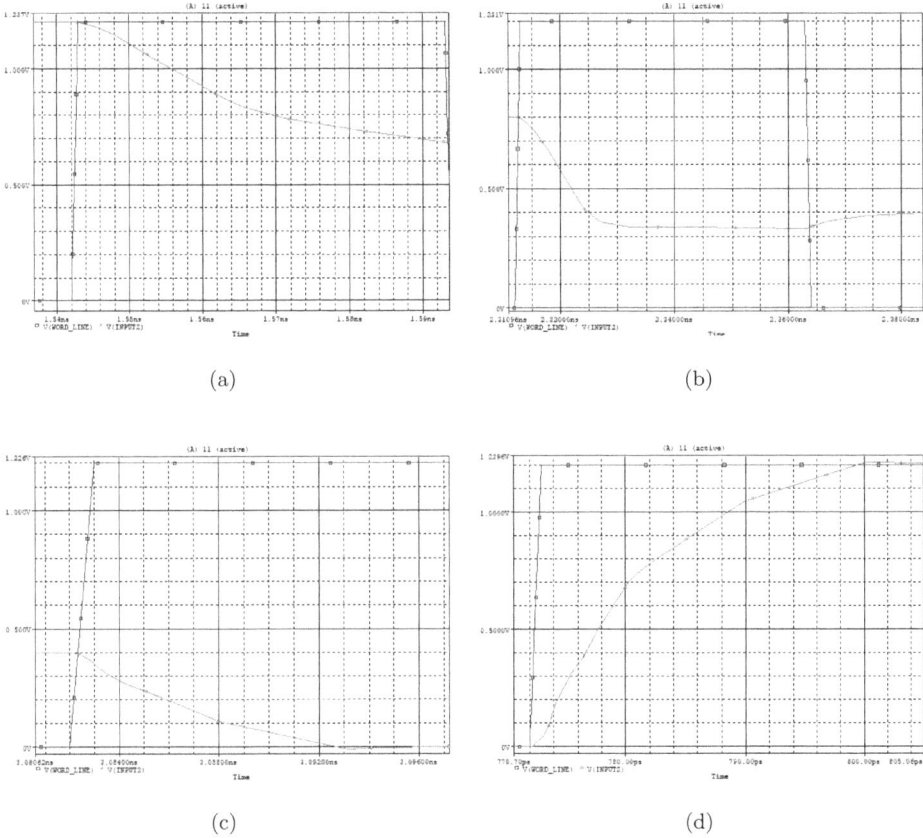

(a)

(b)

(c)

(d)

Fig. 11. The output response for specific input signal transitions.

Table 5. Propagation delay in 4-state/2-bit 9T SRAM-based SWS-FET at 1.2 V.

Transitions of States	3 to 2	2 to 1	1 to 0	0 to 3
Propagation Delay (ps)	13.5	6.9	5	6.7

4.3. *Power dissipation simulation result*

Figure 12 shows the power dissipation in the SWS-FET 9T SRAM at parasitic capacitance of 1 fF under a 1.2 V power supply. In Table 6, the power dissipation values appear to be reasonable and demonstrate a logical trend in accordance with different supply voltage transitions. The transition from 0 V to 1.2 V, with a power dissipation of 4.47 μW, reflects an anticipated increase in power as the voltage rises. Moving from 1.2 V to 0.8 V shows a moderate voltage decrease, resulting in a slightly elevated power dissipation of 9.4 μW. This corresponds to the trend of increased

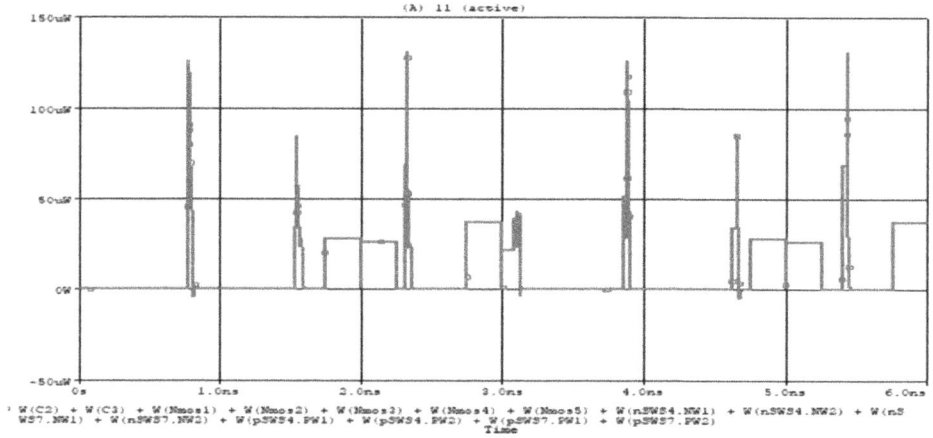

Fig. 12. Power dissipation in the SWS-FET 9T SRAM with a 1 fF parasitic capacitance at 1.2 V supply.

Table 6. Power dissipation in the SWS-FET 9T SRAM with a 1 fF parasitic capacitance at 1.2 V supply.

| Transitions of States | 3 to 2 | 2 to 1 | 1 to 0 | 0 to 3 |
Input Voltages	1.2 V to 0.8 V	0.8 V to 0.4 V	0.4 V to 0 V	0 V to 1.2 V
Av. Power Dissipation (μW)	9.4	15.3	12.3	4.47

power dissipation with smaller voltage changes. During the transition from 0.8 V to 0.4 V, a substantial voltage decrease occurs, leading to higher power dissipation of 15.3 μW. This aligns with expectations for a transition involving a significant voltage drop. Finally, the transition from 0.4 V to 0 V involves a moderate voltage decrease, resulting in a slightly reduced power dissipation of 12.3 μW. This decreasing trend aligns with expectations for a transition with a smaller voltage change. Overall, these observations provide valuable insights into the dynamic behavior of the 9T SRAM during various voltage transitions.

4.4. SWS-FET-based 10 T 2-bit SRAM

Figure 13 shows the circuit schematic of 10T SRAM. This design is expanding upon the 7T, 8T and 9T configurations, and features two cross-coupled SWS-FET-based inverters, ensuring stability through pairs of n-SWS-FET and p-SWS-FET transistors within the cross-coupled inverters. Access control is improved with the inclusion of six nMOS transistors (Nmos1, Nmos2, Nmos3, Nmos4, Nmos5) and the newly integrated Nmos6. These access transistors give smart instructions for both reading and writing functions. During write operations, Nmos1 and Nmos2, responsive to Word_Line inputs, systematically coordinate precisely. Simultaneously,

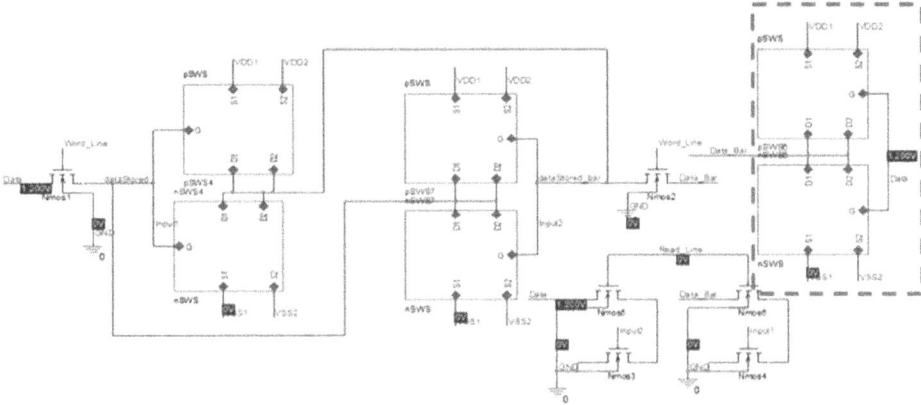

Fig. 13. Schematic of 10T SRAM with two cross-coupled SWS-FET based inverters.

Nmos5 and Nmos6, triggered by the Read_Line input, become conductive during read operations. With Word_Line set high, the consistent connection between Data, Data_Bar, and the cell ensues, enabling fluid reading and writing through the bit lines. Together, with Read_Line activated, Nmos3 and Nmos4 collaborate to establish a robust connection between storage nodes and bitlines, significantly enhancing the efficiency of data retrieval. This refined and expanded configuration in the 10T SRAM not only advances the precision and flexibility of the memory cell but represents a remarkable progress in performance and reliability within memory architecture.

4.5. *Simulation result of* 10 T *SWSFET-based* 2-bit *SRAM*

Figure 14 shows the simulation result for the write operation of 2-bit SRAM based on 10T SWSFET unit cell shown in Fig. 13. The top panel in Fig. 14 shows the

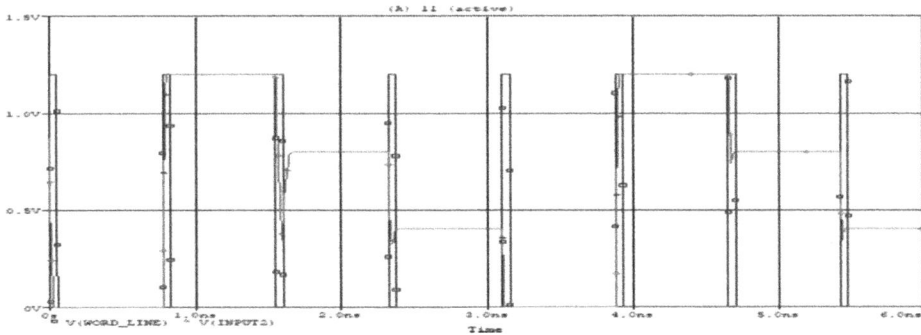

Fig. 14. Simulation result for the write operation of 2-bit SRAM based on 10T SWSFET.

word_line (blue line) and data stored in the memory cell (red line) when wordline input goes high are shown in bottom of the plot. The change in each state for a given input state depicts the storage of each state in this multistate SRAM cell.

4.6. *Propagation delay simulation results*

Figures 15(a)–15(d) show the output response for specific input signal transitions as the output changes from (a) state '3' to state '2', (b) state '2' to state '1', (c) state '1' to state '0', and (d) '0' to '3', with parasitic capacitance of 1 fF at 1.2 V. As shown in Table 7, a comprehensive exploration of signal dynamics within the circuit, representing varied working states corresponding to distinct power levels (ranging from 0 to 3), reveals insightful patterns. The swift transition from a potent 1.2 V to a minimal 0 V occurs remarkably in merely 6.5 ps, showcasing the system's efficient response when moving to the lowest power state. In contrast, the shift from 1.2 V to 0.8 V takes a slightly extended period of 11 ps, indicating subtle behavior during this specific power reduction. The transition from 0.8 V to 0.4 V unfolds

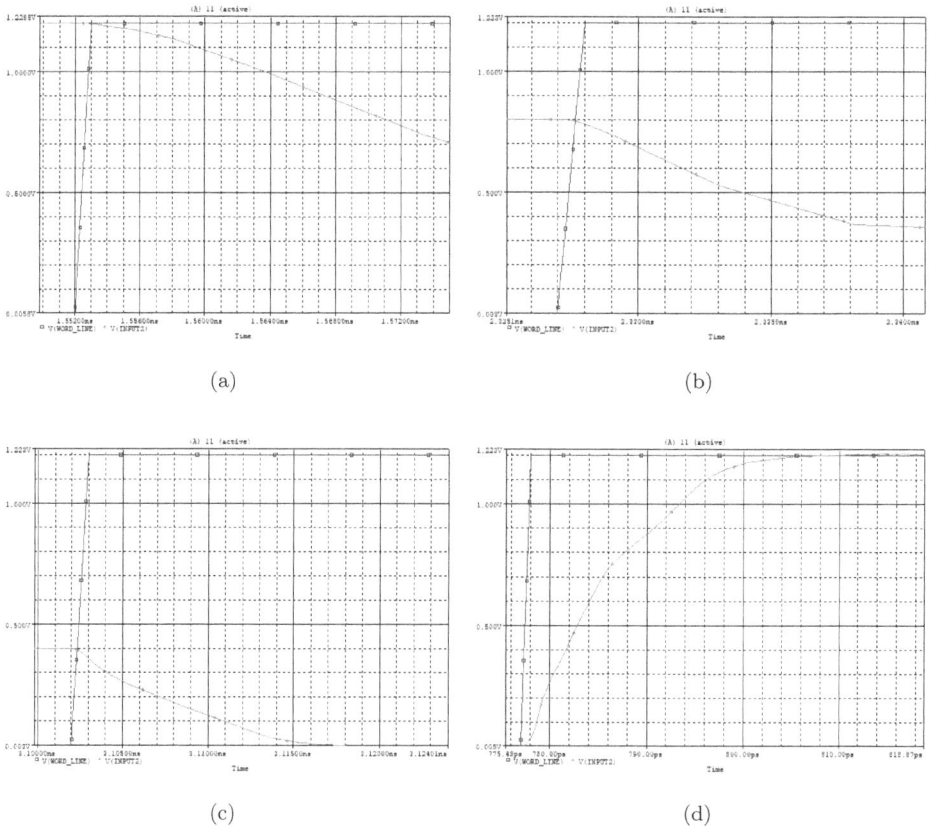

(a)

(b)

(c)

(d)

Fig. 15. The output response for specific input signal transitions.

Table 7. Propagation delay in 4-state/2-bit 10T SRAM-based SWS-FET at 1.2 V.

Transitions of States	3 to 2	2 to 1	1 to 0	0 to 3
Propagation Delay (ps)	11	4.2	4.7	6.5

over a shorter duration of 4.2 ps, emphasizing a relatively smoother shift through intermediate power levels. Finally, the rapid 4.7 ps transition from 0.4 V to 0 V shows the efficiency observed during the initial transition.

4.7. *Power dissipation simulation result*

Figure 16 shows the power dissipation in the SWS-FET 10T SRAM at parasitic capacitance of 1 fF under a 1.2 V power supply. In Table 8, the transition from 0 V to 1.2 V is associated with a power dissipation of 7.81 μW, reflecting the anticipated increase in power as the voltage rises. Subsequently, the transition from 1.2 V to 0.8 V shows an increase in power dissipation to 15.4 μW, aligning with the expected rise in dissipation with smaller voltage decreases. The substantial decrease in voltage from 0.8 V to 0.4 V corresponds to a higher power dissipation of 18.9 μW, consistent with the expected increase in dissipation for significant voltage drops. Finally, the transition from 0.4 V to 0 V yields a slightly higher power dissipation of 21.6 μW, showing the anticipated decrease in transitions associated with smaller changes in

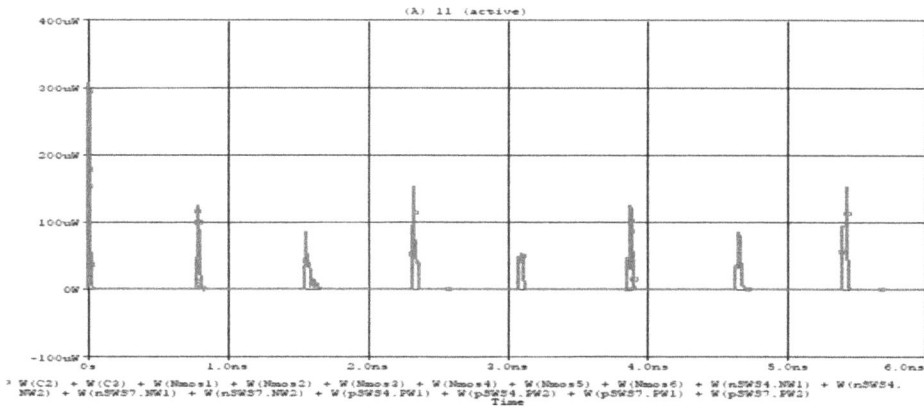

Fig. 16. Power dissipation in the SWS-FET 10T SRAM with a 1 fF parasitic capacitance at 1.2 V supply.

Table 8. Power dissipation in the SWS-FET 10T SRAM with a 1 fF parasitic capacitance at 1.2 V supply.

Transitions of States	3 to 2	2 to 1	1 to 0	0 to 3
Input Voltages	1.2 V to 0.8 V	0.8 V to 0.4 V	0.4 V to 0 V	0 V to 1.2 V
Av. Power Dissipation (μW)	15.4	18.9	21.6	7.81

voltage. Collectively, these calculations reflect a logical behavior of power dissipation during transitions in the 10T SRAM cell.

4.8. Summary of the findings and comparison of 7T–10T 4-states SRAMs

The transformation from 7T to 10T SRAM configurations introduces a notable impact on power dissipation and propagation delay. In terms of propagation delay, the 7T SRAM achieves varied durations, ranging from 5.5 ps during the 1.2 V to 0 V transition to 18.5 ps for the 0.8 V to 0.4 V transition. The 8T SRAM displays consistent swift transitions, with 5.7 ps for the 1.2 V to 0 V, but a slightly longer 13.5 ps for the 1.2 V to 0.8 V transition and 18.5 ps for the 0.8 V to 0.4 V transition. The 9T SRAM indicates increased delays, with 13.5 ps during the 1.2 V to 0.8 V transition. In contrast, the 10T SRAM, featuring an additional transistor, showcases a balance of efficient transitions, with 6.5 ps during the 1.2 V to 0 V shift and slightly longer 11 ps during the 1.2 V to 0.8 V transition.

In the case of power dissipation, the 7T SRAM experiences voltage changes of 6.85 μW when transitioning from 1.2 V to 0 V. The 8T SRAM, with an additional transistor, shows varying dissipation patterns, peaking at 29.6 μW during the 0 V to 1.2 V transition. The 9T SRAM, integrating another transistor, exhibits a mix of moderate and substantial power dissipation, reaching 15.3 μW during the 0.8 V to 0.4 V transition. The 10T SRAM demonstrates increasing power dissipation, hitting 21.6 μW during the 0.4 V to 0 V transition.

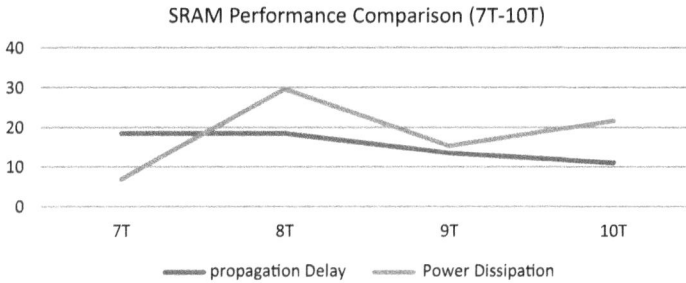

Fig. 17. Comparison of power dissipation and signal delay in different SRAM configurations (7T–10T).

Table 9. Comparison of power dissipation and signal delay in different SRAM configurations (7T–10T).

SRAM Type	Number of Transistors	Propagation Delay (ps)	Power Dissipation (μW)
7T	7	18.5	6.85
8T	8	18.5	29.6
9T	9	13.5	15.3
10T	10	11	21.6

The summary suggests that the increase in the number of transistors impacts both power dissipation and propagation delay, offering a trade-off between efficiency and complexity in SRAM design. Figure 17 and Table 9 provide a simple overview of power dissipation and propagation delay for different SRAM setups (7T–10T).

5. Conclusion

This study presents an in-depth analysis of power dissipation and propagation delay in 2-bit SRAM configurations ranging from 7T to 10T. Utilizing Cadence simulations and real-world 0.18-μm technology considerations, the research explores the design intricacies of SWS-FET SRAMs, highlighting their advantages in speed and power consumption. The analysis reveals varying trends in power dissipation and propagation delay across different SRAM configurations, showcasing the trade-offs between efficiency and complexity. Notably, the increase in the number of transistors from 7T to 10T impacts both power dissipation and propagation delay, emphasizing the need for careful optimization in SRAM design. Overall, this study contributes valuable insights into improving memory circuitry performance using advanced technologies like SWS-FETs.

ORCID

A. Husawi ◉ https://orcid.org/0009-0008-5926-0567

R. H. Gudlavalleti ◉ https://orcid.org/0000-0002-7727-8030

A. Almalki ◉ https://orcid.org/0009-0001-1954-4644

F. C. Jain ◉ https://orcid.org/0000-0003-3961-6665

References

1. F. C. Jain, B. Miller, E. Suarez, P.-Y. Chan, S. Karmakar, F. Al-Amoody, M. Gogna, J. Chandy and E. Heller. Spatial wavefunction-switched (SWS) InGaAs FETs with II–VI gate insulators. *Journal of Electronic Materials* 40, no. 8 (2011): 1717–1726.
2. F. Jain, B. Saman, R. Gudlavalleti, R. Mays, J. Chandy and E. Heller. Low-threshold II–VI lattice-matched SWS-FETs for multivalued low-power logic. *Journal of Electronic Materials* 50, no. 5 (2021): 2618–2629.
3. R. H. Gudlavalleti, B. Saman, R. Mays, E. Heller, J. Chandy and F. Jain. A novel peripheral circuit for SWSFET based multivalued static random-access memory. *International Journal of High Speed Electronics and Systems* 29, no. 01n04 (2020): 2040010.
4. B. Saman, P. Gogna, El-Sayed Hasaneen, J. Chandy, E. Heller and F. C. Jain. Spatial wavefunction switched (SWS) FET SRAM circuits and simulation. In *Microelectronics and Optoelectronics: The 25th Annual Symposium of Connecticut Microelectronics and Optoelectronics Consortium (CMOC 2016)*, pp. 37–48. 2017.
5. A. Almalki, B. Saman, J. Chandy, E. Heller and F. C. Jain. Propagation delay evaluation for spatial wavefunction switched (SWS) FET-based inverter. *International Journal of High Speed Electronics and Systems* 31, no. 01n04 (2022): 2240008.
6. F. Jain, M. Lingalugari, B. Saman, P. Y. Chan, P. Gogna, E. S. Hasaneen, J. Chandy and E. Heller. Multi-state sub-9 nm QDC-SWS FETs for compact memory circuits. In *46th IEEE Semiconductor Interface Specialists Conference (SISC)*, pp. 2–5. 2015.

7. A. Husawi, R. H. Gudlavalleti, B. Saman, A. Almalki, J. Chandy, E. Heller and F. C. Jain. Propagation delay and power dissipation analysis for a 2-bit SRAM using multi-state SWS inverter. *International Journal of High Speed Electronics and Systems* 32, no. 02n04 (2023): 2350023.
8. A. Husawi, B. Saman, A. Almalki, R. Gudlavalleti and F. C. Jain. Power dissipation and cell area: Quaternary logic CMOS inverter vs. four-state SWS-FET inverter. *International Journal of High Speed Electronics and Systems* 31, no. 01n04 (2022): 2240009.

Optimizing TCAD Model and Temperature-Dependent Analysis of Pt/AlN Schottky Barrier Diodes for High-Power and High-Temperature Applications#

Md Maruf Hossain ⓞ*, Showmik Singha ⓞ, Twisha Titirsha ⓞ, Sazia A. Eliza ⓞ
and Syed Kamrul Islam ⓞ

Department of Electrical Engineering and Computer Science
University of Missouri, Columbia, MO 65211, USA
*mh5md@umsystem.edu

This research presents a comprehensive investigation and optimization of the Pt/AlN Schottky Barrier diode (SBD) using technology computer-aided design (TCAD) modeling. The study explores the electrical characteristics of AlN SBDs with various metal contacts, including Aluminum (Al), Silver (Ag), Tungsten (W), Gold (Au), Nickel (Ni), and Platinum (Pt). Through the comparative analyses of different metal/AlN Schottky contacts, the Pt/AlN structure emerges as the most promising due to its superior barrier height and lower leakage current. At 300°K, the diode demonstrates a barrier height of 2.72 V, a nearly ideal leakage current of 0.046 pA, and a breakdown voltage of 363 V. The research extends to examining the temperature-dependent electrical behavior of Pt/AlN Schottky diodes, particularly for high-power and high-temperature applications. Analysis carried out across temperatures ranging from 300°K to 550°K reveals a trend of increasing ON resistance and consistently lower leakage current with rising temperature. Importantly, the study indicates that the impact of temperature on the barrier height and breakdown voltage of the diode is negligible, thus rendering it suitable for high-temperature operation. Leveraging the unique properties of AlN as an ultra-wide bandgap material within the III-V compound semiconductor family, this research provides valuable insights into the potential applications of Pt/AlN Schottky contact. The study highlights that the Pt/AlN Schottky contact is effective not only for high-power, high-temperature SBDs but also as superior metal/semiconductor gate contacts for field-effect transistors (FETs). Their suitability is attributed to their ability to handle high voltages, minimize reverse leakage current, and demonstrate improved thermal stability.

Keywords: Pt/AlN; Schottky barrier diode; TCAD modeling; reverse leakage current; ultra-wide bandgap.

*Corresponding author.
#This chapter appeared previously on the International Journal of High Speed Electronics and Systems. To cite this chapter, please cite the original article as the following: Md M. Hossain, S. Singha, T. Titirsha, S. A. Eliza and S. K. Islam, *Int. J. High Speed Electron. Syst.*, **33**, 2440065 (2024), doi: 10.1142/S0129156424400652.

1. Introduction

In recent years, the escalating demands of electric power, industrial control, consumer electronics, and automotive electronics industries have ignited an unprecedented need for high-performance power semiconductor devices [1,2]. This has led to intensive research focused on wide bandgap semiconductor materials. Among these, aluminum nitride (AlN) has emerged as a promising candidate due to its ultra-wide bandgap (6.2 eV) and exceptionally high breakdown electric field (15.4 MV/cm), setting it apart from conventional materials such as Silicon (Si), as well as later developed alternatives such as Silicon Carbide (SiC) and Gallium Nitride (GaN). Correspondingly, AlN exhibits a better thermal conductivity (340 W/mK) and notably greater Baliga's figure of merit (FOM) [3,4]. Moreover, achieving high-performance semiconductor devices hinges on precise fabrication of AlN layers, controlling thickness, crystal quality, and doping levels. Semiconductor growth technologies such as MOCVD and HVPE are pivotal for producing AlN layers with superior structural and electrical characteristics [4–8]. These parameters stand as a pivotal metric for gauging the suitability of a semiconductor material for power device applications. Hence, AlN emerges as a promising contender for the development of next-generation high-power devices, encompassing SBDs and FETs.

Schottky rectifiers are preferred for their quick switching capability, which is crucial for improving the efficiency of devices such as inductive motor controllers and power supplies, while also reducing on-state losses during operation [9,10]. Unlike bipolar devices, Schottky rectifiers are not hampered by minority-carrier storage effects, ensuring consistent performance [11, 12]. In enhancing SBD performance, vertical SBDs on conducting substrates are favored as they provide distinct advantages over lateral SBDs on insulating substrates. These vertical SBDs can manage higher power levels with full backside ohmic electrodes and facilitate increased current flow through optimal utilization of the conducting area [13–15].

The preference for SBDs arises particularly when high power and faster switching frequencies are required [16]. Given their majority carrier mechanism and lower reverse recovery time, Schottky diodes facilitate achieving faster switching frequencies at high voltage operations. Moreover, AlN SBD holds potential for attaining a higher barrier height and superior reverse breakdown voltage due to higher bandgap and lower electron affinity of AlN. The metal/semiconductor Schottky contact is pivotal not only for Schottky barrier diodes but also for high electron mobility transistor (HEMT) devices. It serves as the gate electrode, influencing crucial parameters such as gate capacitance, gate resistance, and switching speed. Optimizing this contact is vital for enhancing HEMT device performance, including transconductance, cutoff frequency, and power efficiency [17]. Understanding its behavior is crucial for advancing the capabilities of HEMT and tunneling field-effect transistors (TFET) in high-frequency and high-power applications [18, 19]. While researchers have fabricated AlN Schottky diodes with various device geometries analyzing their temperature-dependent current-voltage characteristics, insights remain limited. These studies have unveiled intriguing deviations from the ideal thermionic emission model and negative temperature dependencies in breakdown voltage and leakage current patterns, elucidated through a 2D variable-range hopping conduction model. Despite the limited fabrication outcomes regarding AlN Schottky barrier diodes, the absence of an optimized TCAD model poses a notable challenge. This deficiency impedes or delays research endeavors aimed at conducting performance comparisons under diverse operating conditions. Consequently, this study is dedicated to developing a comprehensive TCAD model for the metal/AlN Schottky barrier diode. The focus extends to capturing temperature-dependent electrical behaviors, including diode forward characteristics, reverse leakage current, and reverse breakdown voltage. Additionally, attention is directed towards exploring the electrical behavior of AlN Schottky diode with varying Schottky metals.

Utilizing this refined TCAD simulation model, researchers strive to enhance their grasp of device physics and precisely predict device performance under various operational scenarios. This TCAD simulation model for the metal/AlN Schottky diode facilitates understanding of the device physics and forecasting of the device behavior. Additionally, it aids in predicting device on-resistance and reverse leakage current at varying operating temperatures. Furthermore, it enables the prediction of the required Schottky metal contact for the AlN diode when specific barrier heights are needed. Ultimately, the optimized TCAD simulation model will serve as a valuable tool for designing AlN Schottky diodes tailored to specific applications within the semiconductor industry.

2. Device Schematic and Energy Band Diagram

The schematic depiction of the cross-sectional geometry of the vertical AlN SBD is depicted in Fig. 1. The structure features a 5-μm full-area back ohmic contact, succeeded by a 50-μm thick n-doped AlN layer, followed by a 5-nm undoped AlN layer. Atop the undoped AlN layer lies a 5-μm metal layer serving as a Schottky contact. The front contact is designated as the anode, while the back contact is termed the cathode. The front portion of the device is passivated by an oxide (SiO_2) layer. Additionally, a 5-μm undoped AlN layer is strategically employed to augment the Schottky barrier width and mitigate tunneling effects, thereby contributing to leakage current reduction.

In optimizing the simulation model, various ideal and non-ideal effects were considered to reconcile analytical and experimental data. Given that the Schottky diodes primarily operate via majority carrier transport mechanisms, the simulation model particularly focuses on these aspects, bypassing the intricate calculations of minority carrier phenomena typical in traditional semiconductor device simulations. The model encompasses several critical considerations. Mobility of carriers, contingent upon their concentration, is integrated into the simulation model to accurately reflect real-world behavior. Additionally, the model incorporates field-dependence of carrier mobility to capture nuanced transport phenomena. Shockley-Read-Hall

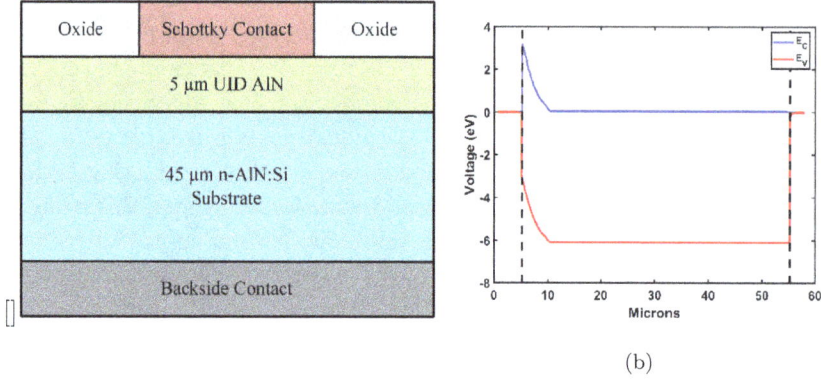

(b)

Fig. 1. (a) Device architecture of Pt/AlN Schottky diode, (b) energy band diagram of proposed Pt/AlN Schottky barrier diode.

(SRH) and Auger recombination mechanisms are accounted for in the model to simulate realistic carrier recombination processes [20–22]. Furthermore, the assumption of a perfect ohmic contact with zero barrier height is presumed for the cathode in the simulation model, facilitating accurate representation of device behavior. Alongside the ideal effects, the non-ideal factors such as bandgap lowering are considered in the simulation model to align simulated outcomes with experimental observations [23, 24]. For reverse characteristics analysis, both tunneling, and thermionic emission effects are considered to accurately capture the dominant transport mechanisms influencing the AlN Schottky rectifier.

Figure 1(a) portrays the energy band structure of the Pt/AlN Schottky rectifier. The energy band diagram is depicted along the Y-axis at the cut line x = 100. Notably, the depicted energy band structure indicates a barrier height of approximately 2.7 eV, a critical parameter influencing device performance and characteristics.

3. Device Physics

3.1. *Forward I–V characteristics*

The forward $I - V$ characteristics were investigated with the thermionic emission model [25, 26]. The current density generated by thermionic emission is represented as

$$J = J_o \left[\exp \left(\frac{eV}{nkT} \right) - 1 \right], \tag{1}$$

$$J_o = A^* T^2 \exp \left(-\frac{e\emptyset_b}{kT} \right), \tag{2}$$

where J represents the current density, J_o denotes the saturation current density, A^* stands for the Richardson constant, T represents the temperature in Kelvin, q is the charge of an electron, $\emptyset_b$ indicates the Schottky barrier height, n is the ideality factor, and k is the Boltzmann constant. The calculation employed a Richardson constant of $38.4\,\mathrm{Acm^{-2}K^{-2}}$ with an effective electron mass of 0.32 times the free electron mass (m_0). After the on voltage the slope of forward characteristics curve would be the diode on resistance.

3.2. *Reverse leakage current*

In SBDs, the reverse leakage current, which flows when the diode is under reverse bias, is a critical parameter influencing device performance. This leakage current can be attributed to various mechanisms, including thermionic emission, tunneling, and field emission processes. Understanding and controlling these mechanisms are essential for optimizing the performance of the SBDs, especially in high-frequency and high-power applications.

Moreover, the presence of a high surface electric field in SBDs can significantly impact the reverse leakage current behavior. Under such conditions, tunneling processes, such as thermionic field emission (TFE) and field emission (FE), become dominant contributors to the reverse leakage current [27]. Analytical models, such as those developed by Murphy and Good, and later refined by Padovani and Stratton, aim to describe and quantify these tunneling processes [28]. However, these models may overlook certain factors, such as the doping effect in semiconductors or the influence of the image force lowering (IFL) phenomenon.

To address these limitations and provide a more comprehensive understanding of reverse leakage characteristics in the SBDs, researchers have developed numerical models that incorporate factors such as doping effects and IFL [9,29]. These models enable accurate prediction and analysis of the total reverse leakage current density (J), allowing for improved device design and optimization.

The total reverse leakage current density (J) is given by

$$J = \frac{A}{k_B} \int_{\varepsilon_{\min}}^{+\infty} \mathcal{T}(\mathcal{E}) \times In\left[1 + \exp\left(-\frac{\epsilon - \varepsilon_{Fm}}{k_B T}\right)\right] d\varepsilon, \tag{3}$$

where $A = 4\pi m^* k_B^2 e/h^3$ is the Richardson constant, E is the electron energy, ε_{Fm} is the Fermi-level energy in metal, and $E_{\min}$ is the minimum energy required for tunneling which is equal to the effective constant potential energy inside the metal. $\mathcal{T}(\mathcal{E})$ represents the probability of transmission across the barrier. Using ε_{Fm} as the zero-energy level, the potential energy distribution of the Schottky barrier under IFL is given by

$$\varepsilon_c(x) = e\emptyset_B - eE_x - \frac{e^2}{16\pi\varepsilon_x x} + \frac{e^2 N_D x^2}{2\varepsilon_s}, \tag{4}$$

where $\emptyset_B$ is the Schottky barrier height, E is the surface electric field, N_D is the net donor concentration, and ε_s is the dielectric constant of AlN ($8.9\varepsilon_o$). The third and the fourth terms in Eq. (4) represent the IFL and doping effects, respectively. IFL rounds and lowers the top of the barrier by $\Delta\emptyset = \sqrt{eE/4\pi\varepsilon_s}$, resulting in $\varepsilon_{c,\max} = e(\emptyset_B - \Delta\emptyset)$ as seen in Fig. 1(a). This barrier-rounding effect must be carefully considered to accurately evaluate the tunneling probability using Eq. (4).

4. TCAD Model Optimization and Simulation Results

In the evaluation of the proposed AlN Schottky barrier diode, the focus was on its forward and reverse characteristics, break-down voltage, and reverse leakage current. The device model was calibrated using experimental data obtained from the Ni/AlN Schottky barrier diode [15]. Subsequently, this optimized model was employed to assess the performance of various metal contacts, including Al, Ag, W, Au, Ni, and Pt. The objective was to understand the relationship between the Schottky barrier height and the work functions of these metals.

Among the metals studied, the Pt/AlN Schottky device structure emerged as the preferred choice due to its superior performance compared to the previously proposed Ni/AlN Schottky rectifiers. This preference was established through a comprehensive analysis of the performance matrix for different metal contacts. Furthermore, exploration was conducted into the temperature-dependent electrical characteristics of the Pt/AlN Schottky diode to assess its suitability for high-temperature operations. These findings contribute valuable insights into the potential applications and performance optimization of Schottky barrier diodes in various electronic devices.

4.1. *Forward characteristics with different Schottky metals*

In Fig. 2, the forward characteristics of the AlN SBD are depicted across various Schottky metals at an operating temperature of 300°K. These metals include aluminum (Al), silver (Ag), tungsten (W), gold (Au), nickel (Ni), and platinum (Pt), all recognized for their applications in Schottky gate contacts or Schottky diodes. The $J-V$ characteristics curve depicted in Fig. 2, illustrates the compatibility between

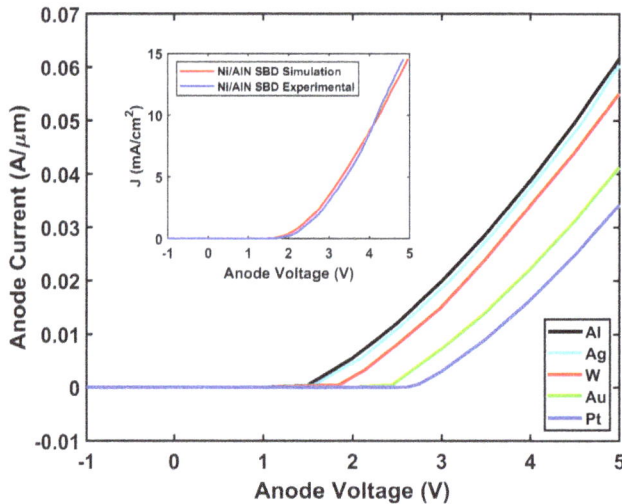

Fig. 2. Forward characteristics of AlN Schottky barrier diode for different Schottky metals.

our TCAD model and experimental data for Ni/AlN SBDs, as reported in [15]. The red line represents the experimental forward $J-V$ curve of the Ni/AlN SBD, while the blue line signifies the TCAD simulation results, showcasing a remarkable alignment between the two datasets. This validation confirms the reliability and accuracy of our TCAD model for simulating AlN SBD behaviors.

Leveraging the optimized TCAD model, the electrical behavior is examined for different Schottky metals. The $I-V$ characteristics in Fig. 2 vividly demonstrate that the Schottky barrier height or diode voltage escalates with increasing metal work functions. Among the six metals studied, aluminum (Al) boasts the lowest work function at 4.26 eV, while platinum (Pt) boasts the highest at 5.6 eV in terms of Schottky barrier height. Consequently, Al/AlN showcases a minimum barrier height of approximately 1.51 V, while Pt/AlN displays the highest barrier height of around 2.65 V. Table 1 represents the correlation between the metal work functions and the Schottky barrier height. Drawing from these insights, we propose a Pt/AlN metal-semiconductor contact. Furthermore, the measurements of the reverse characteristics of Pt/AlN were conducted at an operating temperature of 300°K to examine the reverse breakdown voltage and reverse leakage current of the diode.

Figure 3 illustrates the reverse characteristics of the Pt/AlN Schottky barrier diode. Analysis of the reverse characteristics reveals a reverse breakdown voltage (V_{BR}) of approximately 363 V for the Pt/AlN Schottky rectifiers, rendering the device suitable for high-voltage applications. The high Schottky barrier height at the Pt/AlN interface mitigates the tunneling effect and thermionic emission, contributing to the attainment of a high reverse breakdown voltage and a lower reverse leakage current. The $I-V$ curve within Fig. 3, displayed on a logarithmic scale, indicates a leakage current of approximately 0.046 pA at an operating temperature of 300°K. This leakage current remains consistent within a similar range from 0 V to –100 V reverse voltage, suggesting near-ideal leakage current behavior. Lower off-state leakage current is always preferred in semiconductor devices such as transistors and diodes. This observation suggests that a Pt/AlN Schottky contact could serve as an optimal choice for Schottky gate contacts when necessitating a high threshold voltage and lower off-state gate leakage current for the device.

Table 1. Forward characteristics parameters of AlN SBD with different Schottky metals.

Schottky metal	Metal work-function ($\emptyset_m$)	V_{bi} from simulation
Aluminum (Al)	4.26 eV	1.51 V
Silver (Ag)	4.28 eV	1.5 V
Tungsten (W)	4.55 eV	1.90 V
Gold (Au)	5.10 eV	2.44 V
Platinum (Pt)	**5.65 eV**	**2.72 (Proposed)**

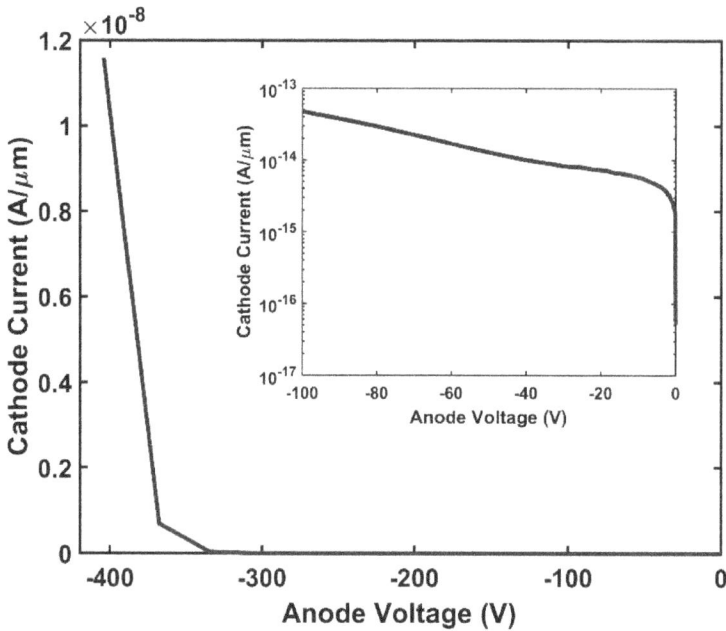

Fig. 3. Reverse characteristics of Pt/AlN Schottky barrier diode at 300°K temperature.

4.2. *Temperature modeling of Pt/AlN Schottky rectifiers*

When considering high-power applications, it is crucial for devices to exhibit consistent behavior even under high temperatures. To address extreme operating conditions, temperature modeling of Pt/AlN was undertaken to observe its behavior across different temperatures. Theoretical principles suggest that elevated temperatures typically exacerbate the ON resistance and leakage current of semiconductor devices. In this study, the forward and the reverse characteristics of Pt/AlN Schottky barrier diodes were examined across various operating temperatures ranging from 300°K to 550°K at 50°K intervals.

Figure 4 illustrates the forward $I-V$ characteristics of the Pt/AlN Schottky diode at different operating temperatures within the specified range. The figure demonstrates that the ON voltage of the device remains relatively consistent across this temperature range. However, the ON resistance of the device, depicted by the slopes of the forward $I-V$ curves after the device switches on, increases with temperature. Table 2 provides a summary of the Pt/AlN Schottky diode's ON resistance at various operating temperatures. This increasing behavior in the ON resistance of the Pt/AlN SBD with increasing temperature aligns with the majority carrier semiconductor device theory.

Figure 5 depicts the reverse leakage current of the Pt/AlN Schottky barrier diode across different operating temperatures. The reverse $I-V$ characteristics of the

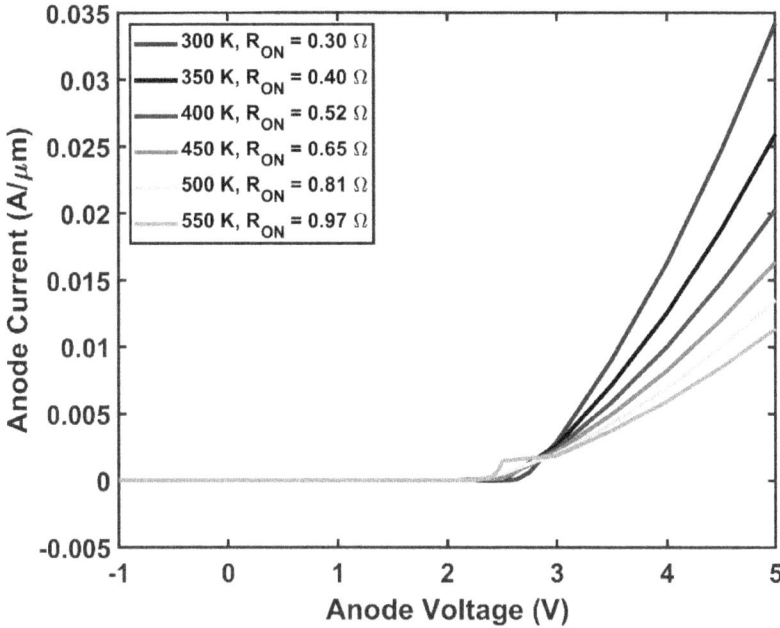

Fig. 4. Forward characteristics of Pt/AlN Schottky barrier diode at different temperatures.

Table 2. Temperature-dependent parameters of Pt/AlN Schottky barrier diodes.

Temperature	Reverse leakage current	On resistance ($\Omega.\mu m^2$)
300° K	0.046 pA	0.30
350° K	0.72 pA	0.41
400° K	2.10 pA	0.52
450° K	5.12 pA	0.65
500° K	7.81 pA	0.81
550° K	11.2 pA	0.97

proposed Schottky contact were examined by varying the voltage from 0 V to –100 V. The plot of anode voltage versus cathode current in Fig. 5 clearly illustrates that the reverse leakage current increases with temperature. At 300°K, the leakage current of the Pt/AlN Schottky rectifier measures around 0.046 pA, while at 550°K, this leakage current escalates to 11.2 pA. Despite the increase in reverse leakage current with temperature, the Pt/AlN Schottky rectifier maintains a relatively low leakage current even at extreme temperatures of 550°K. Based on the simulation data of the reverse leakage current of the Pt/AlN Schottky contact at high temperatures, it is anticipated that the proposed Pt/AlN metal semiconductor contact will exhibit superior performance at elevated temperatures by minimizing leakage current and ensuring reliable device operation under extreme conditions. Table 2 summarizes the

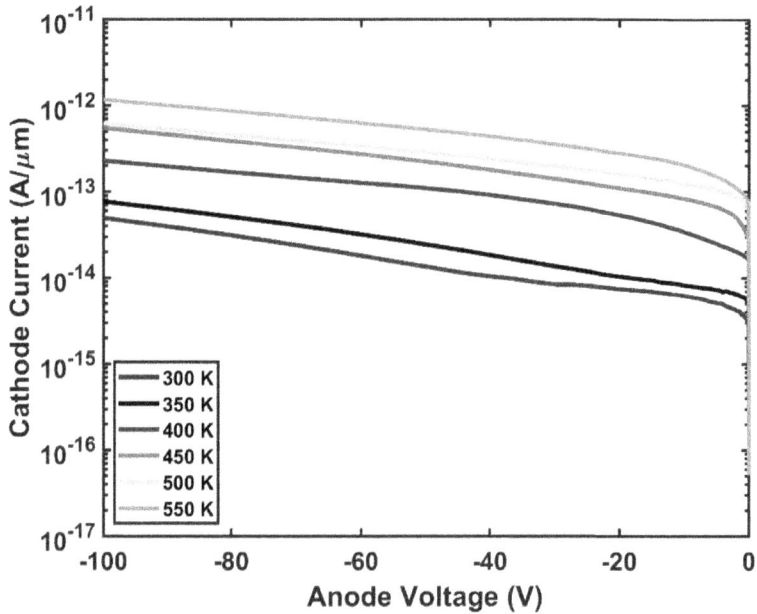

Fig. 5. Reverse leakage characteristics of Pt/AlN Schottky barrier diode at different temperatures.

reverse leakage current behavior at various operating temperatures for the Pt/AlN Schottky barrier diode.

5. Conclusion

This study extensively investigates and optimizes Pt/AlN Schottky Barrier diodes (SBDs) using TCAD modeling. It focuses on exploring the electrical characteristics of AlN Schottky Barrier diodes with various metal contacts, ultimately identifying Pt/AlN as the most promising structure due to its superior barrier height and lower leakage current. Temperature-dependent behaviors show negligible impact on barrier height and breakdown voltage, rendering Pt/AlN SBDs suitable for high-temperature operations. Leveraging ultra-wide bandgap properties of AlN, Pt/AlN Schottky contacts show potential for high-power and high-temperature applications, including as superior metal/semiconductor gate contacts for FETs. Vertical SBDs on conducting substrates are preferred for their quick switching ability and higher power handling capabilities compared to lateral SBDs on insulating substrates. The significance of AlN as a promising semiconductor material for high-performance power devices, including SBDs and FETs, is emphasized along with the importance of precise fabrication techniques. Future work potentially involves developing SPICE model simulations for Pt/AlN Schottky diodes to accurately measure circuit-level performance and investigating radiation effects under extreme operating conditions which crucial for various applications in challenging environments.

ORCID

Md Maruf Hossain ☉ https://orcid.org/0009-0007-0287-8131

Showmik Singha ☉ https://orcid.org/0000-0003-0011-5273

Twisha Titirsha ☉ https://orcid.org/0000-0002-2142-2283

Sazia A. Eliza ☉ https://orcid.org/0009-0000-8706-0791

Syed Kamrul Islam ☉ https://orcid.org/0000-0002-0501-0027

References

1. S. Fujita, "Wide-bandgap semiconductor materials: For their full bloom," *Jpn J Appl Phys*, vol. 54, no. 3, p. 030101, Mar. 2015, doi: 10.7567/JJAP.54.030101.
2. J. Y. Tsao, S. Chowdhury, M. A. Hollis and D. Jena, "Ultrawide-bandgap semiconductors: Research opportunities and challenges," *Adv Electron Mater*, vol. 4, no. 1, Jan. 2018, doi: 10.1002/aelm.201600501.
3. A. L. Hickman, R. Chaudhuri, S. J. Bader, K. Nomoto and L. Li, "Next generation electronics on the ultrawide-bandgap aluminum nitride platform," *Semicond Sci Technol*, vol. 36, no. 4, p. 044001, Apr. 2021, doi: 10.1088/1361-6641/abe5fd.
4. R. Rounds, B. Sarkar, A. Klump and C. Hartmann, "Thermal conductivity of single-crystalline AlN," *Appl Phys Exp*, vol. 11, no. 7, p. 071001, Jul. 2018, doi: 10.7567/APEX.11.071001.
5. W.-J. Lin and J.-C. Chen, "Numerical study of growth rate and purge time in the AlN pulsed MOCVD process," *Crystals (Basel)*, vol. 12, no. 8, p. 1101, Aug. 2022, doi: 10.3390/cryst12081101.
6. H. Fu, X. Huang, H. Chen, Z. Lu and Y. Zhao, "Fabrication and characterization of ultra-wide bandgap AlN-based Schottky diodes on sapphire by MOCVD," *IEEE J Electr Dev Soc*, vol. 5, no. 6, pp. 518–524, Nov. 2017, doi: 10.1109/JEDS.2017.2751554.
7. H. Wu, K. Zhang, C. He, L. He, Q. Wang, W. Zhao, and Z. Chen, "Recent advances in fabricating wurtzite AlN film on (0001)-plane sapphire substrate," *Crystals (Basel)*, vol. 12, no. 1, p. 38, Dec. 2021, doi: 10.3390/cryst12010038.
8. P. Reddy, I. Bryan; Z. Bryan, and W. Guo, "The effect of polarity and surface states on the Fermi level at III-nitride surfaces," *J Appl Phys*, vol. 116, no. 12, Sep. 2014, doi: 10.1063/1.4896377.
9. W. Li, D. Saraswat, Y. Long, K. Nomoto, D. Jena, and H. G. Xing, "Near-ideal reverse leakage current and practical maximum electric field in β-Ga2O3 Schottky barrier diodes," *Appl Phys Lett*, vol. 116, no. 19, May 2020, doi: 10.1063/5.0007715.
10. Y. Yao, R. Gangireddy, J. Kim, K. K. Das, R. F. Davis, and L. M. Porter, "Electrical behavior of β-Ga2O3 Schottky diodes with different Schottky metals," *J Vacuum Sci Technol B, Nanotechnol Microelectron: Mater, Process, Measure, Phenomena*, vol. 35, no. 3, May 2017, doi: 10.1116/1.4980042.
11. G. Gupta, S. Dutta, S. Banerjee, and R. J. E. Hueting, "Minority carrier injection in high-barrier Si-Schottky diodes," *IEEE Trans Electr Dev*, vol. 65, no. 4, pp. 1276–1282, Apr. 2018, doi: 10.1109/TED.2018.2807926.

12. M. A. Green and J. Shewchun, "Minority carrier effects upon the small signal and steady-state properties of Schottky diodes," *Solid State Electr*, vol. 16, no. 10, pp. 1141–1150, Oct. 1973, doi: 10.1016/0038-1101(73)90141-X.

13. T. Pu, U. Younis, H.-C. Chiu, K. Xu, H.-C. Kuo, and X. Liu, "Review of recent progress on vertical GaN-Based PN diodes," *Nanoscale Res Lett*, vol. 16, no. 1, p. 101, Dec. 2021, doi: 10.1186/s11671-021-03554-7.

14. D. Disney, Hui Nie, A. Edwards, D. Bour, H. Shah, and I. C. Kizilyalli, "Vertical power diodes in bulk GaN," in 2013 *25th International Symposium on Power Semiconductor Devices & IC's (ISPSD)*, IEEE, May 2013, pp. 59–62. doi: 10.1109/ISPSD.2013.6694455.

15. D.-H. Kim, S.-J. Min, J.-M. Oh, and S.-M. Koo, "Fabrication and characterization of oxygenated AlN/4H-SiC heterojunction diodes," *Materials*, vol. 13, no. 19, p. 4335, Sep. 2020, doi: 10.3390/ma13194335.

16. H. Fu, I. Baranowski, X. Huang, H. Chen, Zhijian Lu, and J. Montes, "Demonstration of AlN Schottky barrier diodes with blocking voltage over 1 kV," *IEEE Electr Dev Lett*, vol. 38, no. 9, pp. 1286–1289, Sep. 2017, doi: 10.1109/LED.2017.2723603.

17. M. M. Hossain, M. M. H. Shuvo, T. Titirsha, and S. K. Islam, "Modeling of enhancement mode HEMT with Π-gate optimization for high power applications," *Int J High-Speed Electron Syst*, vol. 32, no. 02n04, Dec. 2023, doi: 10.1142/S0129156423500064.

18. Md. M. Hossain, Md. M. Hassan, and Md. M. Rahman Adnan, "Impact of structural modification by spacer layer inclusion on AlGaN/GaN HEMT performance," in 2021 *International Conference on Information and Communication Technology for Sustainable Development (ICICT4SD)*, IEEE, Feb. 2021, pp. 76–81. doi: 10.1109/ICICT4SD50815.2021.9396787.

19. S. Hussain, N. Mustakim, S. Singha, and J. K. Saha, "A Comprehensive study on tunneling field effect transistor using non-local band-to-band tunneling model," *J Phys Conf Ser*, vol. 1432, no. 1, p. 012028, Jan. 2020, doi: 10.1088/1742-6596/1432/1/012028.

20. W. Shockley and W. T. Read, "Statistics of the recombinations of holes and electrons," *Phys Rev*, vol. 87, no. 5, pp. 835–842, Sep. 1952, doi: 10.1103/PhysRev.87.835.

21. R. N. Hall, "Electron-hole recombination in germanium," *Phys Rev*, vol. 87, no. 2, pp. 387–387, Jul. 1952, doi: 10.1103/PhysRev.87.387.

22. M. Farahmand, C. Garetto, E. Bellotti, K. F. Brennan, M. Goano, and E. Ghillino, "Monte Carlo simulation of electron transport in the III-nitride wurtzite phase materials system: Binaries and ternaries," *IEEE Trans Electr Dev*, vol. 48, no. 3, pp. 535–542, Mar. 2001, doi: 10.1109/16.906448.

23. M. P. Lepselter and S. M. Sze, "Silicon Schottky barrier diode with near-ideal I-V characteristics," *Bell Syst Tech J*, vol. 47, no. 2, pp. 195–208, Feb. 1968, doi: 10.1002/j.1538-7305.1968.tb00038.x.

24. S. Chand and J. Kumar, "Effects of barrier height distribution on the behavior of a Schottky diode," *J Appl Phys*, vol. 82, no. 10, pp. 5005–5010, Nov. 1997, doi: 10.1063/1.366370.

25. T. Kinoshita, T. Nagashima, T. Obata, S. Takashima, R. Yamamoto, and R. Togashi, "Fabrication of vertical Schottky barrier diodes on n-type freestanding AlN substrates grown by hydride vapor phase epitaxy," *Appl Phys Exp*, vol. 8, no. 6, p. 061003, Jun. 2015, doi: 10.7567/APEX.8.061003.

26. Q. Zhou, H. Wu, H. Li, and X. Tang, "Barrier inhomogeneity of schottky diode on nonpolar AlN grown by physical vapor transport," *IEEE J Electr Dev Soc*, vol. 7, pp. 662–667, 2019, doi: 10.1109/JEDS.2019.2923204.

27. E. L. Murphy and R. H. Good, "Thermionic emission, field emission, and the transition region," *Phys Rev*, vol. 102, no. 6, pp. 1464–1473, Jun. 1956, doi: 10.1103/PhysRev.102.1464.

28. F. A. Padovani and R. Stratton, "Field and thermionic-field emission in Schottky barriers," *Solid State Electr*, vol. 9, no. 7, pp. 695–707, Jul. 1966, doi: 10.1016/0038-1101(66)90097-9.

29. H. Fu, H. Chen, X. Huang, I. Baranowski, J. Montes, T. Yang, and Y. Zhao "A comparative study on the electrical properties of vertical β-Ga2O3 Schottky barrier diodes on EFG single-crystal substrates," *IEEE Trans Electr Dev*, vol. 65, no. 8, pp. 3507–3513, Aug. 2018, doi: 10.1109/TED.2018.2841904.

https://doi.org/10.1142/9789811297427_0015

Additively Manufactured, Flexible 5G Electronics for MIMO, IoT, Digital Twins, and Smart Cities Applications[#]

Theodore W. Callis [*], Kexin Hu, Hani Al Jamal
and Dr. Manos M. Tentzeris

*School of Electrical and Computer Engineering,
Georgia Institute of Technology, 85 5th St NW,
Atlanta, Georgia, 30308, USA
tcallis7@gatech.edu

This review encompasses additive manufacturing techniques for crafting 5G electronics, showcasing how these methods innovate device creation with novel examples. A wearable phased array device on commonplace 3D printed material is described, with integrated microfluidic cooling channels used for thermal regulation of integrated circuit bulk components. Mechanical and electrical tunability are exemplified in an origami-inspired phased array structure. A 3D printed IoT cube structure shows the flexibility in the number of geometries additively manufactured 5G devices can adhere to. Finally, integrating 3D optical lenses with 5G electronics is shown.

Keywords: 5G; additive manufacturing; inkjet printing; 3D printing; phased array; IoT; origami; RFID.

1. Introduction

A continuing rise in the popularity of wireless electronic devices for applications such as smart cities, supply chain management, healthcare, and asset tracking is taking place. IoT and sensor integration with their respective digital twins is becoming increasingly commonplace. There is a growing need for innovation in the fabrication of these devices and their respective components as well as their performance. Key device metrics include the range, localization, sensitivity, flexibility, cost, geometry customization, and time to prototype. The vast range of quality factors to improve and the increase in the ubiquity of wireless electronics in the modern era underscore the importance of the researches performed.

---

*Corresponding author.
#This chapter appeared previously on the International Journal of High Speed Electronics and Systems. To cite this chapter, please cite the original article as the following: T. W. Callis, K. Hu, H. A. Jamal and M. M. Tentzeris, *Int. J. High Speed Electron. Syst.*, **33**, 2440066 (2024), doi: 10.1142/S0129156424400664.

Inkjet and 3D printing additive manufacturing technologies allow for quick fabrication of 5G electronics in a wide variety of shapes and substrates. Inkjet printing of conductive traces has proven to be an efficient and quick method of patterning substrates for electronic devices [1, 2]. An additional benefit of inkjet printing is the fabrication of 5G flexible devices which provide a plethora of new applications [3]. Combining both inkjet and 3D printing creates more degrees of freedom when designing electronics. The design process has transcended the constraints of rigid substrates and planar designs, and has become swifter and more environmentally friendly process.

2. Additively Manufactured Phased Array Antennas for Wearable, Digital Twin, and Massive MIMO Applications

The evolution of 5G/mm Wave technologies, characterized by their expansive bandwidth and exceptional data rates, holds the promise of revolutionizing various sectors, including IoT systems, smart cities, and health monitoring wearables. This advancement underscores the necessity for compact, intricate System-on-Package (SoP) designs capable of real-time responsiveness, deployable packaging, and high integration levels. These designs are pivotal for digital twins, structural health monitoring of large structures, robotic applications with rigorous bending requirements, and the implementation of conformal intelligent metasurfaces, particularly in the context of Industry 4.0 applications. While flexible hybrid electronics (FHE) are commonly employed in wearable designs due to their flexibility, most operate within low-frequency ranges, such as MHz. Meeting the demands for mm Wave operation, which include reliable performance across a wide operational bandwidth, minimal parasitic losses from interconnects, and the integration of high-gain phased arrays with beamforming capabilities to mitigate high path loss, remains a significant challenge. Additionally, these systems must possess a conformal form factor and exhibit resilience to bending and on-body effects from both RF and mechanical standpoints, thereby enabling the implementation of massively scalable MIMO systems.

To address these challenges, a customizable mm Wave flexible packaging and reconfigurable on-package phased array was proposed for 5G wearable applications, conformal digital twins, and deployable massive MIMO systems [4]. Leveraging 3D printing and inkjet printing technologies, this approach enables the production of highly customized electronic devices at reduced costs. Furthermore, the paper investigates the electrical and mechanical properties of 3D printed and inkjet printed materials over the mm Wave band, characterizing different flexible materials from Fused Deposition Modeling (FDM) printing and Stereolithography (SLA) printing. The study also optimizes the fabrication process of inkjet printing onto 3D printed substrates and demonstrates the reliability of fabricated microstrip line and interconnect samples through extensive cyclic bending tests. Moreover, the paper presents a proof-of-concept demonstration of an on-package phased array with an integrated microfluidic cooling channel, as shown in Fig. 1,

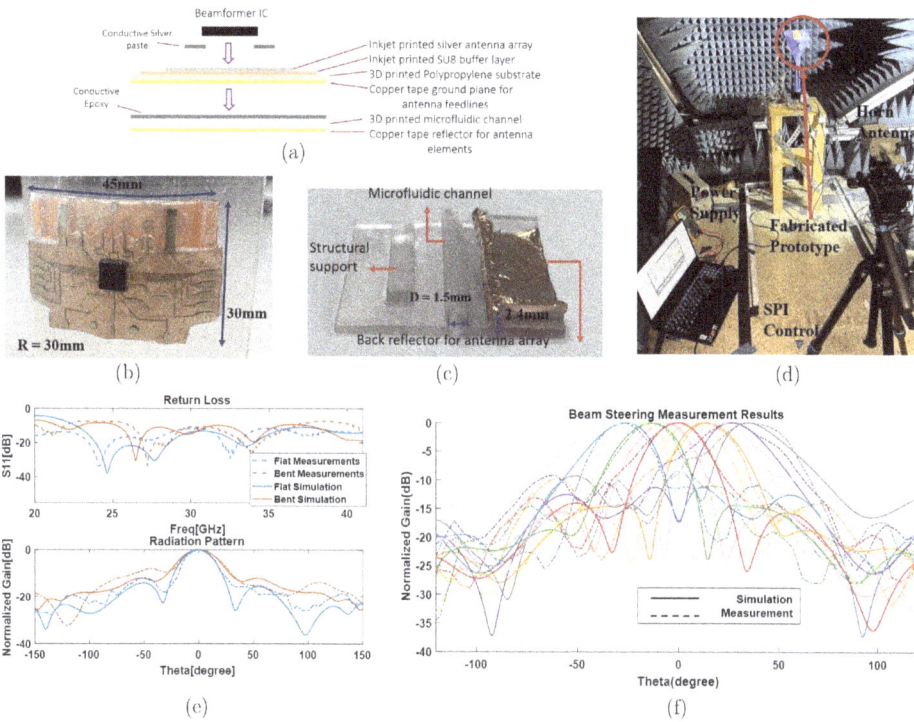

Fig. 1. (a) Cross-sectional schematic of the fabrication process of the proof-on-concept module. (b) Top view of the flexible on-package phased antenna array bent over a curved surface with a 30-mm radius. (c) Side view of the SLA printed microfluidic channel module. (d) Measurement setup. (e) Comparison of simulated and measured S11 values (top) and radiation pattern [H plane] (bottom) for the proof-of-concept prototype in flat and bent with a radius of 30 mm configurations. (f) Beam steering measurement results of the fabricated phased array module prototype with an integrated microfluidic channel [4].

showcasing the reconfigurability and antenna performance of the proposed packaging concept. Overall, this investigation lays the groundwork for next-generation 5G mm Wave broadband FHE, flexible multichip modules (MCM), and phased array on-package modules, offering valuable insights for wearable, smart skin, and implantable applications.

3. Additively Manufactured, Origami-Inspired Phased Array

An additively manufactured, origami-inspired shape-changing, and RFIC-based phased array is presented in [5] for near-limitless radiation pattern reconfigurability in 5G/mm Wave applications. Operating at 28 GHz, this phased array integrates on-structure beamformer RFICs and a flexible feeding network, employing foldable hinge interconnects. By leveraging principles from origami, the art of paperfolding, this array achieves near 360° beamsteering in the azimuth plane, providing

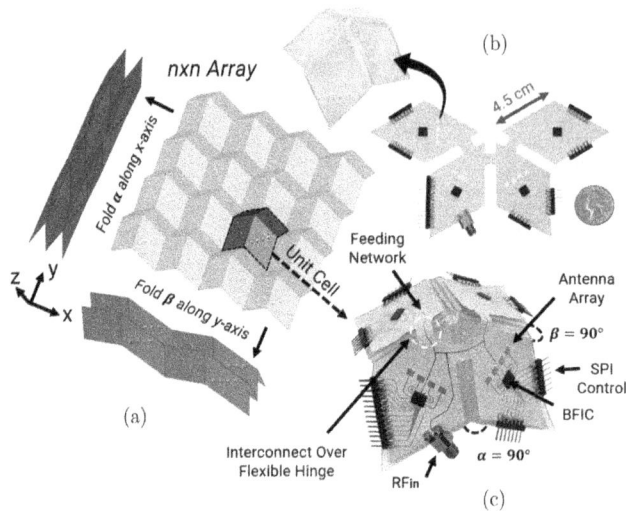

Fig. 2. Origami-inspired, RFIC-based, and fully integrated phased array incorporating foldable "arch" interconnects. (a) nxn array of an "eggbox" origami. (b) Fabrication process. (c) Fabricated unit cell on a flexible 3D printed substrate (Formlabs flexible80A) [5].

reconfigurable radiation patterns crucial for optimizing communication in challenging environments.

Traditionally phased arrays encounter limitations such as restricted angular coverage, narrow beamwidths, bulkiness, and lack of adaptability. In contrast, this origami-inspired design addresses these challenges by offering multi-beam or quasi-isotropic radiation patterns through shape-changing configurations. The utilization of additive manufacturing techniques, including 3D and inkjet printing, enables the fabrication of a lightweight and cost-effective prototype, as shown in Fig. 2.

Central to the design is the origami eggbox structure, foldable along two planes, allowing for various configurations to adapt to different application scenarios. The array achieves seamless integration of beamformer RFICs and a face-to-face feeding network, ensuring stable performance at mm-wave frequencies. In addition, foldable hinge interconnects, known as "arch" hinges, play a pivotal role in enabling reliable performance even under repeated folding.

Measurements validate the array's capabilities, showcasing radiation pattern reconfigurability through two degrees of freedom. This includes simultaneous electrical beamsteering and mechanical shape-change adaptability, shown in Fig. 3. Additionally, the modular tile-based approach permits selective activation of unit cells and faces, offering flexibility in configuring transmission/reception modes and power consumption requirements.

The significance of this origami-inspired phased array extends beyond its technical achievements. It represents a paradigm shift in reconfigurable multi-beam/multifunction phased arrays, demonstrating the potential for enhanced

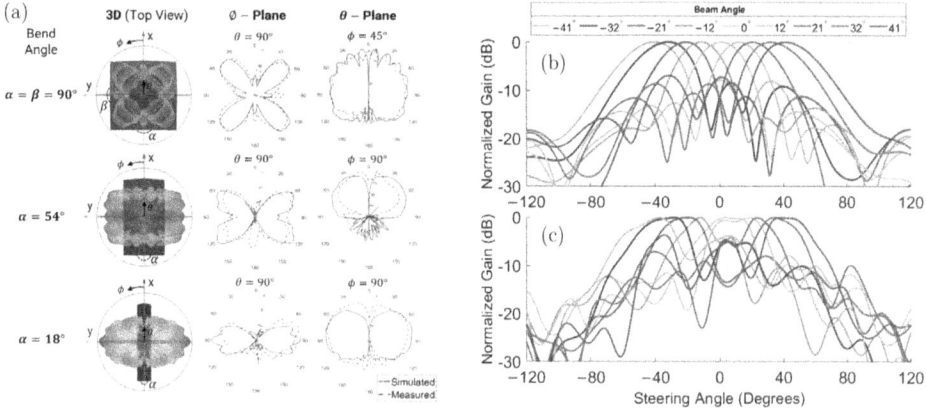

Fig. 3. (a) Simulated and measured radiation patterns for different mechanical bend angles (mechanical reconfigurability). (b) Simulated, (c) measured beamsteering capability with a single eggbox face activated (electrical reconfigurability) [5].

performance, adaptability, and cost-effectiveness in 5G/mm Wave applications. The array's expanded coverage angles, improved beam-steering precision, and radiation pattern reconfigurability pave the way for advancements for simultaneous multi-function phased arrays for smart cities, industry 4.0, and consumer-focused IoT, unlocking new possibilities for high-speed communication in diverse environments.

4. 3D Magic Cube Platform with Fully Integrated Sensors for IoT Applications

A 3D printed $1 \times 1 \times 2$ cm rectangular prism structure was integrated with wireless sensing nodes on each face of the polygon in [6]. The subsystems are designed for seamless integration and different sensor nodes can be incorporated onto each face in a plug-and-play type design. The 3D printed platform has a practical application in wireless sensing networks for smart agriculture, infrastructure, manufacturing, and homes. The additive process allows for rapid prototyping and complete integration of sensing nodes onto a singular IoT device at low cost.

The fabrication process is shown in Fig. 4. All faces of the prism are fabricated together, rather than stitching them together post-fabrication which would take much more time. SU8 photoresist was inkjet printed as a buffer layer to print the silver on. Directly printing silver on 3D printed material is often challenging as these materials are often too hydrophobic and have striation on the surface. After the SU8 buffer layer is printed, a silver nanoparticle ink was inkjet printed atop the SU8 traces. All bulk components were soldered with a silver epoxy paste that is able to cure at room temperature, such that the 3D printed substrate does not melt during the soldering process.

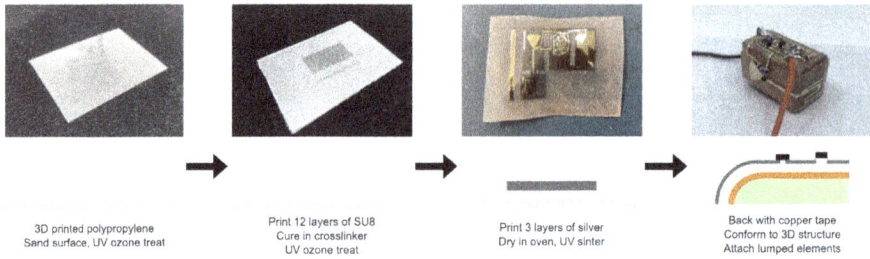

Fig. 4. Fabrication of the cube-like IoT Device as seen in [6].

In this example, a simple thermistor is used to monitor the temperature of the environment. However, this could easily be swapped out with other sensing node architectures for monitoring of different environmental variables. The power module could also be swapped out for a different source. In this case, a solar cell was used, but a wireless energy harvesting architecture, thermoelectric device, or piezoelectric could be swapped out to provide energy. An antenna operating at 5.8 GHz with an omnidirectional pattern was printed on one face for this example but any antenna operating at a frequency band of choice that fits on the face of the structure could be used. The modular capabilities of the IoT device in terms of sensing nodes, antenna architecture, and power source enable ultimate plug-and-play capabilities.

The engineering of the design shown in Fig. 4 was spent mainly on making radial interconnect lines for connections between the faces of the prism. Bending RF transmission lines around corners poses issues such as radiating electromagnetic energy at the bend point. The stackup of the interconnects and measurement of the results can be seen in Fig. 5. The final design incorporated the baseband circuitry

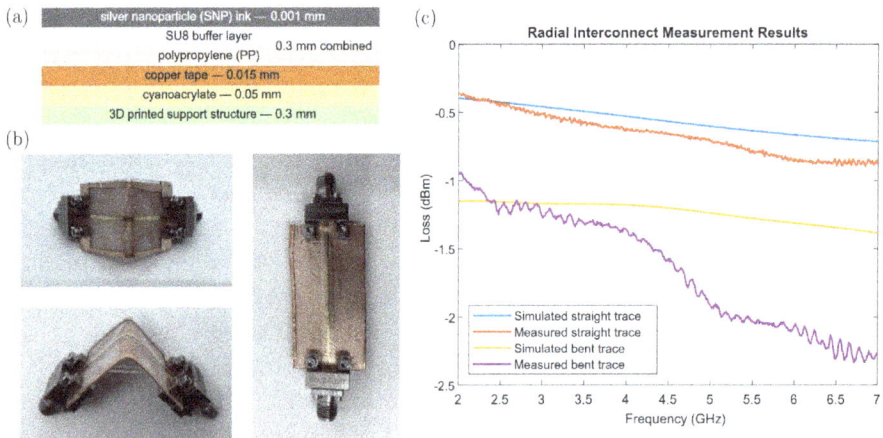

Fig. 5. Radial interconnects between faces of the cube as seen in [6]. (a) The stackup of the interconnect. (b) Bending of the interconnect. (c) Measurement results of the interconnect over frequency.

(a)   (b)

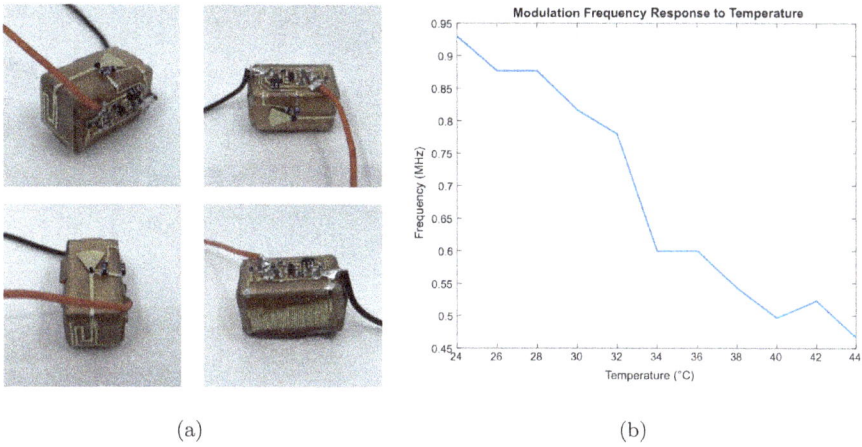

Fig. 6. Final magic cube (a) device and (b) performance [6].

such that temperature response could be frequency modulated back to the reader. The final results are shown in Fig. 6.

5. Camera-Insipired 3D 5G Lens-Based RFID

A retrodirective backscatter system with large angular coverage and extensive range was fabricated utilizing optical lenses in [7]. 3D PTFE lenses were placed in front of an array of elements (or pixels) to focus impinging electromagnetic waves analogous to a camera lens. This 5G camera can be seen in Fig. 7. Cascading multiple lenses were shown to have increased level of magnification compared to the single lens model. The lens diameter was parametrized to achieve a balance of both realized gain and angular coverage in Fig. 8 and a final diameter of 68 mm was chosen. The double lens-based mmID demonstrates $\pm$ 50° of -10 dB of beamwidth, minimal ranging error up to measured distances of 85 m, and differential RCS of -15.7 dBsm in Fig. 9.

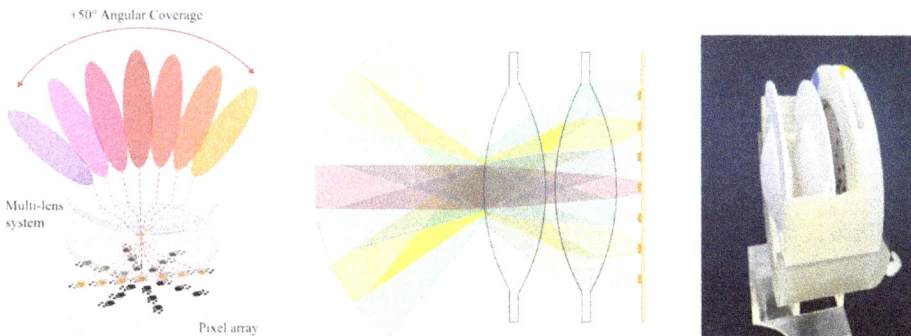

Fig. 7. Operation of 5G lens and working model [7].

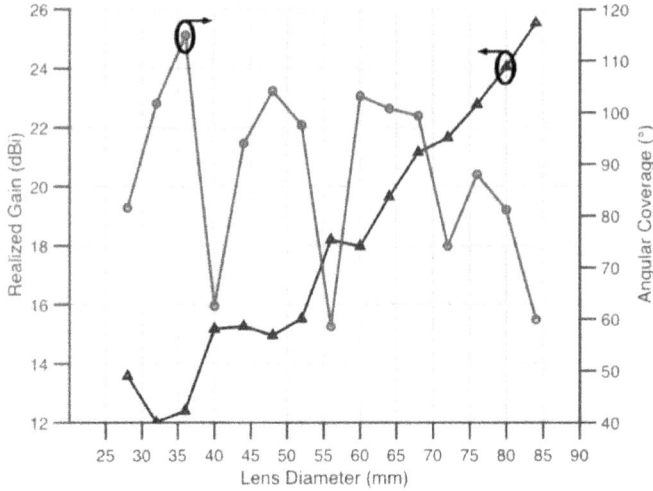

Fig. 8. Variation of lens diameter and effect on realized gain (blue) and angular coverage (red) [7].

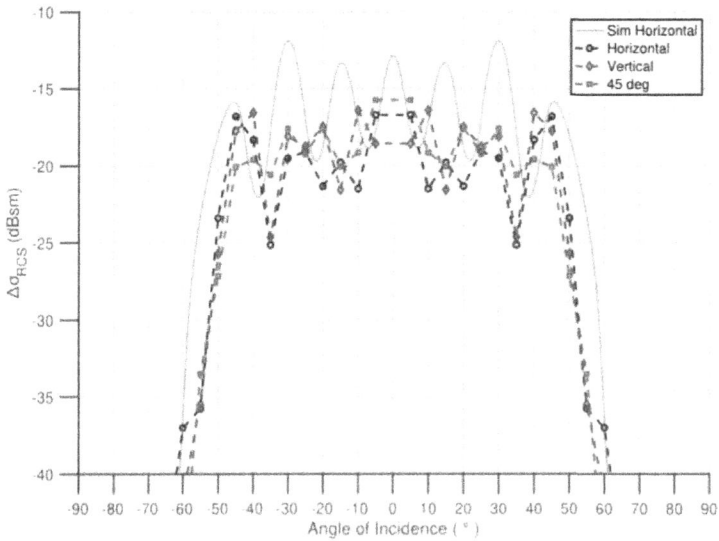

Fig. 9. Differential RCS of the 5G lens mmID at horizontal, vertical, and 45° cuts [7].

6. Conclusion

Additive manufacturing through 3D and inkjet printing is a facile, innovative, and cost-effective method to generate 5G electronics. As shown in this review, additive manufacturing provides a level of geometric freedom that is not possible with standard photolithography processes. Incorporating intricate elements like microfluidics, origami substrates, 3D substrates, and optics-based lenses into the design process facilitates the creation of innovative devices as seen here. Freedom in material choice allows the engineer to choose flexible substrates not often used in standard electronic processes. 3D printed materials offer not only geometric flexibility but also versatility in terms of heat resistance, mechanical stability, and beyond. The ability to select new geometries and materials, coupled with rapid and cost-effective fabrication techniques, endows additive manufacturing with significant potential for developing groundbreaking 5G electronics.

ORCID

Theodore W. Callis ◎ https://orcid.org/0000-0001-5289-2336

Kexin Hu ◎ https://orcid.org/0009-0004-6440-6437

Hani Al Jamal ◎ https://orcid.org/0009-0004-1616-4242

Dr. Manos M. Tentzeris ◎ https://orcid.org/0000-0003-0476-3577

References

1. L. Nayak, S. Mohanty, S. Kumar, S. K. Nayak and A. Ramadoss, "A review on inkjet printing of nanoparticle inks for flexible electronics," *Journal of Materials Chemistry C*, vol. 7, no. 29, pp. 8771–8795, 2019.
2. V. Beedasy and P. J. Smith, "Printed electronics as prepared by inkjet printing," *Materials*, vol. 13, no. 3, Art. no. 3, 2020.
3. A. Nathan *et al.*, "Flexible electronics: The next ubiquitous platform," *Proc. IEEE, Special Centennial Issue* vol. 100, pp. 1486–1517, May 2012.
4. K. Hu, Y. Zhou, S. K. Sitaraman and M. M. Tentzeris, "Additively manufactured flexible on-package phased array antennas for 5G/mmWave wearable and conformal digital twin and massive MIMO applications," *Nature Scientific Reports*, vol. 13, no. 1, p. 12515, 2023.
5. H. Al Jamal, C. Hu, N. Wille and M. M. Tentzeris, "Beyond planar: An additively manufactured, origami inspired, and shape-changing RFIC-based phased array for near-limitless radiation pattern reconfigurability in 5g/mm wave applications," *IEEE Antennas Wireless Propagation Letters.*, 2024. [ACCEPTED].
6. M. E. Holda, C. Lynch and M. M. Tentzeris, "Additively manufactured Magic Cube platforms for fully integrated wireless sensing nodes for internet of things applications," *Scientific Reports*, vol. 13, no. 1, p. 21736, 2023.
7. C. A. Lynch, G. Soto-Valle, J. G. D. Hester and M. M. Tentzeris, "At the intersection between optics and mmwave design: An energy autonomous 5G-enabled multilens-based broadbeam mm ID for "Smart" digital twins applications," *IEEE Transactions on Microwave Theory and Techniques*, vol. 72, no. 4, pp. 2620–2630, 2024.

Nano Thermal Simulation of Graphene Field Effect Transistor Based on Ballistic Diffusive Model#

Faouzi Nasri ⓘ

Center for Research in Microelectronics and Nanotechnology,
Sousse, Tunisia

Laboratory of Thermal Processes,
Research and Technology Center of Energy,
Hammam Lif — Tunisia
nasrifaouzi90@yahoo.fr

Husien Salama ⓘ

Computer Systems Institute, Boston, USA
husien.salama@uconn.edu

Khalifa Ahmed Salama**

Libyan Center for Engineering Research
and Information Technology Bani Walid, Libya
ka5677675@gmail.com

This paper investigates temperature evolution in graphene field-effect transistors. We present a refined Ballistic-diffusive equation model (BDE) tailored for scrutinizing the phonon transport within GNRFET transistors. Furthermore, we examine the influence of the channel length on thermal characteristics. COMSOL Multiphysics is employed to compute the phonon transport and temperature distribution in nano-GNRFET. Our results reveal that the phonon transport anticipated by the suggested BDE model closely corresponds to that derived from the Boltzmann transport equation. The proposed model proficiently gauges the thermal performance of GNRFET, with the outcomes suggesting that the characteristic length of the GNR has a negligible impact on the temperature increase.

Keywords: BDE model; ETC; GNRFET; nano heat conduction; FET devices.

*Corresponding author.
#This chapter appeared previously on the International Journal of High Speed Electronics and Systems. To cite this chapter, please cite the original article as the following: F. Nasri, H. Salama and K. A. Salama, *Int. J. High Speed Electron. Syst.*, **33**, 2440067 (2024), doi: 10.1142/S0129156424400676.

1. Introduction

Significant attention has been directed toward the swift progress of technology nodes and the burgeoning research in advanced nanoelectronics. In recent years, there has been a heightened emphasis on thermal management to enhance heat transport in nanostructured materials and corresponding nanoelectronics [1–5]. Electronic devices based on graphene have garnered significant attention owing to their outstanding electronic properties, including high-doping concentrations and elevated thermal conductivity [6].

While Fourier's law is fundamental to heat transfer theory, dictating that heat conduction occurs in a diffusive regime, it is acknowledged that Fourier's law does not apply to nano heat transport scenarios where the characteristic length is comparable or smaller than the mean free path of the heat carriers [7]. In this study, the Ballistic-diffusive equation (BDE) model has been employed to predict ballistic transport in nano-transistors. We illustrate the influence of the modified ballistic heat conduction represented by ∇qb. To enhance our model, we consider the ballistic nature of the phonon-wall collisions in graphene. A novel ballistic heat condition is proposed within the BDE model. The primary objective of this paper is to investigate the thermal stability of graphene-based transistors, with a specific focus on analyzing the effects of channel length on graphene transistors.

2. Mathematical Model

To capture the quasi-ballistic thermal transport, we have used the enhanced BDE model [8]. The modification in the BDE model is based on adding an equivalent thermal conductivity, a modified phonon relaxation time, and a ballistic heat flux term.

The equivalent thermal conductivity is given as follows [9]:

$$k_{eff}(Kn) = k\left[1 - \frac{2Kn \times tanh(1/2Kn)}{1 + C_B \times tanh(1/2Kn)}\right], \tag{1}$$

where $Kn = \frac{\Lambda}{L}$ is the Knudsen number, Λ is the phonon mean free path, and L is the channel length of the proposed transistors.

The relaxation time related to the phonon scattering is defined as [9]

$$\tau_R = \frac{3 \times k_{eff}}{C \times v^2}. \tag{2}$$

Ballistic heat flux term q_b is calculated by using the following expression [10]:

$$q_b = -k_{eff}\nabla T$$
$$\nabla q_b = -\nabla k_{\text{eff}}\nabla T. \tag{3}$$

Based on the Boltzmann Transport Equation, we can obtain the enhanced BDE model which is rewritten as [10]

$$\tau_R\frac{\partial^2 T(r,t)}{\partial t^2} + \frac{\partial T(r,t)}{\partial t} = \frac{\kappa_{eff}}{C}\nabla\nabla T(r,t) - \frac{\nabla q_b(r,t)}{C} + \frac{Q}{C} + \frac{\tau_R}{C}\frac{\partial Q}{\partial t}. \tag{4}$$

Table 1. The physical properties of the studied architectures [9, 10].

Symbol	V (m s^{-1})	K_0 (Wm^{-1}K^{-1})	C (Jm^{-3}K^{-1})	Λ (nm)	Kn
Si	3000	150	$1.5 \cdot 10^6$	100	10
SiC	13000	490	$2.11 \cdot 10^6$	53	—
HfO$_2$	5900	1.1	$2.6 \cdot 10^6$	0.2	—
GNR	14740	4000	$1.57 \cdot 10^6$	518	—

Table 1 shows the needed constants to generate the proposed model.

2.1.1 *Results and discussion*

This paper focuses on the graphene nanoribbon field-effect transistor (GFET) composed of a single channel. The separations between the source and gate, as well as gate and drain, are set at 50 nm each. Two configurations of the graphene nanoribbon (GNR) Field-Effect Transistor (FET) depicted in Fig. 1 are examined. The given configuration is derived with parameters $L_{ext} = 100$ nm, and $L_g = 10$ nm. To treat the geometry as a 2D transistor, the results are presented at time $t = 50$ ps. The power generation, denoted by $Q = 10^{19}$ (w/m^3), and $T_0 = 300$ K serve as the reference temperature.

Figure 2 shows the temperature evolution in the centerline of the transistor. The temperature profile has the same trend. With $Lc = 50$ nm, we show a rapid increase in the temperature. The increase is caused by the important value of the phonon scattering rate ($1/\tau_b$). At $t = 15$ ps, the temperature reaches 313 K, and the saturation is observed.

The 2-D temperature distribution in the GNRFET at $t = 30$ ps has been presented in Fig. 3. It is very clear that the temperature is maximal in the interface HfO2–GNR. We show that the maximum temperature is 313 K inside the device.

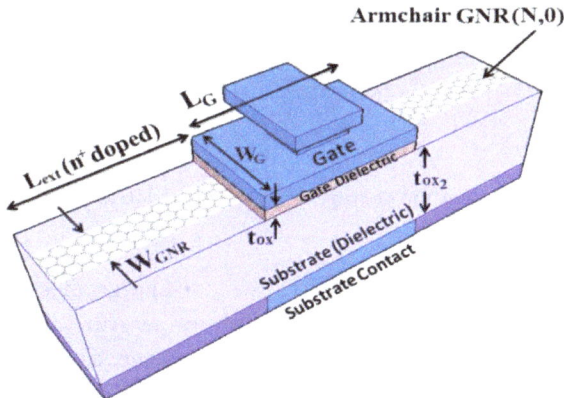

Fig. 1. Schematic geometry of Graphene nanoribbon FET devices [11].

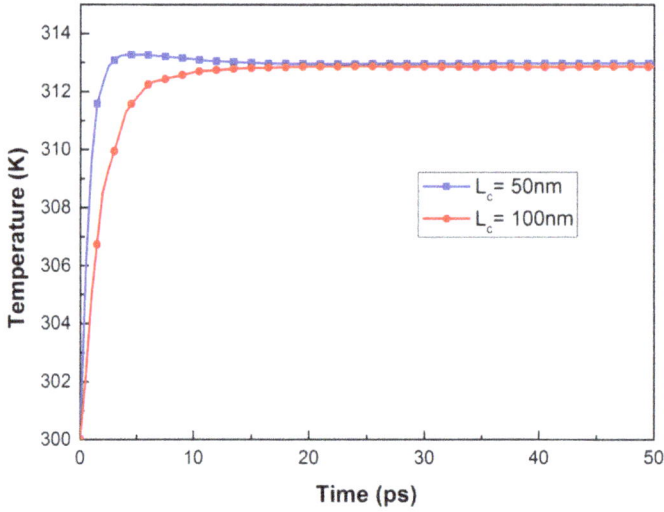

Fig. 2. Temperature evolution in the centerline of the transistor.

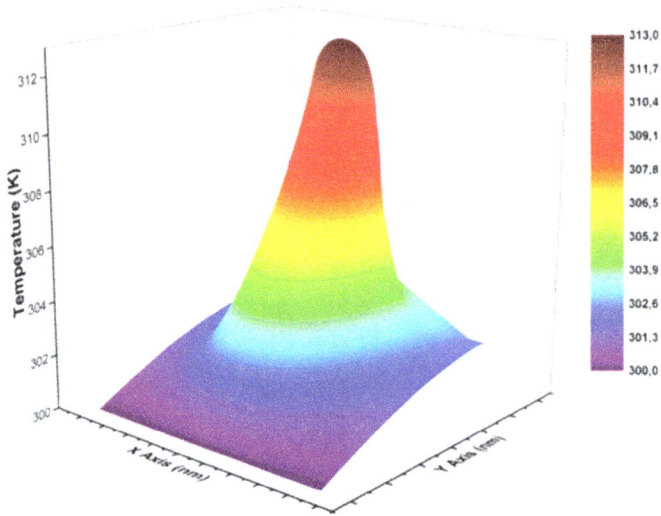

Fig. 3. 2-D temperature distribution in the GNR MOSFET at $t = 30$ ps.

The temperature decreases in the SiC substrate. The gradient temperature is in the channel region. By comparing the gradient temperature along the channel, we found that heat dissipation is important in the heat zone. The thermal conductivity is minimal in the interface scattering.

Figure 4 represents the heat flux profile versus y-direction at $t = 10$ ps with $L_c = 50$ and 100 nm. The heat flux along the y-axis reaches the value 4.32 (10^{12}W/m^2)

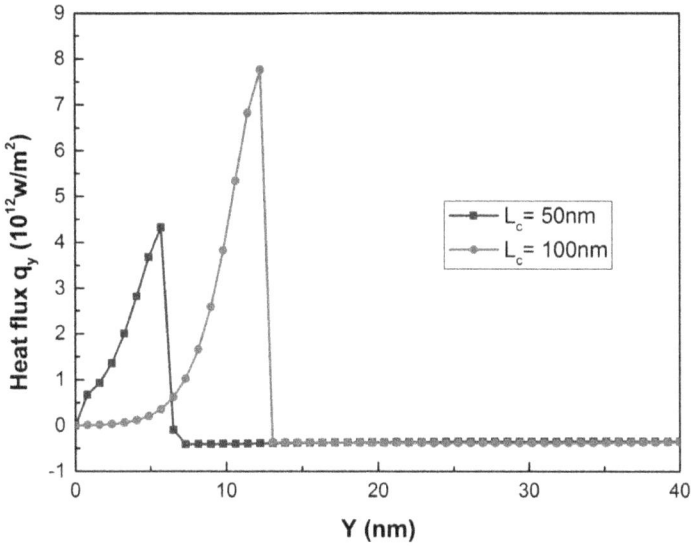

Fig. 4. Heat flux profile versus y-axis with different L_c at $t = 10\,\text{ps}$.

for $L_c = 50\,\text{nm}$ and 7.76 $(10^{12}\ \text{W/m}^2)$ for $L_c = 100\,\text{nm}$. With $L_c = 50\,\text{nm}$, the Knudsen number increases and reaches the value Kn = 10 in the GNR. The heat flux increases in the nano-transistor when the ballistic phonon transport is dominant. In the GNR–SiC interface, the Knudsen number attains Kn = 1 the phonon transport becomes ballistically diffusive. In the interface oxide semiconductor (HfO2–GNR), the Knudsen number decreases, and the phonon transport is purely diffusive [12].

3. Conclusion

In this study, we introduce a modified BDE model for predicting phonon transport in GNRFET devices. Our modification involves incorporating an expression for the Energy Transmission Coefficient (ETC) regarding the Knudsen number, secularity parameter, and nonzero heat flux (∇qb) into the BDE model. This enhancement allows us to establish a new model capable of explaining and predicting the heat transfer process and phonon transport in nano-GNRFET. From a technological standpoint, we observe that the temperature distribution and heat flux profiles in 50 nm and 100 nm GNRFETs are comparable, exhibiting the same maximum values.

ORCID

Faouzi Nasri ◎ https://orcid.org/0000-0002-9170-8279
Husien Salama ◎ https://orcid.org/0000-0002-3452-086X

References

1. J. Wang, X. Mu, M. Sun, The thermal, electrical and thermoelectric properties of graphene nanomaterials, Nanomaterials 9 (2) (2019) 218.
2. D. S. Schulman, A. J. Arnold, S. Das, Contact engineering for 2D materials and devices, Chem. Soc. Rev. 47 (2018) 3037–3058.
3. A. Malhotra, K. Kothari, M. Maldovan, Enhancing thermal transport in layered nano-materials, Sci. Rep. 8 (2018) 1880.
4. J. S. Kang, M. Li, H. Wu, H. Nguyen, Y. Hu, Experimental observation of high thermal conductivity in boron arsenide, Science 361 (2018) 575–578.
5. L. M. Peng, Z. Zhang, S. Wang, Carbon nanotube electronics: Recent advances, *Mater. Today* 17 (9) (2014) 433–442.
6. H. Rezgui, F. Nasri, M. F. Ben Aissa, H. Belmabrouk and A. A. Guizani, Modeling thermal performance of nano-GNRFET transistors using ballistic-diffusive equation, IEEE Trans. Electron Dev. 65 (4) (2018) 1611–1616, doi: 10.1109/TED.2018.2805343.
7. H. Mzoughi, F. Nasri, M. Almoneef, S. Soltani, M. Mbarek and A. Guizani, Investigation of nano-heat-transfer variability of AlGaN/GaN-heterostructure-based high-electron-mobility transistors.," Electronics 13 (1) (2023) 164.
8. F. Nasri, H. Rezgui and M. F. Ben Aissa, Numerical investigation of nano heat transfer process in FET nano devices using Ballistic-diffusive model, 2018 IEEE International Conference on Smart Materials and Spectroscopy (SMS), Hammamet, Tunisia (2018), pp. 12–17, doi: 10.1109/SMS44485.2018.9101397.
9. H. Rezgui, F. Nasri, M. F. Ben Aissa, H. Belmabrouk and A. A. Guizani, Modeling thermal performance of nano-GNRFET transistors using ballistic-diffusive equation, IEEE Trans. Electron Dev. 65 (4) (2018) 1611–1616, doi: 10.1109/TED.2018.2805343.
10. F. Nasri, N. Rekik, H. Bahri, U. Farooq, A. W. N. Hussein, H. Affan and B. Ouari, Temperature effects on electrical response of FinFET transistors in the static regime, IEEE Trans. Electron Dev. **70**(4) (2023) 1595–1600.
11. N. Dinesh Kumar, S. Rajendra Prasad, C. Raja Kumari and C. Dhanunjaya Naidu, Design and analysis of different full adder cells using new technologies, in Favorskaya, M. N., Mekhilef, S., Pandey, R. K., Singh, N. (eds.), *Innovations in Electrical and Electronic Engineering*, Lecture Notes in Electrical Engineering, vol. 661 (Springer, Singapore, 2021). https://doi.org/10.1007/978-981-15-4692-1_45.
12. F. Nasri and H. Salama, Numerical investigation of the electrothermal properties of SOI FinFET transistor, Int. J. High Speed Electronics Syst. 32 (02n04) (2023) 2350020.

Explosives Detection Using Octanethiol Gold Nanoparticle Inkjet Printed Devices[#]

Md Delowar Hossain ⊙*, John Grasso ⊙† and Brian G. Willis ⊙‡

*Chemical and Biomolecular Engineering, University of Connecticut,
191 Auditorium Road, Storrs, CT 06269, USA*
*delowar@uconn.edu
†John.grasso@uconn.edu
‡brian.willis@uconn.edu

This study focuses on implementing 1-octanethiol gold nanoparticles (OT-AuNPs) as chemiresistive sensors and comparing fabrication by inkjet-printing with conventional drop-casting to detect volatile organic compounds (VOCs) and explosives including HMTD, PETN, KClO$_3$, TATP, RDX, TNT, and UN. Inkjet-printing technology potentially offers more controlled deposition of OT-AuNPs and more uniform sensor properties compared to a drop-casting process. Using inkjet-printing, we found that a minimum of 600 OT-AuNPs printed layers with 2.4 pL drop volume, 9 drops per layer, and 60 μm drop spacing was sufficient to reduce chemiresistor device baseline resistances to around 10^3 Ω and obtain more consistent responses. By contrast, 12 μL OT-AuNPs drop-casting yielded device resistances spread over a range of $10^2 - 10^6$ Ω due to uneven deposition. For inkjet-printed devices, higher response magnitudes with lower variability were achieved for explosive vapor detection whereas a larger spread was observed for drop casting. The improved uniformity of baseline resistances and sensor responses indicates inkjet-printed devices were more reproducible and repeatable than drop-casting. In addition, inkjet-printing consumes lower amounts of OT-AuNPs for material and cost savings. The results demonstrate that inkjet-printed devices are promising for use in sensor arrays to fabricate electronic noses for VOCs and explosives detection.

Keywords: Explosives detection; octanethiol sensor; inkjet-printed device.

1. Introduction

Researchers have been investigating electronic noses that can be used in different industries (including chemical, food, beverage, pharmaceutical, cosmetics, and perfume), environment pollution monitoring, quality control, medical diagnostics, safety and security, agricultural sector, space exploration, and explosives

*Corresponding author.
[#]This chapter appeared previously on the International Journal of High Speed Electronics and Systems. To cite this chapter, please cite the original article as the following: Md D. Hossain, J. Grasso and B. G. Willis, *Int. J. High Speed Electron. Syst.*, **33**, 2440068 (2024), doi: 10.1142/S0129156424400688.

detection [1–5]. Electronic noses typically have arrays of sensors with different chemistries that are designed to generate unique response patterns through chemical interactions between sensing materials and chemical vapors. The major challenges are to select suitable sensing materials, deposit selected materials onto the sensor arrays, minimize moisture interference, extract signal from low sample concentrations, and differentiate responses to identify different compounds [6, 7]. To deposit sensing materials on sensor arrays, drop casting, material jetting (inkjet printing technology, pico-drop technology), material extrusion (fused deposition modeling, direct ink writing techniques), screen printing, and other deposition techniques can be used [8–10]. Drop-casting is commonly used due to its simplicity, but adding different chemistries to dense arrays of devices with conventional drop-casting processes is difficult to achieve since the minimum drop-cast volume ($\sim$0.5 μL) spreads out to cover many devices instead of depositing only on targeted devices. Moreover, due to uncontrolled droplet spreading, device reproducibility and uniformity of sensing materials on devices are major challenges for conventional drop-casting techniques [11]. Coffee ring formation is also a challenge for drop casting [12]. As an alternative, material jetting techniques like inkjet printing technology can deposit pico-liter-sized droplets onto specific devices in a more controlled way. Therefore, more reproducible, repeatable, and reliable devices are expected to be produced using inkjet printing compared to conventional drop-casting. In this context, this study was carried out to fabricate chemiresistive sensors made through inkjet-printing and drop-casting of OT-AuNPs and compare performance for detection of VOCs and explosives. OT-AuNPs were chosen because they are readily available and commonly used for making sensors [13, 14].

Explosive detection is important for public safety, national security, demining, and explosion prevention. Explosive detection is difficult due to the low vapor pressures, presence of variable humidity, interference of background contaminants, surface stickiness, and variable vapor composition in real samples [15–17]. Currently, several different methods are used for sensing explosives including canine (K-9) detection, laser-based techniques (laser-induced breakdown spectroscopy (LIBS), Raman spectroscopy), and trace chemical detection (ion mobility spectrometry, mass spectrometer, biosensors, optical transmission, and reflection spectroscopy) [18,19]. These techniques all have some limitations including limited response speed, explosion initiation for some sensitive compounds, large system size, heavy system weight, and high system cost [20]. This investigation is focused on micro-fabricated devices that can detect volatiles emitted by explosive compounds and overcome the shortcomings of current sensor technologies.

Steps to implement OT-AuNPs sensing materials as chemiresistive sensors for detecting chemical compounds including VOCs and explosives are shown in Fig. 1. For chemical detection, OT-AuNPs are deposited onto pre-patterned microfabricated electrodes to make resistive device elements. OT-AuNPs deposition creates particle networks that form conductive bridges between electrodes with

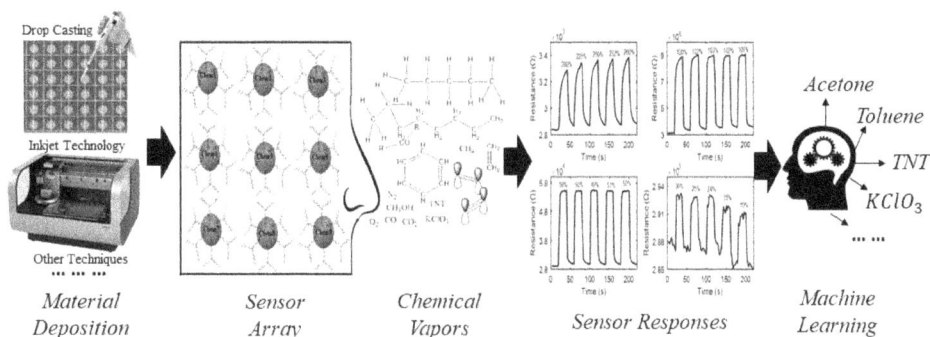

Fig. 1. Steps to fabricate and test chemiresistive sensors.

chemically responsive electrical resistances. The exposure of the sensor devices to vapor analytes induces reversible changes of electrical resistances due to chemical interactions between sensing materials and vapor analytes [21]. The amplitudes of electrical resistance changes depend on the nature of sensing materials and analytes, operating conditions, surface treatments, particle network morphologies, and electrode designs. This study measures reversible electrical resistance changes for arrays of chemiresistor sensors exposed to VOCs and explosive compounds where array data are analyzed by machine learning to identify analytes. The performance and reproducibility of sensor arrays fabricated by drop-casting and inkjet-printing are compared.

2. Materials and Methods

OT-AuNPs (15 nm diameter) were deposited onto micro-fabricated electrodes following drop-casting and inkjet-printing processes. A FujiFilm Dimatix Materials Printer (DMP-2800 series) with 2.4 pL samba jetting module was used for the inkjet printing process, while drop-casting was done by hand using a 0.5 μL pipette. OT-AuNPs — ethanol solutions of 1:9 and 1:99 dilutions were prepared from a stock solution of 3.3 mg/mL OT-AuNP (OD 50) in ethanol and mixed with ultra-pure deionized water with a 2:1 ratio to deposit OT-AuNPs onto devices through drop-casting and inkjet-printing, respectively. The higher dilution for inkjet printing was used to minimize clogging of DMP-2800 samba cartridge nozzles and improve the uniformity of nanoparticle deposition onto devices.

Before depositions, sensor chips were prepared by soaking in Remover PG (KAYAKU Advanced Materials) to strip a protective photoresist layer used in chip dicing. Sensor chips were then rinsed with isopropyl alcohol and dried with N_2 gas. Samples were further cleaned with a UV ozone treatment for 10 minutes followed by oxygen plasma treatment (20 sccm flow, 15 mtorr pressure, 50 sec duration, 80 W power). Sensor surfaces were activated by immersion in 3 v/v % MPTES (3-mercaptopropyl(trimethoxysilane)) toluene solution for 3 hrs, and then baked at

122°C in air for 1 hr. Oxygen plasma treatment removes surface contaminants and increases surface wettability [22]. MPTES immersion introduces silane groups that improve adhesion between the substrate and sensing materials [23].

The sensor chip arrays have a total of 143 devices in four quadrants where 35–36 devices are in each quadrant. Each device has a 250 μm diameter circular electrode design with 2 μm electrode spacing. For drop-casting, OT-AuNPs solution (2 to 30 μL) was applied using a micropipette of 0.5 μL drop volume onto each quadrant of a sensor chip. After drop-casting, sensor chips were annealed at 70°C and – 29 inch Hg in a vacuum oven for 24 hrs to evaporate solvents and make devices functional. In contrast, inkjet-printing used knowledge of the device design to align print patterns with devices. The inkjet printing parameters include drop spacing (DS), number of drops per layer (DPL), and total number of layers (L) per devices (Fig. 2). Depending on the DS and DPL, printed areas may merge or remain distinct. The number of OT-AuNPs deposited layers varied between 20 and 800. Each drop is 2.4 pL of OT-AuNPs solution. After inkjet printing, sensor chips were stored in open air to evaporate solvents.

Saturated vapors of VOCs (hexane, toluene, acetone, and isopropanol) and explosive training aids (hexa-methylene triperoxide diamine (HMTD), penta-erythritol tetra-nitrate (PETN), potassium chlorate (KClO$_3$), tri-acetone tri-peroxide (TATP), royal demolition explosive (RDX), 2,4,6-trinitrotoluene (TNT), and urea nitrate (UN)) were produced for sensor testing. Solvents were acquired from Fisher Chemical and explosive training aids were obtained from Signature Science. To prepare saturated vapors for sampling, separate 5 liter tedlar bags were

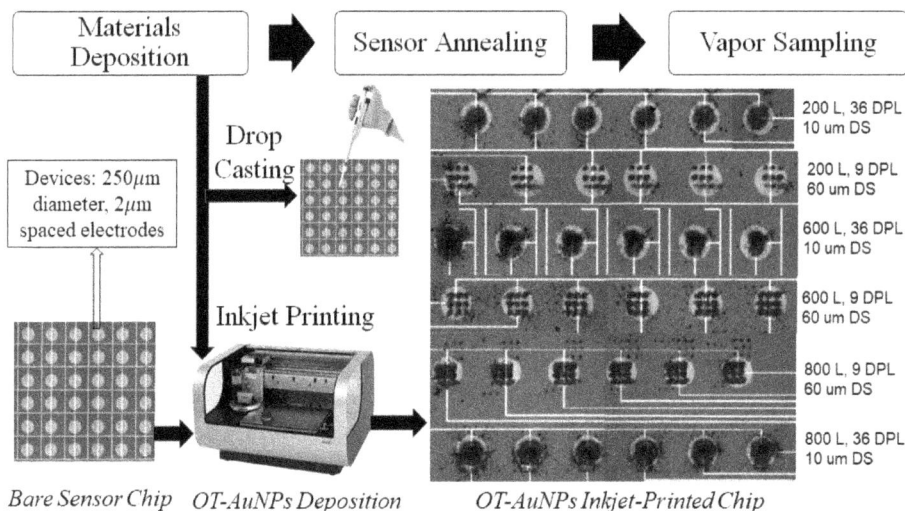

Fig. 2. Experimental procedure to produce OT-AuNPs functionalized devices (L, DPL, DS, and OT denote layer, drops per layer, drop spacing, OT-AuNPs respectively).

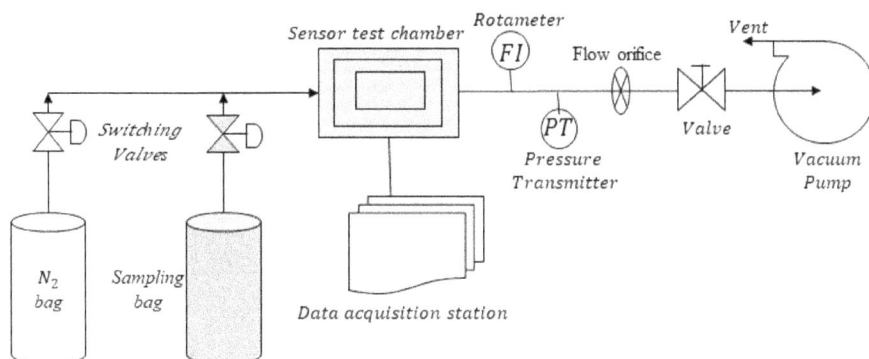

Fig. 3. Experimental setup for vapor sampling.

loaded with small amounts of VOC liquids and explosive training aids and filled to 80–90% capacity with nitrogen gas. The bags were kept at ambient condition for 24 hrs to equilibrate before sensor testing.

Fabricated sensors were wire-bonded to a ceramic chip carrier and inserted into a 3D printed plastic flow cell test chamber to measure VOC and explosive vapor responses as shown in Fig. 3. Saturated vapors with a gas flow rate of 500 ml/min at room temperature and atmospheric pressure were introduced into the flow cell with 20-second purge and pulse cycles during each test run. Nitrogen was used as a purge gas. Due to chemical interactions of analyte vapors with sensor materials, electrical resistances of devices changed reversibly and were collected using a multiplexed data acquisition station with a sampling rate of 60 Hz, which cycled through 140 measurements in 2.3 seconds. Although not used here, collected data can be further processed using machine learning algorithms to analyze and classify the VOCs and explosive chemicals [24].

3. Results and Discussion

3.1. *Resistance scaling of OT-AuNPs*

Electrical resistances of surface-cleaned sensor chips without AuNPs are $>10^8\,\Omega$, which changes after OT-AuNPs deposition. Resistance changes after OT-AuNPs deposition through drop-casting and inkjet-printing are shown in Fig. 4. It is evident that resistances decrease with depositing more OT-AuNPs onto the devices. Drop-cast device resistances decrease from 10^6 to $10^4\,\Omega$ for increasing nanoparticle solution deposition from 2 to 30 μL, whereas inkjet-printed device resistances decline from 10^6 to $10^3\,\Omega$ for increasing deposition layers from 20 to 200. Inkjet-printed devices show a saturation after 400 or more OT-AuNPs layers. Drop-cast results shown in Fig. 4(a) table indicate 100% of device resistances were in the range of 10^2–$10^6\,\Omega$ for 12 μL OT-AuNPs deposition but 20 μL or more produced around 44–50% devices with very low resistances ($<10^2\Omega$) that did not generate sensible responses

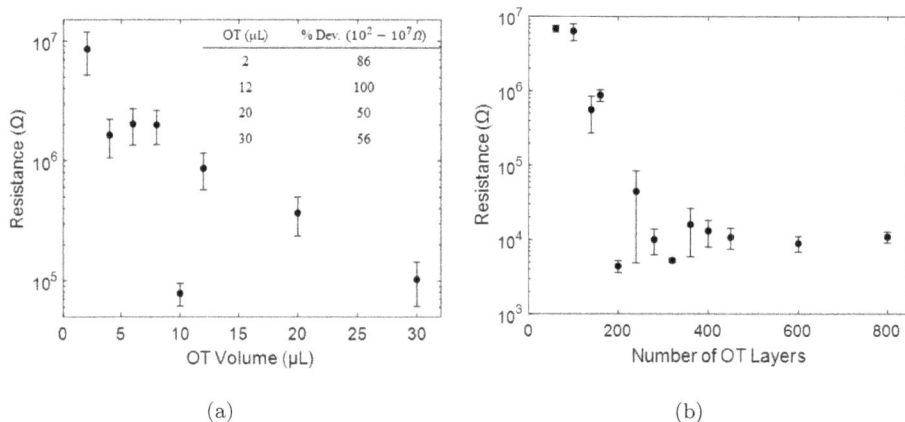

Fig. 4. Resistance scaling of OT-AuNPs (a) drop-cast devices, (b) inkjet-printed devices under dry nitrogen flow (standard error bars reflect the resistance spread of devices with the same OT-AuNPs solution volume).

to vapor analytes. Excessive nanoparticle deposition through drop-casting appears to lower device resistances and create short circuits between devices. For both fabrication methods, resistance declines with added material, but the inkjet method produces more devices in the desired 10^3–$10^4 \, \Omega$ range without causing shorting. To achieve 100% device yield, 200 L or more OT-AuNPs deposition was necessary in inkjet printing, whereas 12 μL AuNPs was used in drop-casting.

3.2. Correlation between resistance and response magnitudes

Figure 5 plots device resistances and response magnitudes ($\Delta R/R_0 * 100$) during toluene vapor sensing. For drop-casting, there is a clear trend towards higher sensitivity with lower resistance, but very low resistances ($\leq 10^2 \, \Omega$) yield poor responses, which are not included on the plot. The data for inkjet-printed sensors are more scattered (Fig. 5(b)), but there is some trend that responses may also increase with decreasing device resistances. For example, inkjet-printed samples of 9 DPL 60 μm DS with 800 layers produced lower resistance and higher response magnitude devices compared to 400 layers. Also, 36 DPL produces higher response magnitudes than 9 DPL for the same number of layers. In other words, for the range 10^3–$10^4 \, \Omega$, response magnitudes increase with increasing mass of OT-AuNPs deposited. Interestingly, for the best drop-cast samples, the response magnitudes are comparable or even better than the best inkjet data, which is an indication of the importance of the particle network morphology for high sensitivity devices. Further study is needed to explain how device resistance and response magnitudes are correlated. We speculate that increasing mass of OT-AuNPs lowers resistance of devices and creates more conductive pathways with improved overall sensitivity, but more study is needed.

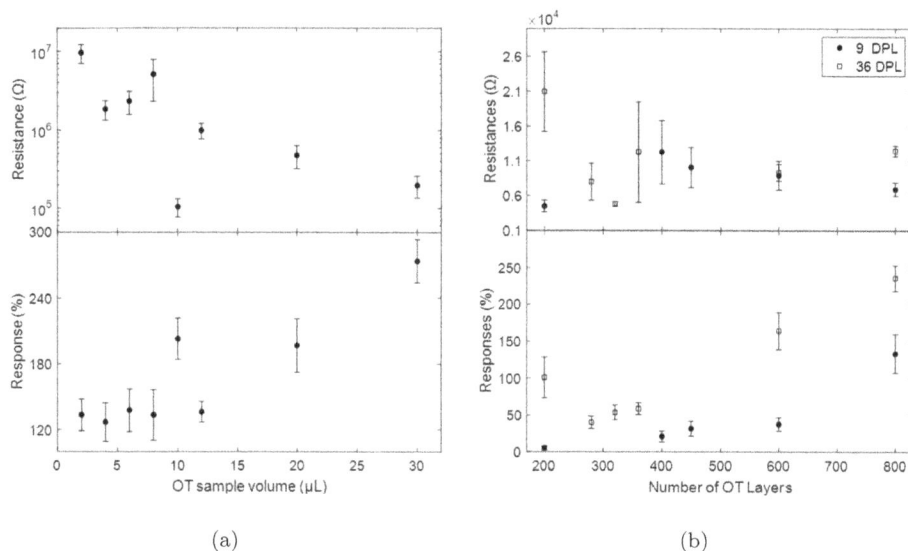

(a)   (b)

Fig. 5. Changes of resistances (R) and responses ($\Delta R/R_0$) of OT-AuNPs (a) drop-cast devices, (b) inkjet-printed devices to toluene vapor (standard error bars reflect the resistances and responses spread of devices with the same OT-AuNPs solution volume).

3.3. *Drop spacing effect on resistance and response magnitudes*

The effects of drop spacing on inkjet-printed devices was investigated by depositing OT-AuNPs on devices using variable spacing over the range 10–60 μm. All droplets impinge on devices, but larger spacing gives distinct droplets whereas smaller spacing produces single regions (see Fig. 2). As shown in Fig. 6, device resistance declines from 10^5 to 10^4 Ω for increasing drop spacing from 10 to 30 μm for the same DPL. Similar resistances were obtained with larger drop spacing 40 and 60 μm using smaller DPL, and thus less mass. The smaller DPL is possible because larger DS produces the effect of resistors in parallel, which decreases the overall device resistance. Similar to the discussion regarding Fig. 5, device response increases with increasing drop spacing, which reinforces the correlation of response sensitivity with device resistance. For 30 μm DS, responses to urea nitrate are near 12% compared with only 10% for the smaller 10 μm DS. However, the sensitivity increases further to 15% for 60 μm DS even though resistance stays nearly the same. Thus, other factors beyond device resistance are also important for optimizing sensitivity.

3.4. *Resistance uniformity: Inkjet versus drop-cast devices*

The distribution of resistances for inkjet-printed and drop-cast devices is presented in Fig. 7. The pie charts show that drop-cast devices are more widely distributed than inkjet-printed devices. For instance, 87% of inkjet-printed and 75% of the best drop-cast device resistances were within the ranges of 10^3–10^4 and 10^5–10^6 Ω,

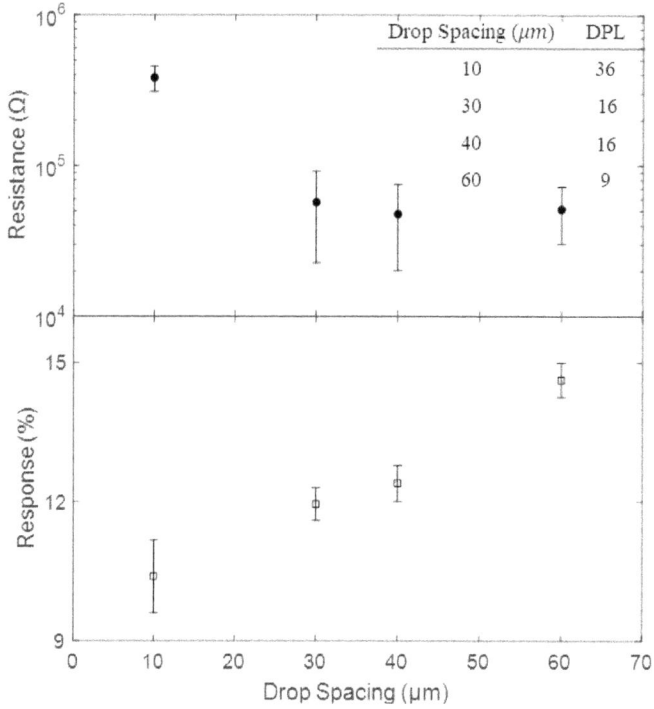

Fig. 6. Drop spacing effect on resistances (R) and responses ($\Delta R/R_0$) of 200 layers OT-AuNPs inkjet-printed devices with urea nitrate saturated vapor introduction (standard error bars reflect the resistances and responses spread of devices with the same drop spacing).

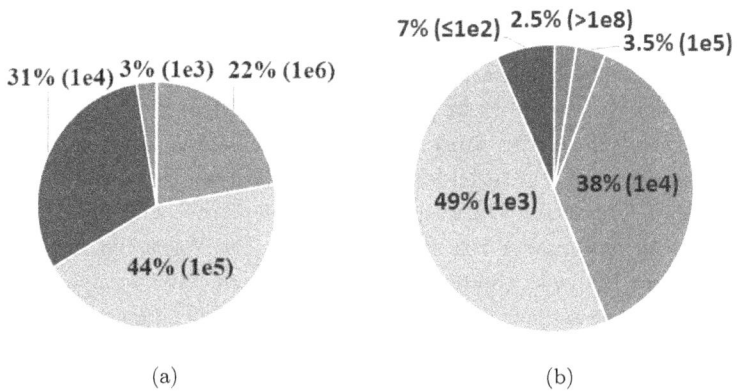

(a)   (b)

Fig. 7. Resistance distribution of OT-AuNPs: (a) drop-cast devices using $12\,\mu L$, (b) inkjet-printed devices.

respectively. Due to the correlation of response magnitudes and device resistances, inkjet-printed devices have more consistent responses due to a tighter distribution of resistances than drop-casting. The larger variation for drop-cast devices is the result of different amounts of OT-AuNPs deposited on different devices. Therefore, inkjet-printed devices are more reproducible with tighter distributions of resistances than drop-cast devices.

3.5. *Responses of OT-AuNPs inkjet and drop-cast devices*

Sensing data show that 800L inkjet-printed devices have higher responses to analytes than $12\,\mu L$ OT-AuNPs drop-cast. Figures 8(a)–8(d) show sensing data for a $12\,\mu L$ AuNPs drop-cast device with responses of 3–4%, 4–5%, 1.6–1.8%, and 3.6–4.0% to HMTD, PETN, TNT, and UN vapors, respectively. In contrast, Figs. 8(e)–8(h) show sensing data for a 800 L, 9 DPL, $60\,\mu m$ DS AuNPs inkjet-printed device with 5–8%, 5–9%, 4–6%, and 5–8% to the same explosives. The different response magnitudes are due to the different chemical interactions between the sensing materials and the analytes.

3.6. *VOC sensing with inkjet and drop-cast devices*

Different VOCs including hexane, toluene, acetone, and isopropanol were used to compare the consistency of different device responses for the two fabrication methods. Device responses at optimum conditions ($12\,\mu L$ for drop-casting and 600 L

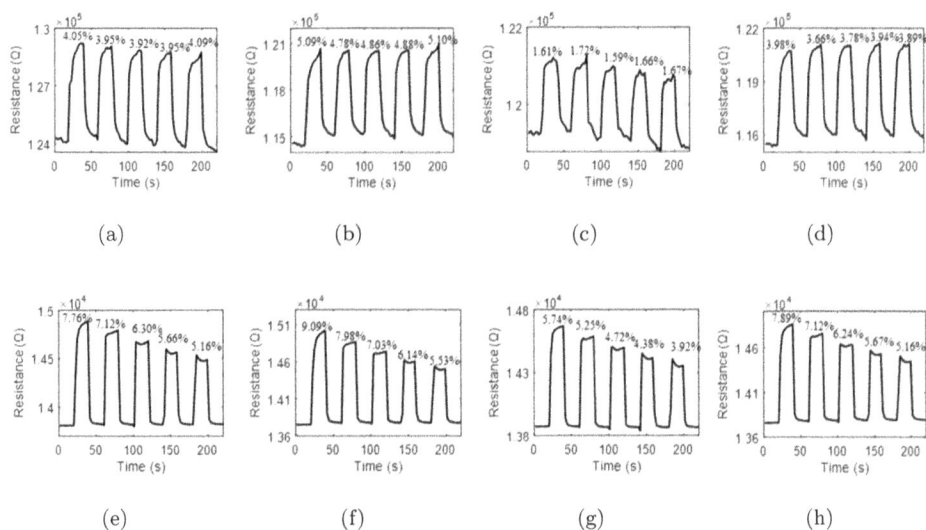

Fig. 8. Responses ($\Delta R/R_o$) of a $12\,\mu L$ OT-AuNPs drop-cast device (a–d) and a 800 L, 9 DPL, $60\,\mu m$ DS inkjet-printed device (e–h) to HMTD (a, e), PETN (b, f), TNT (c, g), and UN (d, h) vapors.

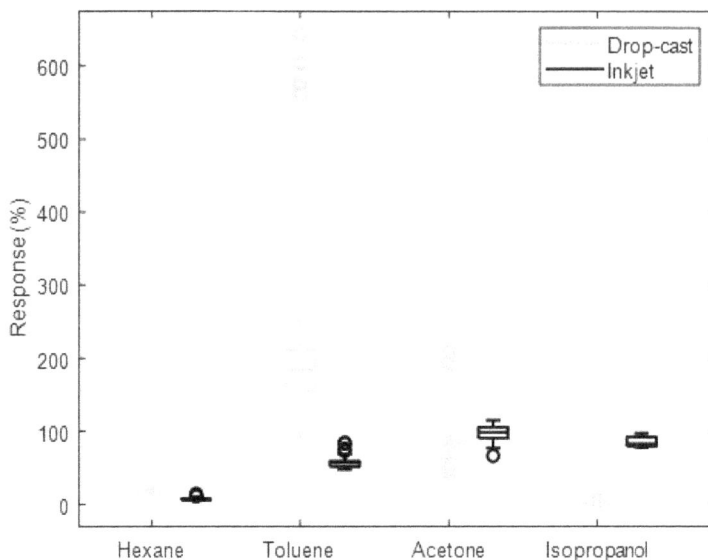

Fig. 9. Spreading of VOC vapor responses in $12\,\mu$L OT-AuNPs drop-cast devices and $600\,$L $9\,$DPL $60\,\mu$m DS inkjet-printed devices.

$9\,$DPL $60\,\mu$m DS for inkjet printing) are shown in Fig. 9. From these configurations, inkjet printing needs $0.285\,$ng AuNP deposition to functionalize a device, whereas drop-casting requires around $73.33\,$ng per device. Figure 9 shows inkjet-printing offers more consistent responses with tighter distribution for different VOCs than drop-casting. For inkjet-printing, 100% devices consistently responded to all the different VOCs, while only 92–94% drop-cast devices responded. Thus, inkjet-printing needs lower mass of OT-AuNPs to obtain higher device yield and more consistent responses than drop-casting.

3.7. *Explosives detection with inkjet and drop-cast devices*

Sensor chips were used to detect explosive vapors including HMTD, PETN, $KClO_3$, TATP, RDX, TNT, and UN, from training aids used for explosive detection dogs. Figure 10 compares the spread of response magnitudes for the two fabrication methods using optimal $12\,\mu$L for drop-cast and $600\,$L $9\,$DPL $60\,\mu$m DS for inkjet-printing. For inkjet printing, each device has approximately $0.285\,$ng OT-AuNPs, whereas approximately 257 times higher mass of AuNPs is required for drop-casting. Figure 10 shows inkjet devices produce similar sensitivity with less variance than drop-cast devices. The different response patterns can be used with machine learning for pattern recognition and detection. The reason for the greater variability of drop-casting is because of the different amounts of materials spread over the arrays. Interestingly, however, many of the drop-cast devices are more sensitive than the inkjet-printed devices. The reason for this sensitivity variation is unclear, but nanoparticle network

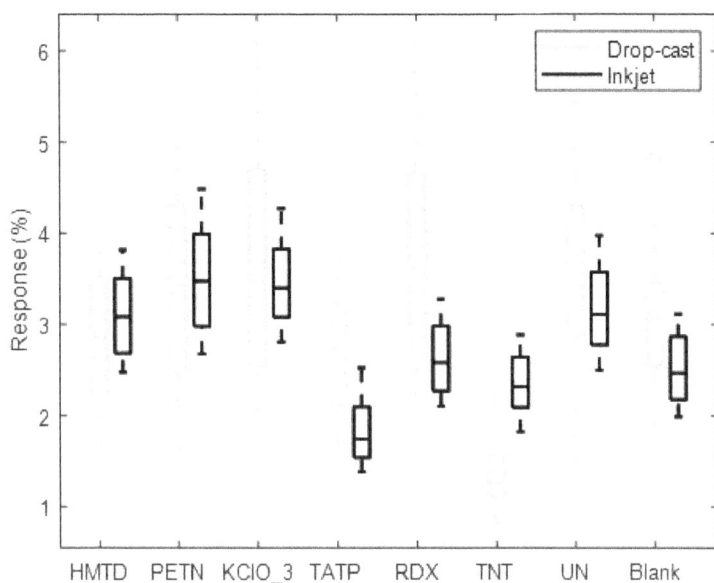

Fig. 10. Spreading of explosive vapor responses of OT-AuNPs drop-cast and inkjet-printed devices.

morphology is the crucial difference between sensors since they all have the same sensor materials. Further optimization of the inkjet process may help to increase sensitivity.

4. Conclusion

In this study, OT-AuNPs were deposited onto microfabricated sensor device arrays to compare conventional drop-casting and inkjet-printing methods for detecting VOCs and explosives. Both methods show a general decrease of device resistance with increasing sensor material deposited, an increase of sensitivity with more OT-AuNPs deposition, and a weak correlation between device sensitivity and resistance. For drop-cast samples, optimum device performance and yield were found for an intermediate droplet volume, while inkjet performance improved for more number of layers and larger drop spacing. Inkjet-printed devices gave more uniform distributions of device resistances and response magnitudes compared to drop-cast devices, and were more reproducible. Inkjet-printed device also showed more consistent responses to the different explosive scents, whereas drop-cast devices had more variability with some devices responding to some analytes but not others. In addition, inkjet printing requires lower OT-AuNPs mass to functionalize devices than drop-casting. The combination of improved device uniformity and reproducibility along with the ability to print micron scale devices makes inkjet-printing very promising for use in sensor arrays to develop electronic noses for VOCs and explosives detection.

Acknowledgment

The authors acknowledge the Office of Naval Research, USA for the financial support (ONR grant # N00014-22-1-2567) to conduct this research. This work was performed in part at the Harvard University Center for Nanoscale Systems (CNS); a member of the National Nanotechnology Coordinated Infrastructure Network (NNCI), which is supported by the National Science Foundation under NSF award no. ECCS-2025158.

ORCID

Md Delowar Hossain ⊚ https://orcid.org/0009-0007-1721-6317

John Grasso ⊚ https://orcid.org/0000-0003-4466-6894

Brian G. Willis ⊚ https://orcid.org/0000-0002-1720-4451

References

1. S. Y. Park, Y. Kim, T. Kim, T. H. Eom, S. Y. Kim and H. W. Jang, *InfoMat* **1**, 289 (2019).
2. Y. Li, K. Yang, Z. He, Z. Liu, J. Lu, D. Zhao, J. Zheng and M. C. Qian, *ACS Omega* **8**, 16356 (2023).
3. J. Li, A. Hannon, G. Yu, L. A. Idziak, A. Sahasrabhojanee, P. Govindarajan, Y. A. Maldonado, K. Ngo, J. P. Abdou, N. Mai and A. J. Ricco, *ACS Sens.* **8**, 2309 (2023).
4. G. Domènech-Gil, N. T. Duc, J. J. Wikner, J. Eriksson, S. N. Paledal, D. Puglisi and D. Bastviken, *Environ. Sci. Technol.* **58**, 352 (2024).
5. R. C. Young, W. J. Buttner, B. R. Linnell and R. Ramesham, *Sens. Actuat. B Chem.* **93**, 7 (2003).
6. G. Shang, D. Dinh, T. Mercer, S. Yan, S. Wang, B. Malaei, J. Luo, S. Lu and C. Zhong, *ACS Sens.* **8**, 1328 (2023).
7. Y. Li, X. Wei, Y. Zhou, J. Wang and R. You, *Microsys. Nanoengg.* **129**, 1 (2023).
8. S. Zhou, Y. Zhao, Y. Xun, Z. Wei, Y. Yang, W. Yan and J. Ding, *Chem. Rev.* **124**, 3608 (2024).
9. E. Chow, J. Herrmann, C. S. Barton, B. Raguse and L. Wieczorek, *Analy. Chimi. Acta* **632**, 135 (2009).
10. N. Zhao, M. Chiesa, H. Sirringhaus, Y. Li, Y. Wu and B. Ong, *J. Appli. Phy.* **101**, 064513 (2007).
11. A. K. S. Kumar, Y. Zhang, D. Li and R. G. Compton, *Elec.Chemi. Commun.* **121**, 106867 (2020).
12. A. K. S. Kumar, Y. Zhang, D. Li, R. G. Compton, *Electrochem. Commun.* **121**, 106867 (2020).
13. H. Wohltjenand A. W. Snow, *Anal. Chem.* **70**, 2856 (1998).
14. J. W. Grate, D. A. Nelson and R. Skaggs, *Anal. Chem.* **75**, 1868 (2003).
15. R. G. Ewing, M. J. Waltman, D. A. Atkinson, J. W. Grate and P. J. Hotchkiss, *Trends Anal. Chem.* **42**, 35 (2013).
16. A. Fainberg, *Science* **255**, 1531 (1992).
17. K. J. Albert and D. R. Walt, *Anal. Chem.* **72**, 1947 (2000).
18. National Research Council, *Existing and Potential Standoff Explosives Detection Techniques* (National Academies Press, Washington DC, 2004).
19. K. C. To, S. Ben-Jaber and I. P. Parkin, *ACS Nano* **14**, 10804 (2020).

20. L. Senesac and T. G. Thundat, *Mater. Today* **11**, 28 (2008).
21. V. K. Khanna, *Nanosensors: Physical, Chemical, and Biological* (CRC Press, Florida, 2012).
22. Y. Milyutin, M. Abud-Hawa, V. Kloper-Weidenfeld, Elias Mansour, Y. Y. Broza, G. Shani and H. Haick, *Nat. Proto.* **16**, 2968 (2021).
23. A. C. Boden, M. Bhave, L. Cipolla, P. Kingshott, *App. S. Sci.* **602**, 154282 (2022).
24. T. Gao, A. Oliveira, C. Zhang, Y. Wang, J. Zhao and B. G. Willis, *Electrochem. Soc.* **66**, 3339 (2020).

The Static Noise Margin (SNM) of Quaternary SRAM using Quantum SWS-FET#

B. Saman ⊕

Department of Electrical Engineering, Taif University,
P.O. BOX 888 - 21974 Al-Hawiyah, Taif, Kingdom of Saudi Arabia
saman@tu.edu.sa

E. Heller

Synopsys Inc. 400 Executive Boulevard,
Ossining, NY 10562, USA
vankheller@gmail.com

F. C. Jain*

Department of Electrical and Computer Engineering,
University of Connecticut, 371, Fairfield Way,
U-2157, Storrs, CT, 06269-2157, USA
*faquir.jain@uconn.edu

Static random-access memory (SRAM) is an essential component in the architecture of modern microprocessors and VLSI circuits. The problems of high power consumption, large area, circuit complexity, and data stability against noise are among the most important indicators of performance and obstacles to the current use of SRAM. Ternary, quaternary, and higher-order logic (MLV) systems have shown the potential in overcoming these limitations in increasing the information density compared to the traditional binary system. The quantum dot channel field-effect transistor (QDC-FET) and quantum well Spatial Wavefunction Switched field-effect transistor (SWS-FET) are a new alternative with multiple operating states, low power consumption, and smaller footprints. This work presents a new four-state SRAM design that uses SWS-FET and compares it with Voltage-Mode CMOS Quaternary logic design. In addition, this work studies the noise margin in the memory circuit of the quadrilateral logic system and its effect on data stability. Furthermore, this study shows the reliability of quaternary SRAM design by evaluating the impact of errors.

Keywords: SWSFETs; SWS-FET; 2-bit SRAM; SNM; MVL; VLSI.

*Corresponding author.
#This chapter appeared previously on the International Journal of High Speed Electronics and Systems. To cite this chapter, please cite the original article as the following: B. Saman, E. Heller and F. C. Jain, *Int. J. High Speed Electron. Syst.*, **33**, 2440069 (2024), doi: 10.1142/S012915642440069X.

1. Introduction

In Quantum Spatial Wavefunction-Switched Field-Effect Transistors (SWSFET) or Quantum well FETs (QW FETs), carriers (electrons or holes) are confined in a narrow region of a semiconductor material. Quantum well or Quantum Dot structures typically consist of two or more thin layers of semiconductor material sandwiched between barrier layers material, forming a potential well in which carriers are confined in two dimensions, this confinement controls the flow of carriers through the device [1, 2].

Each quantum well transport channel can be addressed and its transconductance can be modulated by applying gate voltage of the device. This enables precise control over the flow of current in vertically stacked Quantum Wells of QDots, making QW FETs in Multi-state logic, operating at high speed and low power. Using a two-channel SWS-FET provides encoding of 4 states or 2-bit logic as (00), (01), (10) and (11) [3, 4]. Recently, multi-channel SWS-FETs have been reported for multi-bit CMOS logic, SRAMs, D-flip-flop [5].

2. SWS-FET Structures

Figure 1(a) shows the structures of two well n-channel SWS-FET, the device has upper Si well (W1) and lower Si well (W2) sandwiched between two SiGe barriers

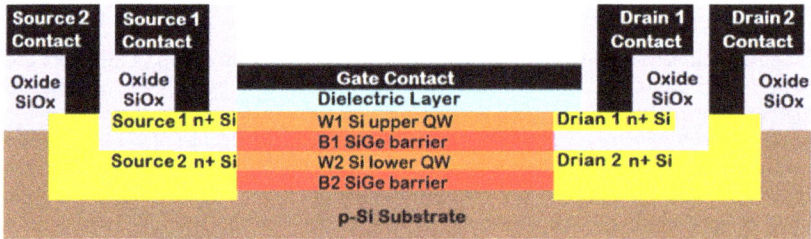

(a)

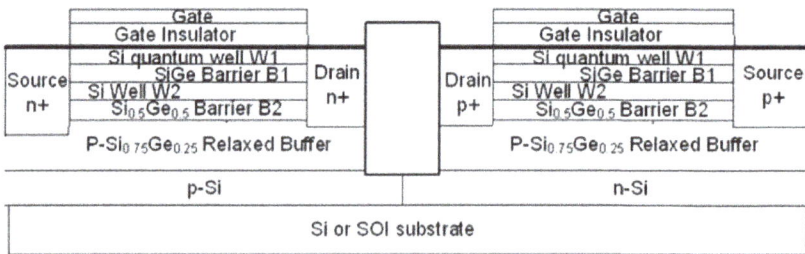

(b)

Fig. 1. (a) Two QWs Si/SiGe n-channel-SWS-FET cross-sectional. (b) Structures of two well n-channel and p-channel SWS-FET.

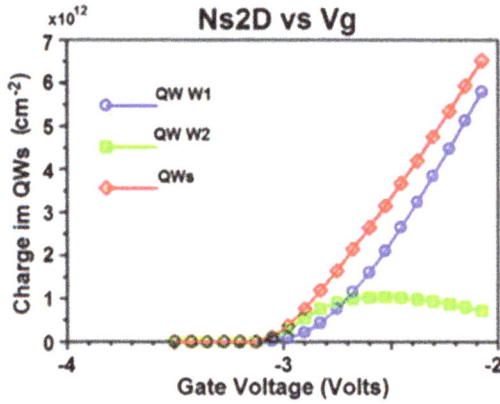

Fig. 2. Charge-densities simulation QWs n-channel SWS-FET.

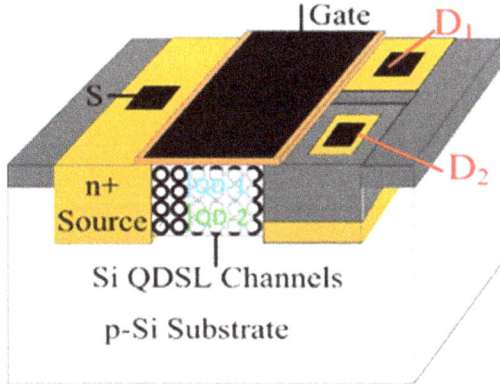

Fig. 3. Fabricated n-channel Quantum Dot SWS-FETs twin-drain device [6].

layers. The structures of two well n-channel and p-channel SWS-FET are shown in Fig. 1(b). Besides, Fig. 2 illustrates the charge-densities simulation of the two QWs n-channel SWS-FET.

Figure 3 illustrates the fabricated n-SWS-FET device, this device has four Quantum Dot (QD) layers and two drains (D1-deep or upper, D2-shallow or lower), the four quantum dots serving as the transport channel where the two lower QD1 layers are connected to lower drain D_2 and upper two QD2 layers are interfaced to upper drain D_1 [6]. ID-VD Wavefunction characteristics have been experimentally observed in Fig. 4.

3. SWS-FET SRAM

A typical brainy *SRAM circuit* consists of memory cell, address decoders, column multiplexers, sense amplifiers, I/Os, and control circuitry. A schematic of 6T SRAM

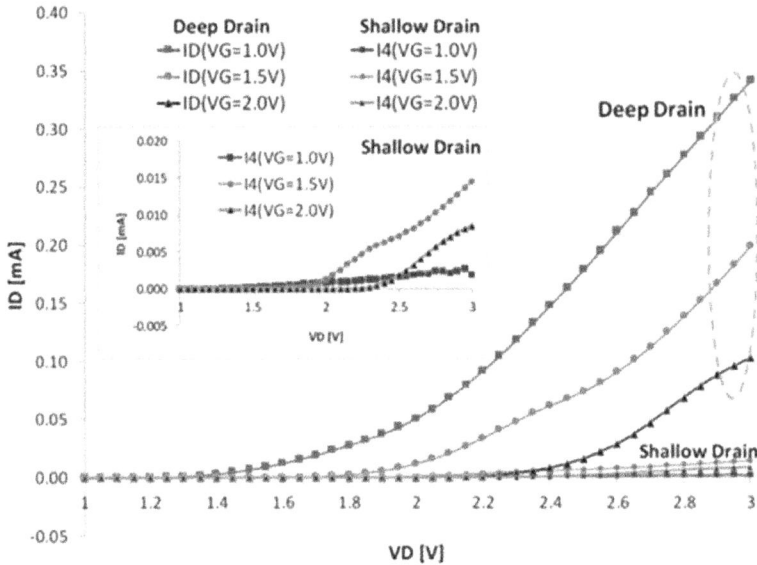

Fig. 4. Experimental ID-VD characteristics of Fabricated n-channel QD SWS-FETs.

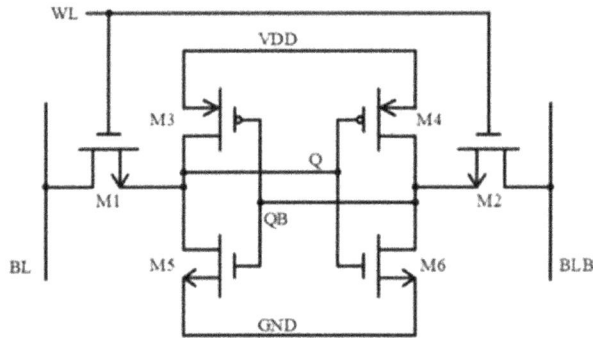

Fig. 5. Schematic diagram of 6T binary SRAM cell.

memory cell is shown in Fig. 5. As long as the cell is connected to a supply voltage VDD store data (Q) maintained by cross-coupled inverters [7].

Since SWS-FET is a promising device for WLV, the four-state (quaternary level or 2-bit) inverter circuit and its simulation are shown in Figs. 6 and 7, respectively, where VDD1 = 1.2 V, VDD2 = 0.8 V, VSS2 = 0.4 V, and VSS1 = 0 as ground. In voltage mode, the quadruple levels are determined as shown in Table 1. The four-state SWS-FET inverter is modeled using the Cadence, the model of the inverter is based on 180 nm technologies as shown by dedicated studies [3, 7].

In the context of quaternary SRAM, Fig. 8 shows the schematic of 6T quaternary SRAM cell, the cell consists of 4 SWS-FET as coupled quaternary inverter, and two

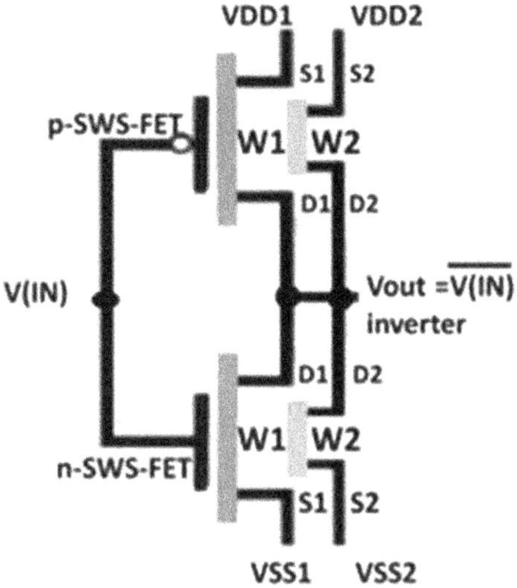

Fig. 6. Schematic diagram of quaternary inverter using SWS-FET.

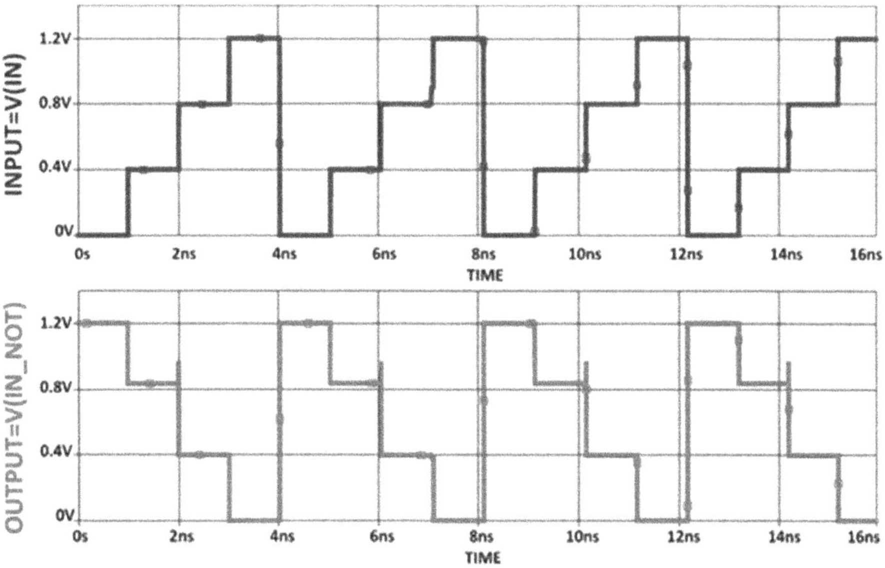

Fig. 7. Simulation of quaternary inverter using SWS-FET.

Table 1. Quaternary voltage level.

P Logic	Voltage	VDD Ration
0	0.0 V	0
1	0.4 V	VDD/3
2	0.8 V	VDD*2/3
3	1.2 V	VDD

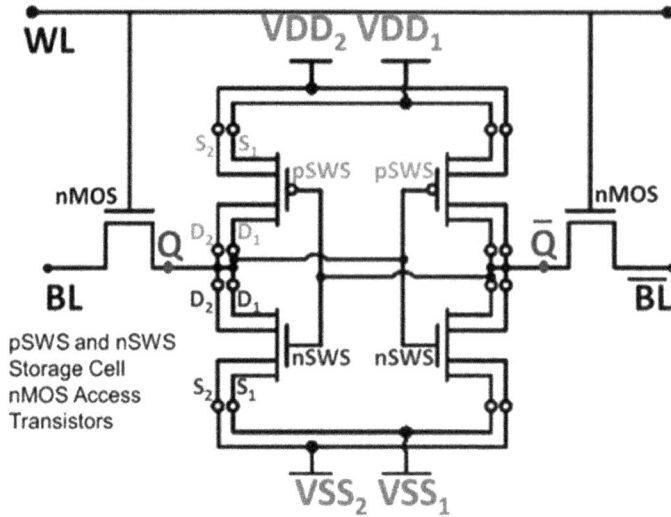

Fig. 8. Simulation of the proposed quaternary SRAM circuit using SWS-FET.

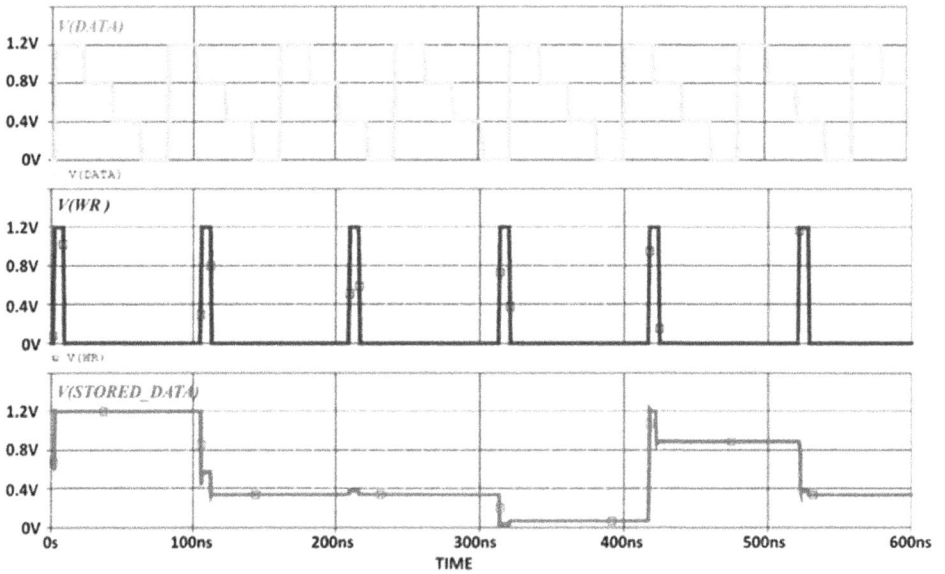

Fig. 9. Simulation of the proposed SRAM circuit using SWS-FET.

access nMOS, here bit is stored as quaternary logic (3, 2, 1, or 0) and control bit line operates binary logic (1 or 0).

Figure 9 illustrates the simulation waveform of the proposed quaternary SRAM circuit using SWS-FET. The first signal is represented Bit line "Din-green", whereas the second signal is represented word line "WL"-blue. The SRAM cell internal node

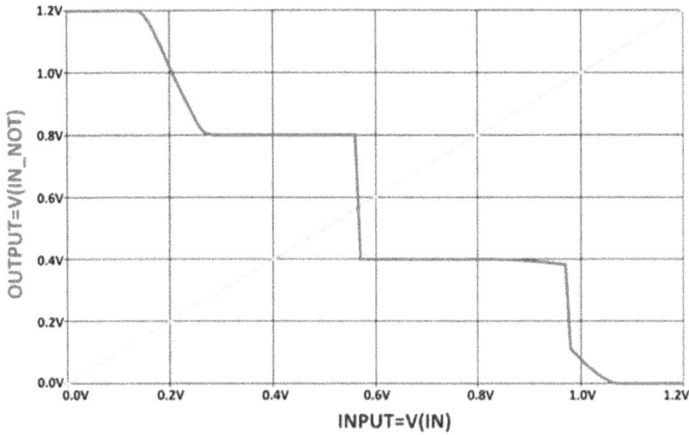

Fig. 10. VTC of the proposed quaternary inverter logic using SWS-FET.

is denoted by "Q-red". Figure 10 gives the inverter voltage transfer curve (VTC) of the proposed quaternary inverter logic of SWS-FET-based SRAM which clearly shows that the device holds four distinct output voltage levels for a large range of input voltage, this allows the couple inverter in the SRAM cell to keep data (hold) during storage operation.

4. Static Noise Margin (SNM) of SRAM Cell

Static Noise Margin (SNM) at a storage node (Q) indicates the effect of DC noise voltage for a SRAM cell, the SNM is defined as the highest value of DC noise voltage necessary to disturb the internal storage node (Q) and flip/change its stored data

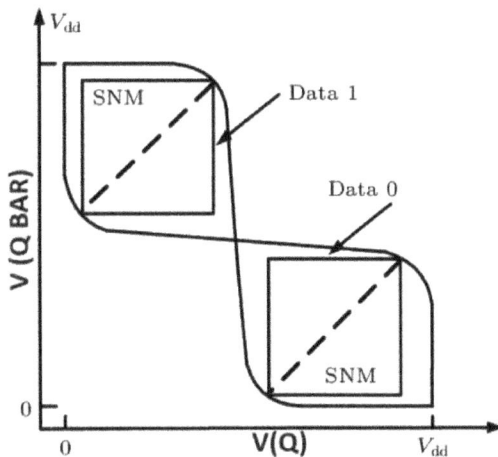

Fig. 11. Butterfly curves for SNM-SRAM cell.

state. Seeing that the inverter supply and date voltage level determine the static noise margin, a decrease in supply voltage for a low-power SRAM results in decrease in noise margin.

There are two ways to determine the SNM of SRAM cell noise curve (N-curve) and butterfly curve [8]. This work used the butterfly curve method, which can be obtained by plotting the VCT of one inverter V(Q) and superimposing the inverse VTC curve of another inverter V(Q BAR) on a single graph. This graph is called the butterfly curve as shown in Fig. 11. The noise margin is then determined by drawing the largest square between the two VTC curves. The SNM is calculated by measuring the length of the diagonal of the biggest square in butterfly curve that can fit inside the butterfly curves.

Based on quaternary SRAM, the butterfly curve of binary logic SRAM has two squares "=2*(2-1)" where the butterfly curve of quaternary logic SRAM has six squares "=2*(4-1)". The SNM of quaternary SRAM is obtained by first finding the maximum squares and measuring the diagonal for each square, the smallest diagonal is SNM. Figure 12 shows the butterfly curve of SWS-FET quaternary inverter, the SNMs quaternary inverter SRAM is 185 mV as shown in Table 2.

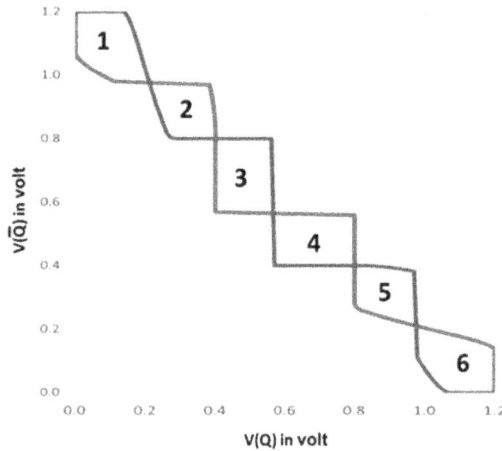

Fig. 12. Butterfly curve of the proposed quaternary SRAM using SWS-FET.

Table 2. Quaternary SNM of the proposed quaternary SRAM using SWS-FET.

Area of Square	Diagonal of maximum square
1	0.196
2	0.185
3	0.227
4	0.245
5	0.199
6	0.203

5. Conclusions

Quaternary (2-Bit) SRAM can be implemented using quantum well SWS-FET which produces four states in their transfer characteristics. A circuit model is used to simulate a four-state state inverter which is the basic building block multi-valued logic (MVL) SRAM circuit design. In this study, a quaternary logic SRAM based on SWS-FET 180nm technologies was realized by using coupled quaternary inverter, the results show that the proposed quaternary logic 6T SRAM cell exhibits non-destructive read and reliable write operations for all the four states. Finally, the Static Noise Margin (SNM) has been estimated, it has been observed that SNM is between 185 to 245 mV. In addition, the proposed quaternary SWS-FET SRAM cell exhibits improved results in terms of transistor count, MVL implementation, and lower circuit complexity, as the reduction in transistor count is 66.67%.

Acknowledgment

The researchers would like to acknowledge the Deanship of Scientific Research, Taif University for funding this work.

ORCID

B. Saman ◎ https://orcid.org/0000-0001-6917-5763

References

1. Jain, F., Lingalugari, M., Saman, B., Chan, P. Y., Gogna, P., Hasaneen, E. S., & Heller, E. (2015, December). Multi-state sub-9 nm QDC-SWS FETs for compact memory circuits. In 46th IEEE Semiconductor Interface Specialists Conference (SISC) (pp. 2–5).
2. Jain, F. C., Miller, B., Suarez, E., Chan, P.-Y., Karmakar, S., Al-Amoody, F., Gogna, M., Chandy, J., & Heller, E. (2011). Spatial wavefunction-switched (SWS) InGaAs FETs with II–VI gate insulators. Journal of Electronic Materials (Vol. 40, Issue 8, pp. 1717–1726).
3. Jain, F., Saman, B., Gudlavalleti, R. H., Chandy, J., & Heller, E. (2018). Multi-State 2-Bit CMOS logic using n- and p- quantum well channel spatial wavefunction switched (SWS) FETs. International Journal of High Speed Electronics and Systems (Vol. 27, Issue 03n04, p. 1840020).
4. Saman, B., Mirdha, P., Lingalugari, M., Gogna, P., Jain, F. C., Hasaneen, E.-S., & Heller, E. (2015). Logic gates design and simulation using spatial wavefunction switched (SWS) FETs. International Journal of High Speed Electronics and Systems (Vol. 24, Issue 03n04, p. 1550008).
5. Jain, F., Saman, B., Gudlavalleti, R. H., Mays, R., Chandy, J., & Heller, E. (2019). Multi-Bit SRAMs, registers, and logic using quantum well channel SWS-FETs for low-power, high-speed computing. International Journal of High Speed Electronics and Systems (Vol. 28, Issue 03n04, p. 1940024).
6. Jain, F., Lingalugari, M., Kondo, J., Mirdha, P., Suarez, E., Chandy, J., & Heller, E. (2016). Quantum dot channel (QDC) FETs with wraparound II–VI gate insulators: Numerical simulations. Journal of Electronic Materials (Vol. 45, Issue 11, pp. 5663–5670).

7. Birla, S., Singh, R. K., & Pattanaik, M. (2011). Stability and leakage analysis of a novel PP based 9T SRAM cell using N curve at deep submicron technology for multimedia applications. Circuits and Systems (Vol. 02, Issue 04, pp. 274–280).

8. Grossar, E., Stucchi, M., Maex, K., & Dehaene, W. (2006). Read stability and write-ability analysis of SRAM cells for nanometer technologies. IEEE Journal of Solid-State Circuits (Vol. 41, Issue 11, pp. 2577–2588).

Fabrication and Characterization of Four-State Inverter Utilizing Quantum Dot Gate Field-Effect Transistors (QDGFETS)#

Bilal Khan ◎*, Roman Mays ◎*, Raja Gudlavalleti ◎*,
Evan Heller ◎† and Faquir Jain ◎*,‡

*Department of Electrical Engineering, University of Connecticut,
352 Mansfield Rd, Storrs, CT 06269, United States

†Synopsis Corporation, NY, USA
‡faquir.jain@uconn.edu

This paper presents the experimental results of nMOS quantum dot gate field-effect transistor (QDG-FET) based four-state inverter fabricated and tested with Si/SiO_2 and Ge/GeO_2 quantum dots. The site-specific self-assembly of SiOx-cladded Si and GeOx-cladded Ge quantum dot layers in the gate region implements both the driver and load FETs in enhancement nMOS inverters. A four-state inverter will allow the reduction of FET count in logic block in microprocessors.

Keywords: QDGFET; multi-state; multi-bit; inverters.

1. Introduction

Quantum dot gate (QDG) FETs exhibiting three-state I-V characteristics have been reported using two layers of thin barrier SiOx-cladded Si quantum dots in the gate region [1]. This has been followed by reports of four-state QDG-FETs and enhancement mode inverters using self-assembled Si-Ge quantum dots [2]. This paper presents new results presented in our recent paper [3]. Moore's law states that the number of transistors doubles every two years [4], and reducing transistor size sub 5 nm presents challenges.

‡Corresponding author.
#This chapter appeared previously on the International Journal of High Speed Electronics and Systems. To cite this chapter, please cite the original article as the following: B. Khan, R. Mays, R. Gudlavalleti, E. Heller and F. Jain, *Int. J. High Speed Electron. Syst.*, **33**, 2440070 (2024), doi: 10.1142/S0129156424400706.

2. QDG-FET Comprising of 2 Layers of SiOx-Si and 2 Layers of GeOx-Ge QDs

A quantum dot (QD) is a semiconductor structure that is confined in all three dimensions. A QD has two main layers, a small semiconductor core such as Si or Ge surrounded by a cladding layer of either SiOx or GeOx, respectively. The cladding layer acts as a barrier layer which helps to trap charge inside the quantum dots (Fig. 1). When a number of quantum dots are placed next to each other they form a quantum dot superlattice (QDSL). The formation of a QDSL results in mini energy bands which allows the FET device to exhibit multiple states [1].

Three-state inverters using SiOx-Si and GeOx-Ge quantum dot gate FETs have been reported [1,2]. We are reporting the QDG-FET-based inverter employing both SiOx-Si QDs and GeOx-Ge Si QDs in the gate region that resulted in four states for the inverter. In this device, we have two layers of cladded Si QD layers over tunnel gate oxide and two additional layers of Ge QDs are self-assembled on Si QDs (Fig. 2). Every two layers of QDs provide a new threshold voltage which can be used to program the inverter at different states. The QDs are assembled using a colloidal solution over the gate region via a self-assembly method in which QDs are

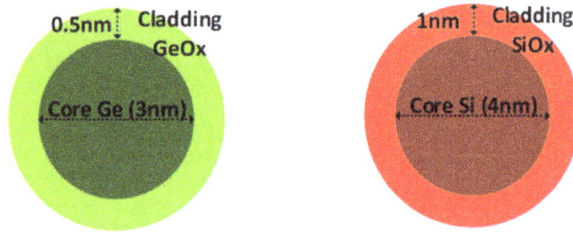

Fig. 1. Quantum Dot cross Section (Green: Germanium Red: Silicon)

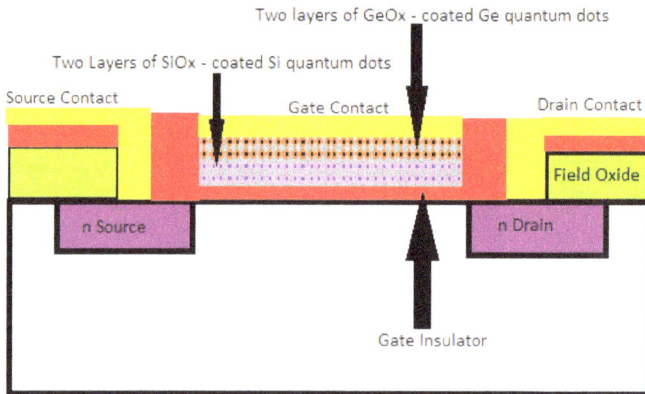

Fig. 2. QDG-FET cross section.

deposited on the p-Si channel region over the tunnel gate oxide. The Si sample was placed in the colloidal solution for 3 minutes after which an annealing process was preformed, this was done for every two layers of QDs.

3. Theory

QDG-FET and a regular transistor have similar operations other than a small number of adjustments. The quantum dots on the gate region allow multiple states in between the on and off operational states. With the addition of intermediate states, there are also threshold voltage shifts, this is calculated with the following equation:

$$
\begin{aligned}
\Delta V_{\mathrm{TH}} &= -\frac{q}{C_{ox}} \int_0^{x_g} \frac{x\rho(x)}{x_g} dx \\
&= -\frac{q}{C_{ox}} \left[\sum \frac{x_{\mathrm{QD1}} n_1 N_{\mathrm{QD1}}}{x_g} + \sum \frac{x_{\mathrm{QD2}} n_2 N_{\mathrm{QD1}}}{x_g} \right].
\end{aligned} \tag{1}
$$

The distance between the core of the first layer of quantum dots (the layer on the gate insulator) and the second layer (the layer furthest from the gate insulator) to the gate contact are represented by X_{QD1} and X_{QD2}, respectively. NQD1 is the charge on the cladding of the first layer of quantum dots, and NQD2 is the charge on the cladding layer of the second layer of quantum dots. X_g is the distance of the interface from gate terminal to the quantum dots, and q is the charge density.

Equation (2) shows the output voltage of an inverter. In the equation below, the output voltage of the QDG-FET inverter is derived and can be used before physically fabricating the device.

$$
V_{\mathrm{OUT}} = \frac{[\beta_R V_{\mathrm{DT}} - V_{1B}] \pm \sqrt{V_{1B}^2 - 2\beta_R V_{\mathrm{DT}} V_{1B} - 2\beta_R V_{\mathrm{DT}}^2}}{2 + \beta_R}. \tag{2}
$$

In the equations above

$$
\beta_R = \frac{\beta_{\mathrm{TOP}}}{\beta_{\mathrm{Bottom}}}
$$

$$
V_{\mathrm{DT}} = V_{\mathrm{DD}} - V_{Teff\ \mathrm{Top}}
$$

$$
V_{1B} = V_{\mathrm{Input}} - V_{Teff\ \mathrm{Bottom}}. \tag{3}
$$

β_R is the ratio of the width and length of the pullup transistor over the pulldown transistor (W/L). V_{DT} is the difference between V_{DD} and the effective threshold of the load transistor, and V_{IB} is the difference between the input voltage and the effective threshold of the driver transistor.

The intermediate states are states in the devices where the devices aren't fully on yet voltage continues to flow in a steady/constant flow. These states are thought to be the third and fourth states of the devices being neither fully operational nor fully off states, yet the devices are still allowing voltage flow.

4. Fabrication

These inverters were fabricated using four-level mask sets. First level of the mask sets open the source and drain regions to create the n$^+$ wells. For diffusion and drive-in diffusion, the process utilized infinite phosphorous sources at 1000° C. Second level in the mask set open the gates; once the gates were opened, a layer of 40 Å of dry oxide was grown utilizing a dry thermal oxidation method. After the oxide was grown, quantum dots were assembled using a bottom-up self-assembly process. The third level of the mask set opened the source and drain contact holes. After opening the source drain contact holes, aluminum was evaporated on the sample at a thickness of 2000 Å. The last mask removed the unnecessary metal so that all the devices can be accessed individually. Once the interconnect was opened and the devices were no longer shorted, testing was performed on the device. In the inverter, the top transistor has a gate width-to-length ratio of 5 microns to 1 micron. The bottom transistor has a gate width by length ratio of 15 microns to 1 micron.

5. Results

Experimental data of Vout-Vin for a 4-state inverter in Fig. 3. This inverter has 3 different threshold voltages. For ON state, Intermediate State 1, and Intermediate State 2. This 4-state inverter has both germanium quantum dots and silicon dots. This device consists of Si and Ge quantum dots which are 4 individual states that can be observed. The ID VD characteristics of the pulldown FET incorporating Si and Ge dot layers are exhibited in Fig. 4, and ID-VG characteristics of the driver FET in the inverter in Fig. 5. The Si dots were deposited first because the temperature for annealing the Si dots (750° C) is much higher than the annealing temperature for Ge dots (425° C). The pH levels for the Si dots and Ge dots are

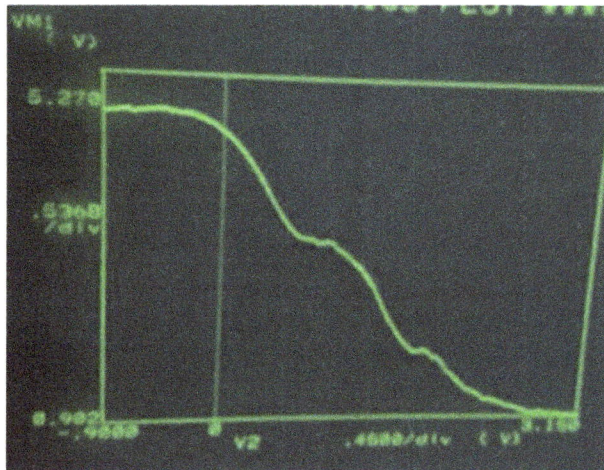

Fig. 3. 4-State Inverter.

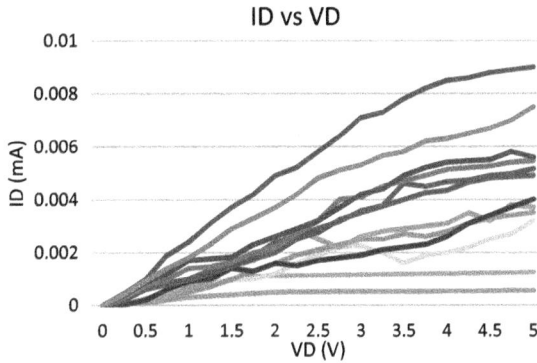

Fig. 4. ID-VD characteristics of the pulldown FET incorporating Si and Ge dot layers.

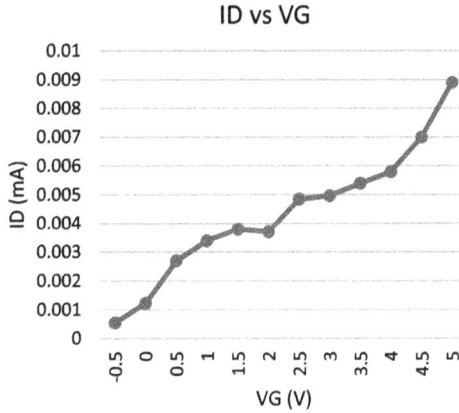

Fig. 5. ID-VG characteristics of the driver FET in the inverter of Fig. 2.

different as well where Ge dots have a pH of 3.3 and Si dots have a pH of 4.6. The annealing temperatures were $425°$ C and $750°$ C, respectively, for Ge and Si QDs.

6. Conclusion

Formation of a Si quantum dot superlattice (QDSL) and Ge QDSL in the gate region results in a four-state inverter. QD-NVRAMs have been experimentally demonstrated with a fast erase/write time [5]. The integration of QDG-FETs with QD-NVRAMs has been envisioned [6]. Finally, Moore's law is prolonged by using extra states exhibited in quantum dot gate (QDG)-FETs.

ORCID

Bilal Khan ⊙ https://orcid.org/0009-0007-5889-1028

Roman Mays ◉ https://orcid.org/0000-0002-3047-6056

Raja Gudlavalleti ◉ https://orcid.org/0000-0002-7727-8030

Evan Heller ◉ https://orcid.org/0009-0005-4405-7089

Faquir Jain ◉ https://orcid.org/0000-0003-3961-6665

References

1. F. C. Jain, E. Heller, S. Karmakar and J. Chandy, "Device and circuit modeling using novel 3-state quantum dot gate FETs," *2007 ISDRS*, 2007, pp. 1–2, doi: 10.1109/IS-DRS.2007.4422254.
2. M. Lingalugari *et al.*, "Novel multi-state quantum dot gate FETs using SiO2 and lattice-matched ZnS-ZnMgS-ZnS as gate insulators," *Journal of Electronic Materials*, 42, pp. 3156–3163, 2013.
3. B. Khan, R. Mays, R. Gudlavalleti and F. C. Jain, "Fabrication and characterization of nMOS inverters utilizing quantum dot gate field effect transistor (QDGFET) for SRAM device," *IJHSES*, 31, no. 01n04, pp. 2240010(1–14), 2022.
4. C. A. Mack, "Fifty years of Moore's law," *IEEE Transactions on Semiconductor Manufacturing*, 24, pp. 202–207, 2011.
5. M. Lingalugari, P.-Y. Chan, E. K. Heller, J. Chandy and F. C. Jain, "Quantum dot floating gate nonvolatile random access memory using quantum dot channel for faster erasing," *Electronic Letters*, 54, p. 36, 2018.
6. F. Jain, R. Gudlavalleti, R. Mays, B. Saman, J. Chandy and E. Heller, "Integrating QD-NVRAMs and QDC-SWS FET based logic for multi-bit computing," *IJHSES*, 31, pp. 2240020–1, 2022.
7. S. Karmakar, J. A. Chandy, M. Gogna *et al.*, "Fabrication and circuit modeling of NMOS inverter based on quantum dot gate field-effect transistors," *Journal of Electronic Materials*, 41, 2184–2192, 2012. https://doi.org/10.1007/s11664-012-2116-4

GNN-Based Reverse Engineering Protection Using Obfuscation Through Electromagnetic Interference[#]

William Stark [*,†], Lei Wang [‡] and Shuai Chen [§]

*Department of Electrical and Computer Engineering,
University of Connecticut, Storrs CT 06269,
Tolland County, USA*

**Current Address is 1315 Washington Boulevard Apartment 520,
Stamford, CT, 06902, Fairfield County, USA*
[†]*william.stark_jr@uconn.edu*
[‡]*lei.3.wang@uconn.edu*
[§]*shuai.chen@uconn.edu*

Yuebo Luo [⊚]

*School of Computing, University of Connecticut,
Storrs, CT 06269, Tolland County, USA*
yuebo.luo@uconn.edu

To combat PCB design reverse engineering and protect intellectual property, a method of obfuscation is developed. Noise generating traces are placed alongside the true circuit, separated by false non-conductive vias visually indistinguishable from regular ones. If the false vias are copied as real conductive vias in illegitimate replications, signals will be distorted, leading to poor performance. The deliberate placement of the generators separated by false vias is tested using a machine learning algorithm (MLA) employing a signal integrity database to determine the most vulnerable traces in the PCB based on their physical properties and interconnections.

Keywords: Reverse engineering; obfuscation; machine learning.

1. Introduction

Printed Circuit Board (PCB) designs are a staple of electronic hardware, with an expansive list of products containing applications of PCB technology. While many circuit types follow standard designs, the actual arrangement of components on the board and combination of circuits are what make PCB designs unique, and as a

---

[†]Corresponding author.
[#]This chapter appeared previously on the International Journal of High Speed Electronics and Systems. To cite this chapter, please cite the original article as the following: W. Stark, L. Wang, S. Chen and Y. Luo, *Int. J. High Speed Electron. Syst.*, **33**, 2440071 (2024), doi: 10.1142/S0129156424400718.

result, potential targets for industrial espionage through reverse engineering. Theft and illegal recreation of PCB designs are extremely common in the industry, with about 90% of PCB designers having reported duplication of intellectual property. The pervasiveness of design theft can be attributed to the ease with which PCBs can be reverse engineered, often through visual analysis. The reaction to this trend has been an incentive to invest in reverse engineering protection measures. The investigation presented in this paper approaches this issue by using transformable electronics to obfuscate and degrade circuit operation, with the arrangement of transformable properties optimized through a machine learning algorithm.

2. Design Obfuscation Strategy

Reverse engineering a PCB can be performed in the absence of design diagrams through both destructive and non-destructive means, beyond a surface-level visual analysis of a product example. While some may choose to use destructive reverse engineering, which involves physically scraping off the outer layers to reveal the design of the interior layers, non-destructive methods that take X-ray photos of the inner layers can be a challenge to account for. Once photos of each layer can be obtained, advanced prediction algorithms can be employed to determine component placement and recreate the circuit, all while keeping the functional example of the PCB intact for testing. If the circuit design of the inner layers is compromised using visual analysis, then visually obfuscating the circuit can confuse an algorithm and prevent it from determining the true design. The inclusion of non-conductive magnesium-oxide via interconnects placed between traces can visually create multiple incorrect network connections (see Fig. 1). This would create enough combinations of possible circuit configurations that would be computationally complex and therefore unrealistically solvable. This solution is simple and inexpensive to accomplish [1].

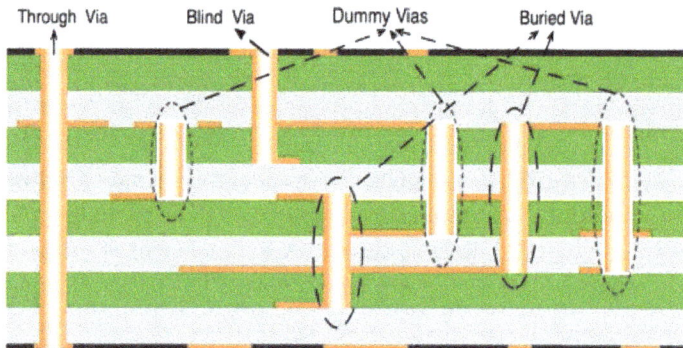

Fig. 1. Insertion of transformable and faux vias.

Meanwhile, destructive means of reverse engineering can be applied, exposing the inner layers of the PCB [3]. For this scenario, transformative electronics in the usage of pure magnesium vias would alter the functioning circuit, preventing a working example from being tested. The magnesium vias would react with the atmosphere upon exposure, turning the once conductive vias into the same magnesium-oxide vias that are used to throw off the appearance of the circuit, making it computationally impossible to determine the true circuit visually and preventing the design from functioning correctly for the attacker. Replacing typical copper vias using magnesium can provide a similarly conductive via material with a resistivity of $44.7\,n\Omega{\cdot}m$ [2].

Transformative electronics can be further by intentionally introducing performance degradation into the design on top of blocking off circuit functionality. When transformative electronics take effect, traces of a circuit can undergo a modification that causes an increase in electromagnetic interference (EMI) if other intentionally placed traces can be used as noise generators. This can subtly invalidate signals containing sensitive information, protecting the data from attackers while allowing the design to function normally without influence from the noise generator traces when used legitimately.

3. Database Selection

Various machine learning studies similar to this one have needed a method to simulate physical circuit models. The SI/PI-Database database project produced by the Institut für Theoretische Elektrotechnik [4] was created with the intention of constructing machine learning models for signal integrity. This database contains several PCB models with numerous physical parameters defined for generating simulation results. Varying the parameters in a physics-based simulation was used to generate frequency domain results in the form of s-parameter data. The specific model selected for this investigation was the 10 cavity PCB with two 5×5 via arrays, with the dataset containing 1500 combinations of physical parameter variation data points and s-parameter results, amounting to 57.2 gigabytes of data. The physical parameters that were simulated include:

- Plane Thickness
- Loss Tangent
- Relative Permittivity
- Antipad Radius
- Cavity Height
- Pad Radius
- Via Pitch
- Array Distance
- Trace Width
- Via Radius

This data will be assigned to the circuit nodes in the neural network as the node features and varied through the training cycles.

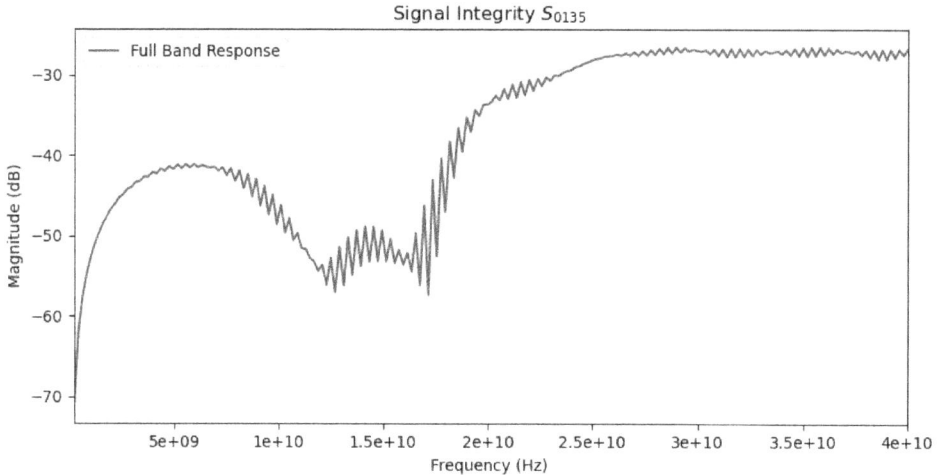

Fig. 2. Signal integrity simulation results from Ref. [4] between ports s_{01} and s_{35}.

The database simulation results are in the form of s-parameters, representing the signal integrity represented as vectors between two points on the physical PCB. S-parameter values were generated for all combinations of simulated 68 ports in the design, swept between 1 and 40 GHz. The model will be interested in the integrity between ports on the bottom of the via that are linked by a trace, such as s_{01} to s_{35} (see Fig. 2). A frequency of interest is needed to consider what the SI is at a base frequency. Many communications products use 5.24 GHz, so this model will take particular interest in the SI at that frequency, given the combination of physical properties.

4. Graph Neural Networks

The implementation of obfuscation tactics is an ongoing investigation determining the most effective placement for the alternative vias. In this case, the generator pairs will be placed at points in the PCB that are naturally most susceptible to noise and can be most easily degraded. The methodology for predicting ideal noise source placement will represent a PCB circuit (see Fig. 3) as a graph neural network (GNN) (see Fig. 4). The circuit will be interpreted as an interconnection of electrical nodes, or traces, and the components between each node. This will then correspond to the network nodes in the model, with their edge connections corresponding to components. Each node in the circuit has physical characteristics that determine the distortion a signal would experience in propagation, or the signal integrity (SI) of the node. The sparse connections in a circuit when interpreted as a network make the GNN model more efficient than typical dense neural networks.

The popularity of using GNN models is currently growing, with more libraries being created to build GNN type models from data sets. The GNN module of

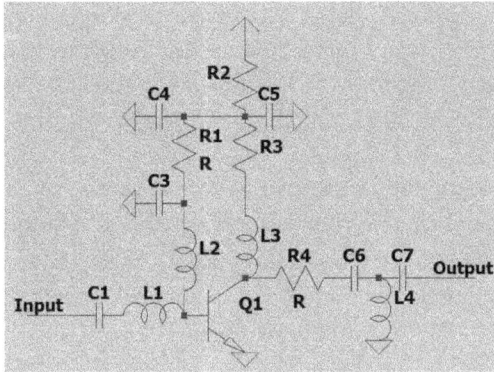

Fig. 3. Basic low noise amplifier (LNA) circuit to test database model.

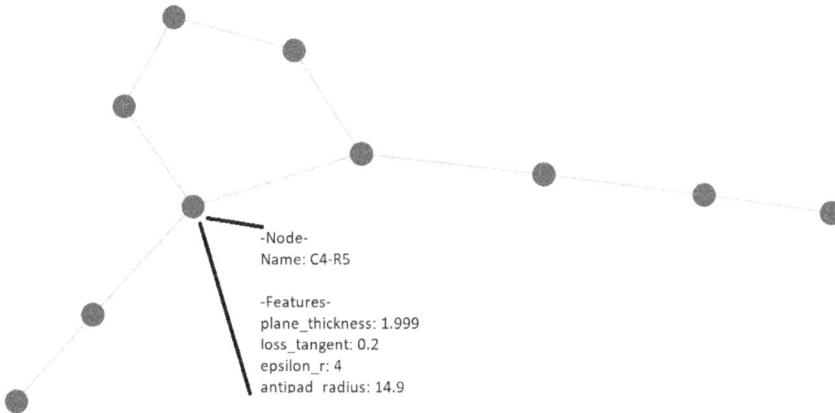

-Node-
Name: C4-R5

-Features-
plane_thickness: 1.999
loss_tangent: 0.2
epsilon_r: 4
antipad radius: 14.9

Fig. 4. LNA circuit converted into a GNN representation.

PyTorch will be used in this case to build the model. First, a selection of two simulation ports will be chosen that share a trace between them in the via array. Each port combination has all 1500 of its data points stored in separate files per port combination. The corresponding s-parameter data, stored as separate files per physical property combination that contain every data point for every port combination, will be read for the desired port combination and simulation results at 5.24 GHz. This will be combined all into one data set in PyTorch. Data training masks will be applied so the trainer can have both the training and validation data.

The netlist of the circuit will be parsed into an adjacency matrix, representing the electrical nodes and component edges. This creates the links between components and electrical nodes of the circuit. The nodes are assigned feature parameter names corresponding with the data set. This will allow the training model data distributor to assign data points to the different nodes in the circuit. The number of data points assigned to the circuit will equal the number of nodes. Convolution layers are then applied to the now created model, which will facilitate the training process converging on a model set. This model will start the experiment using two graph convolution network layers and training results will determine alterations to the model layers.

The training criterion will be defined as the s-parameter data field in the node feature list. An "Adam" style optimizer will be tested. The model is then trained by capturing the loss from the training data points. Gradients to determine the new round of weights are computed and the cycle repeats for each training epoch until a loss percentage is determined. The model will be tested with different convolution layers applied as well as different optimizers to obtain the best possible loss, making the model fit within a tolerable amount. Once the model is trained on the data set, new physical properties can be introduced to the same circuit design based on real specified design parameters, with the model able to extrapolate the SI of each node. This will give an indicator of which nodes in a circuit design would be subjected to the strongest SI, and therefore be good candidates for receiving transformative generator pairs.

5. Prototype Model Program with PyTorch

Testing out the methodology as described in previous sections was conducted in a PyTorch program. The directory for the parameter data file is read into an instance of a class containing the functions to load the data to PyTorch tensors. For now, the program reads in each tensor component from separate sources, as the information needed from both the data set and the circuit files are split apart as such. Node features from each of the 1500 data points were read into entries as the "x" data, while edge features, in this case the name label of each component in the circuit, were placed in a separate tensor. The adjacency matrix was pulled in and converted into a two-column array, with one column containing the index of the source node of each edge and the other containing the target node index. The circuit design offers the simplicity of having every node connected sequentially via a single component, with only one component connection between out of sequence nodes. Finally, the signal integrity magnitude was read into a "y" tensor, with each entry corresponding in index to each line of the physical parameter data. For this first phase of testing, the signal integrity between port s_{01} and s_{35} was used.

With the data loaded into these tensors, the tensors are then put in a Data object. The data object is then saved to a ".pf" formatted file, allowing the program to process the data only once, or if changes are made to the data source files. The PyTorch created data set file is then reloaded, skipping the processing phase. What is left is a file containing a PyTorch data set with 1500 "x" data entries available to the nodes, each of which contains the 12 features from the SI/PI data set, A tensor of sources and targets between node indexes, a tensor of a single edge feature (stored as tokenized values due to the edge name feature being a string), and the magnitudes

```
Torch version: 2.1.2+cpu
Torch geometric version: 2.4.0

Processing...
100%|              | 1500/1500 [02:33<00:00,  9.78it/s]
Done!
Graph properties
==============================================================
Number of nodes: 10
Number of edges: 10
Average node degree: 1.00
Contains isolated nodes: False
Contains self-loops: False
Is undirected: False
Data(x=[1, 12], edge_index=[2, 10], edge_attr=[10], y=[1], num_nodes=10)
Number of features:  12
Edge Index:  tensor([[0, 1],
        [1, 2],
        [2, 3],
        [2, 6],
        [3, 4],
        [4, 5],
        [5, 6],
        [6, 7],
        [7, 8],
        [8, 9]])
Node Features (0th Entry):  tensor([[0.0000e+00, 1.9999e+00, 2.0000e-01, 4.0000e+00, 1.4900e+01, 9.3500e+00,
        8.9700e+00, 5.5480e+01, 7.6400e+02, 5.5800e+00, 5.6700e+00, 0.0000e+00]])
Edge Features: (Tokenized) tensor([0., 1., 2., 3., 4., 5., 6., 7., 8., 9.], dtype=torch.float64)
Tensor of Signal Integrity Values (0th Entry):  tensor([0.5317], dtype=torch.float64)
```

Fig. 5. Output of data set processing into PyTorch and statistics of data set.

(in dB) of the signal integrity between port s_{01} and s_{35} at 5.24 GHz for each node feature set (see Fig. 5).

The data set can be visualized once again now that the edge set is included. NetworkX is a popular library used to visualize graphs in machine learning. This data set will show a graph of 10 nodes connected with 10 edges (one node is considered the input and another the output of the circuit). Names of nodes and edges from the original circuit are applied as labels (see Fig. 6).

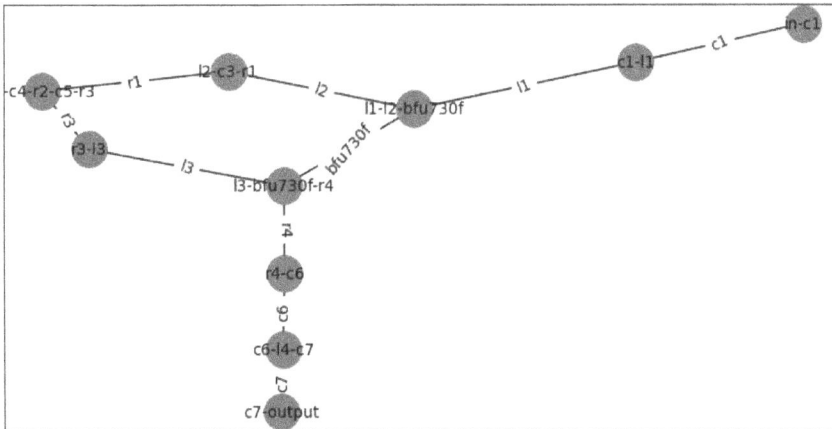

Fig. 6. Graphing the data file edge set with NetworkX.

```
Training Set:  Data(x=[1, 12], edge_index=[2, 10], edge_attr=[10], y=[1], num_nodes=10)
Length:  450
Validation Set:  Data(x=[1, 12], edge_index=[2, 10], edge_attr=[10], y=[1], num_nodes=10)
Length:  1050
```

Fig. 7. Results of separating data set into training and validation groups.

6. Model Training and Considerations

To handle training and validation, the data set is split into these two categories. A 30/70 split scheme is applied, giving the validation set 450 entries and the training set 1050 entries respectively. The sets will be batched in groups of 50 for processing (see Fig. 7). Using the PyTorch supplied randomization generator, seeded with a value of 42, the data set can be split into these two groups without any bias. This ensures the model can thoroughly determine a relationship between the node features and the signal integrity in the training set, and then perform with high accuracy on the validation set.

The Graph Convolution Network class was created initially with two layers, funneling the size from the number of node features down to the single SI value parameter. The feature data and edge list are fed to the first layer. In the forwarding section, the ReLU activation function is applied to the evaluation of the first layer, which is then accepted as the feature data in the second convolution layer. The final result is a logarithmic softmax computation of the second layer output. An "Adam" optimizer is specified with a learning rate of 0.01 and a weight decay of 0.0005. The loss is computed for 200 epochs (see Fig. 8).

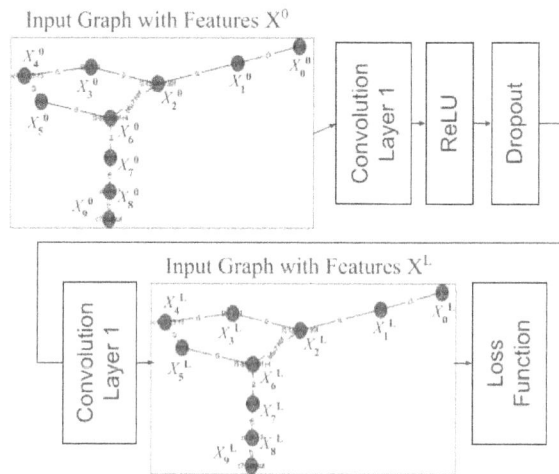

Fig. 8. Diagram of GNN training pipeline with convolution layers and activation function.

With a basic test program established and a data set file created, some details would need investigation to compare results. Certainly, there is a demand for experimentation in regard to the arrangement of convolution layers, activation functions, loss functions and batching. Increasing the number of epochs may provide the necessary amount of convergence on a more desirable loss, without needlessly consuming time and computation resources for every test. There are many more combinations of the 68 ports in the database that the data set could be built from when considering the signal integrity through an array trace. These other combinations provide additional sets of s-parameter data that can be used to evaluate the training model.

7. Construction of a Circuit Analysis Tool

The framework to apply a signal integrity data set to a graph set representation of a circuit can be expanded to work for any provided circuit. Any circuit can be read into the program and converted into a graph representation, whereupon feature data are applied to construct a machine learning model. Applying the same data set to the graph structure of the new circuit will allow the program to then accept real build properties of the circuit design and predict the signal integrity of each trace.

For different circuit designs, the SI/PI database may be more useful if parameters are included or excluded. A different PCB may want to exclude the cavity height parameter if that physical property would be irrelevant to the overall design. Being able to remove properties from the created data set would help limit any overfitting that would result from the training, which would yield poor results when evaluating with custom design parameters. Some circuits may be concerned with other frequencies besides the one sampled for this experiment. Having the option to pull other frequencies from the s-parameter data would be a valuable feature to have in such a tool for use with any PCB design.

The program could be expanded to allow for custom performance comparisons. Once a data set is created from a circuit adjacency matrix, different parameters in the training and validation could be entered to achieve more desirable results, which will depend on the intrinsic circuit design. For example, a circuit might see better validation results for different optimizer parameters, or a different training and validation set split. In a circuit analysis tool, making these customizations available for importing a new circuit would improve flexibility.

Several program components are required to make this process flexible. For now, the model program was tested with one specific circuit design in mind. In order to convert a circuit design of any size or complexity to a graph network, a parsing tool will be needed. SPICE files of a circuit must be read, extracting every component and set of connections from the node list. Assuming the provided format proves to be compatible with the program, any designer could then use the tool to evaluate their PCB for signal integrity, feeding it along the pipeline process of adding magnesium oxide separated noise generator traces.

8. Conclusion

A method to protect PCB designs from being reverse engineered is developed, following principles of circuit obfuscation. This method will utilize the placement of transformative and non-conductive material vias incorporated into the circuit design to provide automatic and effective protection through apparent circuit possibility complexity. A particular application of this via placement has been tested using augmented generator traces to inflict EMI into functional signals. Testing of a machine learning algorithm is underway to predict the optimal placement of generators at points where a PCB would be most susceptible to distortion. This learning model will be in the style of a GNN using the interpretation of circuit nodes as the network nodes and connecting components as the adjacency edges. This model is developed using a physics-based data set to assign physical properties to the supposed circuit nodes. The model is then trained using s-parameter simulation results to predict the signal integrity of each node in the circuit based on the assigned physical properties. This training can then be applied to predefined combinations of physical properties desired for a circuit design to predict the signal integrity of each node, and consequently which node would be most susceptible to EMI distortion.

ORCID

William Stark ⊙ https://orcid.org/0009-0005-7846-9268

Lei Wang ⊙ https://orcid.org/0009-0009-6653-9950

Shuai Chen ⊙ https://orcid.org/0009-0003-0973-1782

Yuebo Luo ⊙ https://orcid.org/0009-0005-1738-886X

References

1. S. Chen and L. Wang. "Transformable Electronics Implantation in ROM for Anti-Reverse Engineering", International Journal of High Speed Electronics and Systems, 2019.
2. S. Chen *et al.* "Chip-level anti-reverse engineering using trans-formable interconnects", IEEE International Symposium on Defect and Fault Tolerance in VLSI and Nanotechnology Systems (DFTS), 2015.
3. J. Grand. "Printed circuit board deconstruction techniques", 8th USENIX Workshop on Offensive Technologies (WOOT 14), 2014.
4. M. Schierholz, A. Sanchez-Masis, A. Carmona-Cruz, X. Duan, K. Roy, C. Yang, R. Rimolo-Donadio and C. Schuster, "SI/PI-database of PCB based interconnects for machine learning applications," IEEE Access, vol. 9, pp. 34 423–34 432, 2021.

A Brief Review of Electronic and Magnetic Structure of TIF₃[#]

Gayanath W. Fernando ◎*, Donal Sheets ◎*, Jason Hancock ◎*,
Arthur Ernst ◎†,‡ and R. Matthias Geilhufe ◎§,¶

*Department of Physics, University of Connecticut,
Storrs, CT 06269, USA

†Max Planck Institute of Microstructure Physics,
Weinberg 2, D-06120 Halle, Germany

‡Institute for Theoretical Physics, Johannes Kepler University,
Altenberger Strasse 69, 4040 Linz, Austria

§Department of Physics, Chalmers University of Technology,
412 96 Göteborg, Sweden
¶matthias.geilhufe@chalmers.se

Materials with perovskite structure are known to exhibit fascinating physical properties such as high-temperature superconductivity, negative thermal expansion (NTE) and colossal magnetoresistance. However, transition metal trifluoride perovskites are less well studied compared to their oxide counterparts though they display marked differences such as NTE behavior in ScF3. Doping of such MF3 perovskites has been the focus of the experimental work of Morelock *et al.* [1] which provides a comprehensive structural study of the material class $Sc1-x Ti_x F_3$. As shown in Fig. 1, there is a structural phase transition assumed to be tied to tilting of corner sharing octahedrons in this class of crystal structures which is believed to have an electrostatic dipolar origin, seen for example even in AlF₃ [2]. However, the insulating and magnetic properties of TiF₃ are closely related to the transition metal $3d$-electrons. The extra valence d-electron that Ti carries compared to Sc [1,3] gives rise to unusual electronic and magnetic properties.

Keywords: Halide Perovskite; ab initio modelling; double-exchange Hubbard model; infrared reflectivity; noncollinear magnetism; spin liquid; correlated electrons.

In our opinion, the magnetic and insulating properties of TiF₃ have not been well understood amidst several controversial claims [4–6]. In the following, we summarize our published work [7,8] on electronic properties and magnetic frustration utilizing

¶Corresponding author.
[#]This chapter appeared previously on the International Journal of High Speed Electronics and Systems. To cite this chapter, please cite the original article as the following: G. W. Fernando, D. Sheets, J. Hancock, A. Ernst and R. Matthias Geilhufe, *Int. J. High Speed Electron. Syst.*, **33**, 2440072 (2024), doi: 10.1142/S012915642440072X.

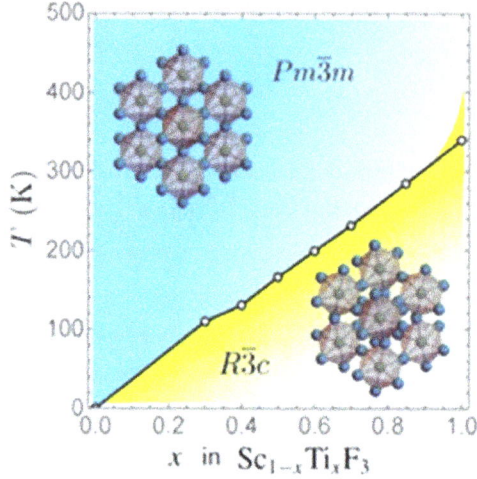

Fig. 1. Structural (cubic-rhombohedral) phase diagram showing temperature vs Ti-doping x in $Sc_{1-x}Ti_xF_3$.

frequency-dependent far-infrared reflectivity measurements and theoretical modeling of TiF_3. Our calculations indicate the possibility of a frustrated magnetic ground state which is insulating in agreement with experiment. In addition, this work explains the absence of a clear antiferromagnetic transition. At high temperature, TiF_3 is cubic with each Ti ion containing an unpaired electron occupying a t2g orbital. However, below 350 K, it can lower its free energy by a rhombohedral distortion, which lifts the degeneracy of the t2g band (see Fig. 1). The following is a list of salient features and conclusions from our study pertaining to TiF_3.

(1) We conclude that TiF_3 is an insulator based on reflectivity measurements [7] having moderate electron correlations. When explicit Coulomb correlations are not considered in theoretical work, an itinerant metallic state is predicted [6] contradicting our experimental results.

(2) First principles VASP calculations using constrained density functional theory [9–11] clearly show the development of an energy gap at the Fermi level in the density of states at moderate Coulomb correlation U (about 3 eV) values with a Hund coupling J_H of about 0.2 eV. A large U value (around 8 eV) as used in [4] appears to be unnecessary and unphysical for Ti.

(3) In addition, noncollinear magnetic order is present in the ground state as opposed to what was predicted in [4]. This was evident from noncollinear VASP as well as a double-exchange Hubbard model-based theoretical work [8].

(4) Recent susceptibility measurements [7] show indications of some kind of a magnetic transition below 10 K hinting toward some complex magnetic transition. Calculated magnon spectrum also indicates possible incommensurate magnetic order or frustration [8]. It would be interesting to see how this order changes with Ti doping along the series shown in Fig. 1.

ORCID

Gayanath W. Fernando ⊕ https://orcid.org/0000-0002-2882-199X

Donal Sheets ⊕ https://orcid.org/0000-0001-8598-9361

Jason Hancock ⊕ https://orcid.org/0000-0003-1101-8962

Arthur Ernst ⊕ https://orcid.org/0000-0003-2842-3032

R. Matthias Geilhufe ⊕ https://orcid.org/0000-0001-9285-0165

References

1. C. R. Morelock, L. C. Gallington and A. P. Wilkinson, Evolution of negative thermal expansion and phase transitions in $Sc1-xTixF_3$, Chemistry of Materials **26**, 1936 (2014).
2. P. B. Allen, Y.-R. Chen, S. Chaudhuri and C. P. Grey, Octahedral tilt instability of ReO3-type crystals, Physical Review B **73**, 172102 (2006).
3. H. Sowa and H. Ahsbahs, Pressure-induced octahedron strain in VF3-type compounds, Acta Crystallographica Section B **54**, 578 (1998).
4. V. Perebeinos and T. Vogt, Jahn-Teller transition in TiF_3 investigated using density-functional theory, Physical Review B **69**, 115102 (2004).
5. S. Mattsson and B. Paulus, Density functional theory calculations of structural, electronic, and magnetic properties of the 3d metal trifluorides MF3 (M = Ti − Ni) in the solid state, Journal of Computational Chemistry **40**, 1190 (2019).
6. F. Qin, X. Wang, L. Hu, N. Jia, Z. Gao, U. Aydemir, J. Chen, X. Ding and J. Sun, Switch of thermal expansions triggered by itinerant electrons in isostructural metal trifluorides, Inorganic Chemistry **61**, 21004 (2022).
7. D. Sheets, K. Lyszak, M. Jain, J. Hancock, G. W. Fernando, I. Sochnikov, J. Franklin and R. M. Geilhufe, Mott insulating low thermal expansion perovskite TiF_3, Physical Review B **108**, 235140 (2023).
8. G. W. Fernando, D. Sheets, J. Hancock, A. Ernst and R. M. Geilhufe, Correlation driven magnetic frustration and insulating behavior of TiF_3, Physica Status Solidi **18**, 2300330 (2023).
9. G. Kresse and D. Joubert, From ultrasoft pseudopotentials to the projector augmented-wave method, Physical Review B **59**, 1758 (1999).
10. J. P. Perdew, K. Burke and M. Ernzerhof, Generalized gradient approximation made simple, Physical Review Letters **77**, 3865 (1996).
11. P.-W. Ma and S. L. Dudarev, Constrained density functional for noncollinear magnetism, Physical Review B **91**, 054420 (2015).

Design and Simulation of 4-State SRAMs Using 4-State Quantum Dot Gate (QDG) FETs[#]

B. Khan [⊚]*, B. Saman [⊚]†, A. Almalki [⊚]*, R. H. Gudlavalleti [⊚]*,§,
J. Chandy [⊚]*, E. Heller [⊚]‡ and F. C. Jain [⊚]*,¶

*Department of Electrical and Computer Engineering,
University of Connecticut, Storrs, CT, USA

†Department of Electrical Engineering, College of Engineering,
Taif University, Saudi Arabia

‡Synopsys Inc., Ossining, NY 10562, USA

§Biorasis Inc., Storrs, CT, USA
¶faquir.jain@uconn.edu

This paper describes fabrication of Quantum Dot Gate n-FETs using SiOx-cladded Si quantum dot self-assembled on the tunnel gate oxide. Experimental I-V characteristics exhibiting 4-states are presented. Simulation is presented for the operation of viable 4-state SRAMs using QDG-FETs.

Keywords: QDGFET; SRAM; ABM model.

1. Introduction

Our group has fabricated quantum dot gate FETs and inverters [1–3]. A recently fabricated FET exhibiting multiple thresholds or intermediate states is shown in Fig. 1 [4]. In parallel, Jain *et al.* [5] have demonstrated inversion charge transferring from lower quantum dot channel to upper quantum dot channel in Quantum spatial wavefunction switched (SWS) FETs.

2. Simulation of 4-State QDG-FET-Based SRAMs

Figure 2 shows the schematic diagram of a QDG-FET-based SRAM cell that uses the traditional six-transistor (6T) architecture. In this case, the QDG-FET is used in place of the driver transistors to give the 4-state outputs for the provided DC input

¶Corresponding author.
[#]This chapter appeared previously on the International Journal of High Speed Electronics and Systems. To cite this chapter, please cite the original article as the following: B. Khan, B. Saman A. Almalki, R. H. Gudlavalleti, J. Chandy, E. Heller and F. C. Jain, *Int. J. High Speed Electron. Syst.*, **33**, 2440073 (2024), doi: 10.1142/S0129156424400731.

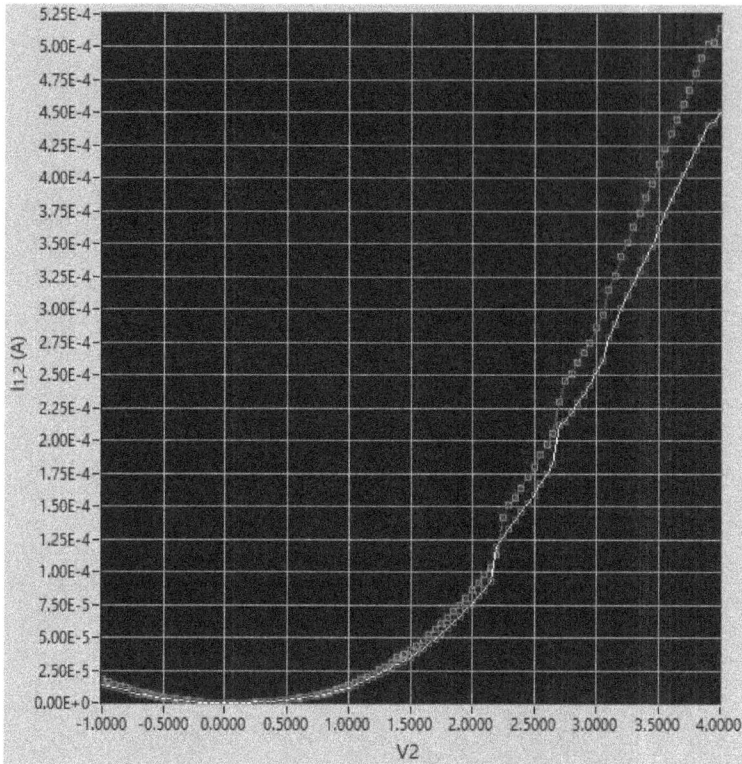

Fig. 1. Quantum dot gate n-FETs with self-assembled SiOx-cladded Si quantum dot on tunnel gate oxide.

voltage. The two inverters in the SRAM cell use P-MOS as pull-up FETs. These 4-state outputs can be thought of as 2-bit binary output or as four digital states. A 50% reduction in cell area can be obtained by using QDG-FETs in conjunction with the conventional 6T SRAM architecture.

2.1. *Analog behavioral model*

The QDG-FET model is based on the combination of the Analog Behavioral Model (ABM) and the Berkeley Short-channel IGFET BSIM Model for 180 nm technology. Figure 3 shows Cadence simulations of input-output characteristics of QDG-FET-based inverters. Simulations of the four states of Data connected to BL and Data_bar connected to BL_bar are shown in Fig. 4.

Figure 5 shows the simulation result for the write operation of 2-bit SRAM cell based on 6T QDG-FET. The top panel shows 4-state input data (green) connected to the bit line (BL). The middle panel shows the word line's pulse voltages at different states (blue). Finally, the QDG-FET-based SRAM's stored data is illustrated in the bottom panel (red) showing the storage of each state in the SRAM cell.

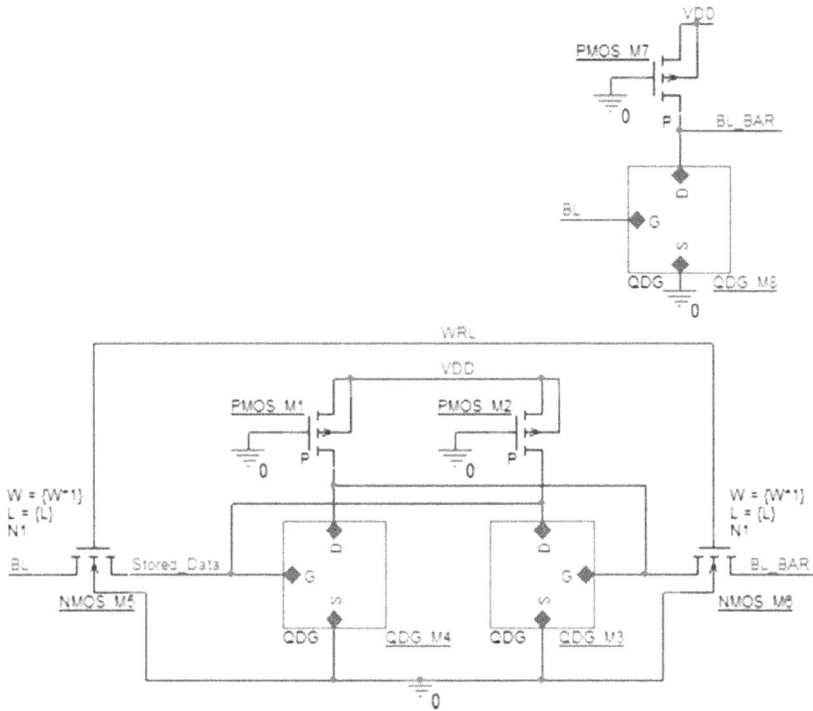

Fig. 2. Schematic of a 4-state SRAM using cross-coupled inverters integrating QDG-FET as pull-down device and P-MOS as pull-up FET.

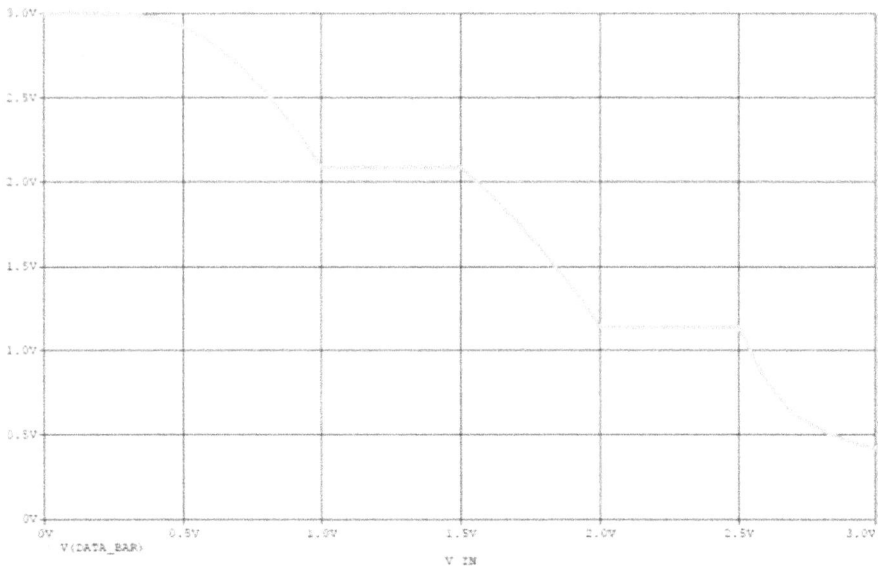

Fig. 3. Vo-Vin simulation based on intermediate states thresholds.

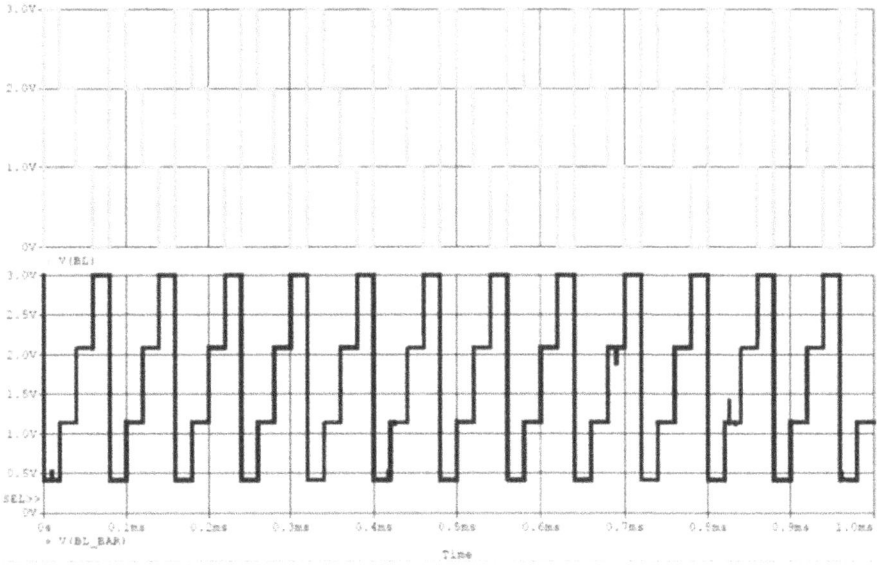

Fig. 4. Simulation of data connected to BL (top panel) and Data-bar connected to BL_bar.

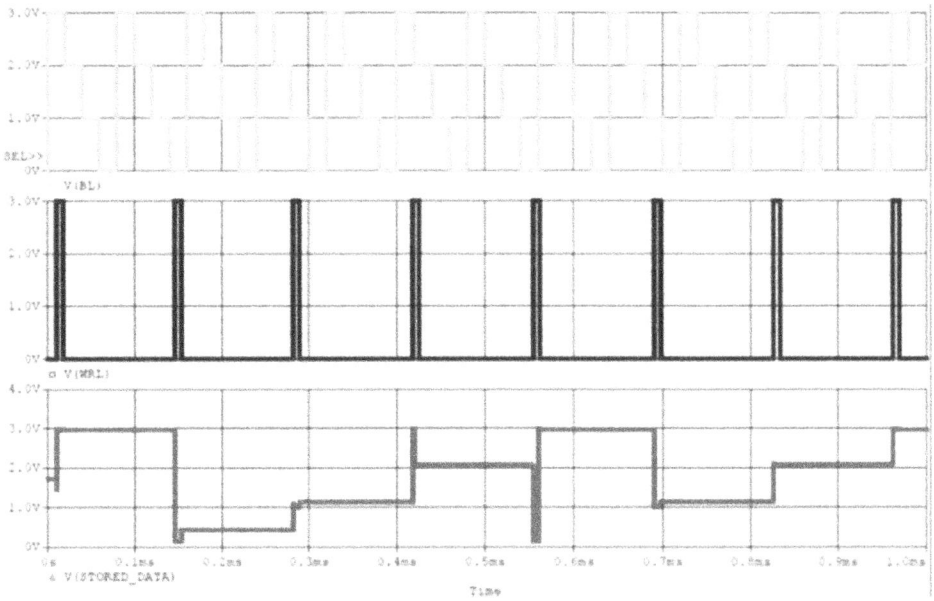

Fig. 5. SRAM Simulation results of data (green), Word line (blue), data stored in the memory cell (red).

ORCID

B. Khan ◎ https://orcid.org/0009-0007-5889-1028

B. Saman ◎ https://orcid.org/0000-0001-6917-5763

A. Almalki ◎ https://orcid.org/0009-0001-1954-4644

R. H. Gudlavalleti ◎ https://orcid.org/0000-0002-7727-8030

J. Chandy ◎ https://orcid.org/0000-0003-3449-3205

E. Heller ◎ https://orcid.org/0009-0005-4405-7089

F. C. Jain ◎ https://orcid.org/0000-0003-3961-6665

References

1. F. C. Jain, E. Heller, S. Karmakar and J. Chandy, "Device and circuit modeling using novel 3-state quantum dot gate FETs," 2007 ISDRS, 2007, pp. 1–2, doi: 10.1109/ISDRS.2007.4422254.
2. M. Lingalugari *et al.*, "Novel multi-state quantum dot gate FETs using SiO2 and lattice-matched ZnS-ZnMgS-ZnS as gate insulators," Journal of Electronic Materials, 42, 3156–3163, 2013.
3. B. Khan., R. Mays, R. Gudlavalleti, and F. C. Jain, "Fabrication and characterization of nMOS inverters utilizing quantum dot gate field effect transistor (QDGFET) for SRAM device," IJHSES, 31, no. 01n04, (March 2022).
4. A. Almalki and F. Jain (private communication).
5. F. Jain, M. Lingalugari, B. Saman, P.-Y. Chan, P. Gogna, E.-S. Hasaneen, J. Chandy, and E. Heller, "Multi-state sub-9 nm QDC-SWS FETs for compact memory circuits," 46th IEEE Semiconductor Interface Specialists Conference (SISC), Atlanta (VA), December 2–5, 2015.

Design of BTO-Based Compact Electro-Optic Modulator[#]

Chengxing He [ORCID], Mohan Shen [ORCID] and Hong X. Tang [ORCID]*

*Department of Electrical Engineering, Yale University,
New Haven, CT 06520, USA*
hong.tang@yale.edu

We present a compact electro-optic (EO) modulator design exploiting the strong Pockels effect of Barium Titanate ($BaTiO_3$ or BTO). This proposed structure, using parallel-plate electrodes tightly sandwiching the EO media, could achieve $V_\pi L$ value as low as $35\,V\cdot\mu m$, a significant reduction from current photonic-integrated Pockels EO modulators with $V_\pi L$ value around $1\,V\cdot cm$. Compared to plasmonic EO modulators, this proposed structure offers much lower optical loss. The small footprint and low-loss properties of this modulator design allow for its future embodiment in photonic-integrated CMOS circuits.

Keywords: Electro-optic (EO) modulator; Barium Titanate (BTO).

1. Introduction

Electro-optic (EO) modulators and switches are essential components for a variety of photonic integrated circuit (PIC) enabled applications, such as optical interconnects [1], microwave photonics [2–4] and on-chip frequency combs [5,6]. The rapid development of PICs calls for EO modulators with smaller footprints and lower loss to further push the performance and scale of PICs. Currently, EO modulation in photonics is mainly realized in two ways. One way is by exploiting free carrier dispersion effects [7] in silicon-based photonics [8–10], and the current state-of-the-art MZI modulators of this type have footprints of ∼1 cm [7], greatly limiting their scalability. Another way of achieving EO modulation is through the employment of Pockels effects in nonlinear crystals [4,6,11–13] or organic nonlinear materials [14–17].

---

*Corresponding author.
[#]This chapter appeared previously on the International Journal of High Speed Electronics and Systems. To cite this chapter, please cite the original article as the following: C. He, M. Shen and H. X. Tang, *Int. J. High Speed Electron. Syst.*, **33**, 2440074 (2024), doi: 10.1142/S0129156424400743.

Traditionally, nonlinear material-based integrated EO modulators are configured as silicon or silicon nitride [18] photonic waveguides straddled by electrodes, placed on top of or cladded by thin films with nonlinear properties. The current state-of-the-art EO modulators have $V_\pi L$ values of around 1 V·cm, also greatly limiting their scalability [12, 13, 19, 20]. On the other hand, plasmonic waveguides made from nonlinear materials sandwiched by electrodes can demonstrate $V_\pi L$ as low as ~60 V·μm. However, this comes at the expense of ~1 dB/μm propagation loss [17], preventing them from large-scale applications where a series of modulations need to be performed.

Progress in deposition [20, 21]/growth [12, 19, 22], bonding [12], etching [23, 24] of nonlinear materials, as well as reliable ion-slicing processes [25–29], prompt a reevaluation of the structure of integrated EO modulators based on nonlinear materials. In this letter, we propose an alternative configuration consisting of electrodes directly sandwiching a photonic waveguide that supports a vertically rather than horizontally elongated propagation mode. With Barium titanate (BaTiO$_3$, BTO) as the nonlinear material, this proposed EO modulator can achieve $V_\pi L$ as low as 35 V·μm and a propagation loss of 0.012 dB/μm. This modulator design paves the way for scalable integration of EO modulators in compact photonic and electronic circuits.

2. Device Concept

The key to achieving small footprint EO modulators is improving EO modulation efficiency. Modulators with high EO modulation efficiency require shorter modulator lengths to achieve the same phase shift as their less efficient counterparts, hence decreased footprint. To enhance the EO modulation efficiency, in addition to using materials with high nonlinear coefficients, our proposed device design improves upon traditional integrated EO modulators in two aspects. One is to increase the modal overlap between propagation mode and nonlinear material, and the other is to shorten the distance between electrodes, providing a stronger electric field given the same applied voltage. In designing a modulator geometry that fulfills both criteria, we derive inspiration from the widely adopted FinFET transistors [30] in the semiconductor industry. Specifically, instead of applying a thin metal film for the planar control of the electric field within nonlinear materials, the electrodes in this proposed modulator extend vertically. This vertical extension facilitates stronger electric fields and enhanced precision in controlling the electric fields within the nonlinear material. The schematic of the modulator is shown in Fig. 1.

The refractive index contrast between nonlinear material (BTO) [31] ($n_{BTO,e}$ = 2.25 at 1550 nm) and that of silica [32] or gold [33] (n_{SiO2} = 1.44, n_{Au} = 0.52 at 1550 nm) provides confinement for propagation modes. In a high confinement configuration, the modal overlap between propagation mode and nonlinear material can be close to 100%. To shorten the distance between electrodes, contrary to the

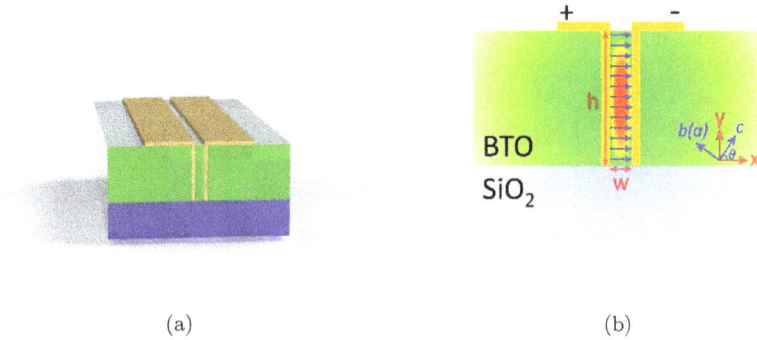

Fig. 1. (a) Schematics of the vertical mode modulator. (b) Cross-sectional view of the modulator, showing the optical field mode and the electric field direction, and illustrating the x, y axis as well as c axis offset angle θ established in the paper.

usual configuration of photonic waveguides, here the propagation mode expands vertically, meaning the horizontal width of the propagation mode can be squeezed below the wavelength. This allows for an order-of-magnitude increase in electric field strength between electrodes compared to more conventional EO modulators, where the distance between electrodes is usually a few microns [12, 13, 19, 20].

While plasmonic waveguides also boast sub-wavelength size and enhancement of electric field strength [17, 34], our photonic waveguide design offers better loss performance. Contrary to plasmonic waveguide which supports a TE mode, i.e. polarization of propagation mode parallel to electric field direction, our photonic waveguide supports a TM mode, i.e. polarization of propagation mode perpendicular to electric field direction. In this configuration, most of the optical energy in a high confinement photonic waveguide is centered in the nonlinear material region, rather than at the more lossy metal-dielectric interface. For large-scale PIC applications involving a series of EO modulations [35], our design allows for series integration of multiple EO modulators before signal amplification is needed.

To achieve the highest EO modulation efficiency, we propose to exploit BTO as our nonlinear material. BTO has one of the largest electro-optic coefficients of all materials [36] at $r_{42} = r_{51} = 1300 \pm 150 \,\mathrm{pm/V}$, $r_{33} = 113 \pm 5 \,\mathrm{pm/V}$ and $r_{13} = 11.2 \pm 1.3 \,\mathrm{pm/V}$. To best utilize the large r_{42} and r_{51} coefficients for EO modulation, the polarization of light traveling within the modulator need to offset from the c axis of BTO crystal. In the following section, the dependence of EO modulation efficiency on the offset angle will be discussed, and an optimal offset angle will be derived. The performance metrics of the EO modulator based on polycrystalline BTO thin film, where the c axis of BTO has a random distribution with no preferred alignment, will also be examined.

3. Optimization of BTO Thin Film Orientation

The change of relative permittivity (at 1550 nm) of BTO crystal when an external electric field is applied can be derived via the reduced tensor [37]:

$$
\begin{pmatrix}
\Delta(1/\epsilon)_1 \\
\Delta(1/\epsilon)_2 \\
\Delta(1/\epsilon)_3 \\
\Delta(1/\epsilon)_4 \\
\Delta(1/\epsilon)_5 \\
\Delta(1/\epsilon)_6
\end{pmatrix}
=
\begin{pmatrix}
0 & 0 & r_{13} \\
0 & 0 & r_{13} \\
0 & 0 & r_{33} \\
0 & r_{42} & 0 \\
r_{51} & 0 & 0 \\
0 & 0 & 0
\end{pmatrix}
\cdot
\begin{pmatrix}
E_a \\
E_b \\
E_c
\end{pmatrix}
$$

where E_a, E_b, E_c are the modulation electric field strengths projected onto the a, b, c axis of BTO crystal. Establish a coordinate system for the EO modulator, with the polarization axis of the TM mode represented as the y axis and the axis of the electric field between the electrodes as the x axis, as shown in Fig. 1(b). Suppose the c axis of the BTO crystal is offset from the previously defined x axis by angle θ, then without loss of generality due to the rotational symmetry in ab plane, one can write $E_a = 0$, $E_b = -E_x \cdot \sin\theta + E_y \cdot \cos\theta$, $E_c = E_x \cdot \cos\theta + E_y \cdot \sin\theta$.

The relative permittivity tensor ϵ and electric-field induced change in relative permittivity tensor $\Delta\epsilon$ are:

$$
\epsilon =
\begin{pmatrix}
\epsilon_{11} & 0 & 0 \\
0 & \epsilon_{22} & 0 \\
0 & 0 & \epsilon_{33}
\end{pmatrix}
$$

$$
\Delta\epsilon =
\begin{pmatrix}
\epsilon_{11}^2 \cdot \Delta(1/\epsilon)_1 & \epsilon_{11} \cdot \epsilon_{22} \cdot \Delta(1/\epsilon)_6 & \epsilon_{11} \cdot \epsilon_{33} \cdot \Delta(1/\epsilon)_5 \\
\epsilon_{11} \cdot \epsilon_{22} \cdot \Delta(1/\epsilon)_6 & \epsilon_{22}^2 \cdot \Delta(1/\epsilon)_2 & \epsilon_{22} \cdot \epsilon_{33} \cdot \Delta(1/\epsilon)_4 \\
\epsilon_{11} \cdot \epsilon_{33} \cdot \Delta(1/\epsilon)_5 & \epsilon_{33} \cdot \epsilon_{22} \cdot \Delta(1/\epsilon)_4 & \epsilon_{33}^2 \cdot \Delta(1/\epsilon)_3
\end{pmatrix}
$$

where $\epsilon_{11} = \epsilon_{22} = n_o^2 = 5.29$ and $\epsilon_{33} = n_e^2 = 5.06$ are the relative permittivity at 1550 nm or 193.4 THz. From this, the change in effective refractive index for the propagation mode can be related to the relative permittivity tensor and change in relative permittivity tensor by:

$$
\Delta n = \frac{n_{eff}}{2} \cdot \frac{(O_a, O_b, O_c)\Delta\epsilon(O_a, O_b, O_c)^H}{(O_a, O_b, O_c)\epsilon(O_a, O_b, O_c)^H}
$$

where O_a, O_b, O_c is the optical field strength projected onto the a, b, c axis of BTO crystal. Following the same transformation rules as modulation electric field, $O_a = 0$, $O_b = -O_x \cdot \sin\theta + O_y \cdot \cos\theta$, $O_c = O_x \cdot \cos\theta + O_y \cdot \sin\theta$.

Given the orientation of the ferroelectric domains within BTO is aligned in one direction as opposed to either direction along the c axis, the $V_\pi L$ value of the

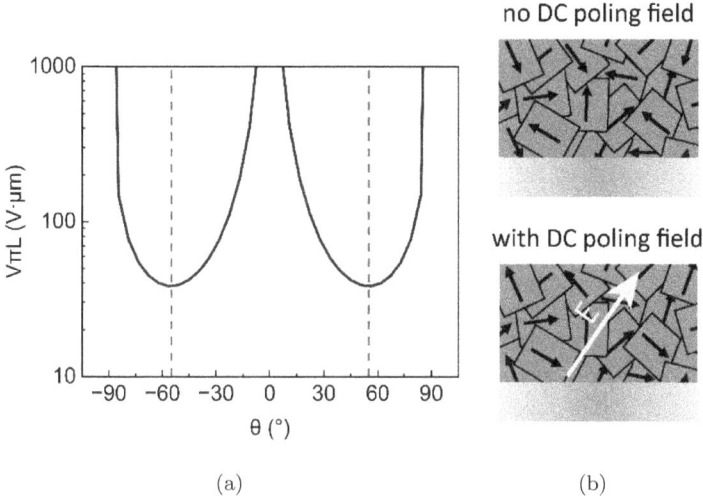

Fig. 2. (a) Dependence of VπL on offset angle θ for modulators with single-crystal BTO sections measuring 1.5 μm tall and 500 nm wide, with the optimal offset angle at 55° for highest EO modulation efficiency. (b) In polycrystalline BTO film, DC poling field induces local BTO crystalline regions to preferentially align with the orientation exhibiting a positive component parallel to the external electric field.

modulator exhibits a dependence on θ. In a simplified model of EO modulator supporting a TM mode where $E_y = 0$ and $O_x = 0$, $V_\pi L$ reaches a minimum value when $\theta = 55°$, as shown in Fig. 2(a). Simulations conducted on modulators operating at a wavelength of 1550 nm, comprising parallel plate electrodes encapsulating a single crystal BTO as shown in Fig. 1(b) with $h = 1.5\,\mu m$ and $w = 500$ nm, and BTO c axis offset from x axis by variable θ, suggests that for this particular geometry, the lowest value of $V_\pi L$ is 35 V·μm, which is achieved when θ is 55°. This result closely aligns with the value predicted by the simple analytical model.

In polycrystalline BTO with mixed alignment, an external DC poling field in the direction of the optimal offset angle can induce local BTO crystalline regions to preferentially align in the orientation exhibiting a positive component parallel to the external electric field, rather than the one exhibiting a negative component, as shown in Fig. 2(b) [38]. Consequently, the change in effective refractive index for polycrystalline BTO is determined by the weighted average of the changes in effective refractive index across all orientations with a positive component parallel to the external electric field:

$$\Delta n_{poly} = \int_0^{\pi/2} \Delta n(\theta)\sin\theta d\theta.$$

In the simplified model introduced above where $E_y = 0$ and $O_x = 0$, $\Delta n_{poly} = 0.65\Delta n(55°)$. i.e. the $V_\pi L$ for polycrystalline BTO film is 1.54 times the $V_\pi L$ for single-crystal BTO film with optimal offset angle.

4. Optimization of Modulator Geometry

In addition to the orientation of the BTO thin film, the geometry of the modulator, or more specifically the height h and width w of the cross-section, also profoundly impacts the performance metrics of the EO modulator. Having established an optimal BTO c axis offset angle at $\theta = 55°$ in the last section, in this section, the orientation of the BTO c axis is kept constant at the optimal angle. In Fig. 3, we show how h and w affects the modal overlap between propagation mode and BTO (quantified by $\Gamma = \frac{\iint_{\text{BTO}} O(x,y)\epsilon_{\text{BTO}}O^*(x,y)dxdy}{\iint O(x,y)\epsilon(x,y)O^*(x,y)dxdy}$), the propagation loss (α), VπL, and the total loss of a π phase shift modulator working at 1 V applied voltage ($\alpha \cdot$ VπL).

As for polycrystalline BTO films, the approximation $E_y = 0$ and $O_x = 0$ used to derive $\Delta n_{poly} = 0.65\Delta n(55°)$ remains reasonable for the simulations in this section, and the refractive index for ordinary ($n_o = 2.3$) and extraordinary ray ($n_e = 2.25$) in BTO is reasonably close. Hence, within good approximations, the modulator performance metrics for polycrystalline films share the same Γ and α as the single-crystal BTO film at optimal offset angle, while V$_\pi$L and $\alpha\cdot$V$_\pi$L gain a factor of 1.54. In the following paragraphs, only the values for single-crystal BTO film at optimal offset angle are quoted.

Figure 3(a) shows that for a fixed w of 500 nm, as h increases from 0.5 μm to 1.5 μm, a length comparable to the wavelength of propagating light at 1550 nm, Γ gradually increases from $\sim$93% to $\sim$98%. This is because when the cross-sectional area of BTO is smaller than that of the propagation mode, the BTO section could not offer high confinement, and the propagation mode has to extend beyond the BTO region into the substrate and cladding material regions, resulting in a low Γ. As h becomes larger than wavelength, the BTO section is able to provide high confinement and a further increase in h results in diminishing returns in Γ, remaining around 98% as h increases from 1.5 μm to 3 μm. This trend in Γ as a function of h is reflected in V$_\pi$L, where it drops from 58 V·μm to 36 V·μm as h increases from 0.5 μm to 1.5 μm, but only further drops to 35 V·μm as h is further increased to 3 μm.

The propagation loss α is directly related to the modal overlap between the propagation mode and gold electrodes, where light experiences a high loss of $\sim$60 dB/μm [33]. When the propagation mode is in the low confinement regime, increase in h results in more overlap between the propagation mode and gold electrodes, hence an increase in α, as the comparison between $h = 0.5 \mu$m and 1 μm data points suggests. However, as h becomes high enough for the BTO/gold electrode structures to provide high confinement of the propagation mode, with further increases in h, the profile of the propagation mode becomes more extended in the vertical direction while more squeezed in the horizontal direction, resulting in a decrease of the overlap factor and therefore α, as the comparison between $h = 1 \mu$m and 1.5 μm data points suggest. As h increases even further, the horizontal size of the propagation mode does not decrease further, resulting in a negligible change in α.

The parameter of $\alpha \cdot V_\pi L$ is the total loss of a π phase shift modulator at 1 V applied voltage. When a π phase shift modulator is designed with higher operation voltages in mind, the length required for a π phase shift is inversely decreased, and therefore inversely less propagation loss through the modulator. Generally reflecting the trends of α and $V_\pi L$, $\alpha \cdot V_\pi L$ drops significantly from 0.9 V·dB to 0.4 V·dB as h increases from 0.5 μm to 1.5 μm, and has negligible change as h increases from 1.5 μm to 3 μm.

Figure 3(c) demonstrates that for a fixed h of 3 μm, as w increases from 400 nm to 1000 nm, modal overlap Γ shows an increase from ~97% to ~100%. This increase

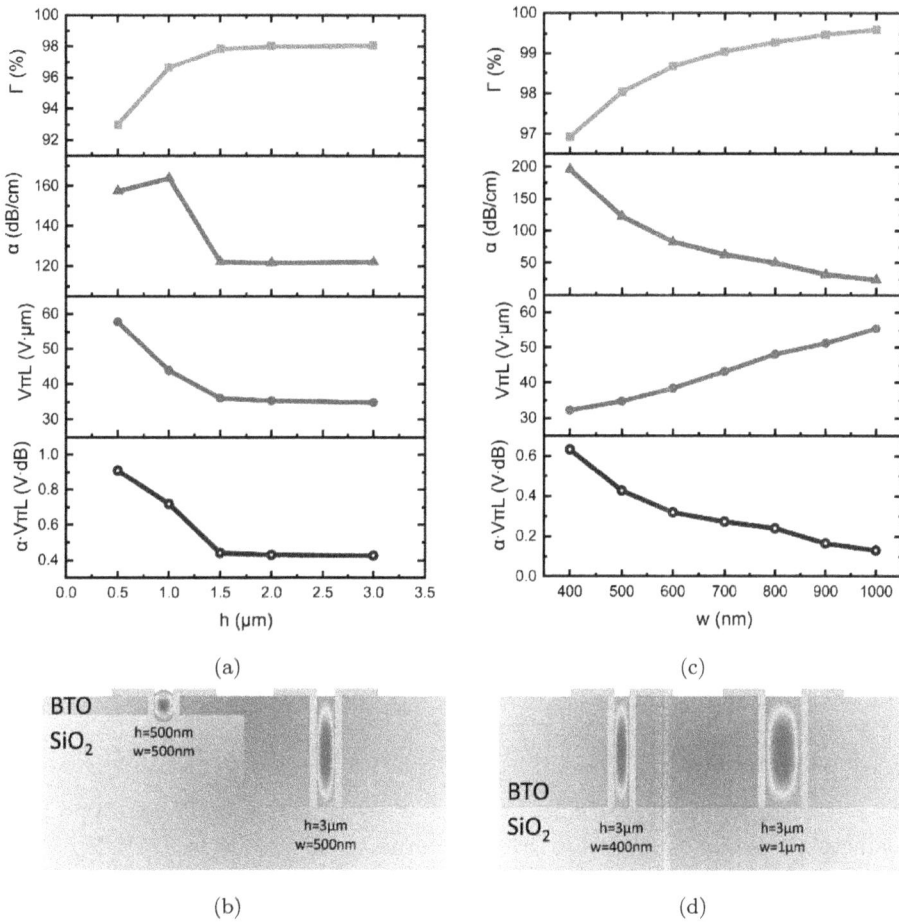

Fig. 3. Change in modal overlap between propagation mode and BTO Γ, propagation loss α, $V\pi$L, and total loss of a π phase shift modulator working at 1 V applied voltage $\alpha \cdot V\pi$L, as (a) the height h of the BTO section is scanned from 0.5 μm to 3 μm with a constant width w of 500 nm, the propagation modes of the extreme cases shown in (b), or (c) the width w of the BTO section is scanned from 400 nm to 1000 nm with a constant height h of 5 μm, the propagation modes of the extreme cases shown in (d).

in modal overlap with BTO comes from the decrease of modal overlap with lossy gold electrodes, which helps the decrease in propagation loss α from $200\,\text{dB/cm}$ to $23\,\text{dB/cm}$. Increase in w with the same applied voltage comes at the price of inversely reduced electric field strength, hence decreased EO modulation efficiency which results in increased $V_\pi L$ from $32\,\text{V}\cdot\mu\text{m}$ to $55\,\text{V}\cdot\mu\text{m}$ as w increases from $400\,\text{nm}$ to $1000\,\text{nm}$. The combined effects of change in w on α and $V_\pi L$ results in a decrease in $\alpha \cdot V_\pi L$ from $0.63\,\text{V}\cdot\text{dB}$ to $0.13\,\text{V}\cdot\text{dB}$ as w increases from $400\,\text{nm}$ to $1000\,\text{nm}$.

Considering the tradeoff between the modulation efficiency and loss of the EO modulator, we identify an optimized modulator geometry to be $h = 1.5\,\mu\text{m}$ and $w = 500\,\text{nm}$. Given a typical modulation voltage of $3\,\text{V}$ V_{pp} [17], a π phase modulator with this geometry would have a length of $12\,\mu\text{m}$ and bear a propagation loss of $0.15\,\text{dB}$. As for a polycrystalline BTO film-based modulator, this geometry would boast a length of $19\,\mu\text{m}$ and bear a propagation loss of $0.23\,\text{dB}$. These two performance metrics can be further halved if a push-pull MZM configuration is employed. With these performance metrics, the optical modulator structure proposed in this paper boasts footprint as compact as plasmonic modulators, while providing $10\,\text{dB}$ less loss. The aforementioned simulation results provide a guideline for further geometry optimization to meet specific application requirements.

5. Conclusion

In conclusion, we propose a novel configuration of EO modulator featuring BTO as the waveguide core. With our recommended geometry and typical modulation voltage of $3\,\text{V}$ V_{pp}, the modulator would have a small footpint of $12\,\mu\text{m}$ by $500\,\text{nm}$ when single-crystal BTO thin film with optimal offset angle is utilized, or $19\,\mu\text{m}$ by $500\,\text{nm}$ when polycrystalline BTO thin film is utilized, and bear a propagation loss of only $0.15\,\text{dB}$ for single-crystal film or $0.23\,\text{dB}$ for polycrystalline film, rivaling the state of art EO modulators in terms of footprint while providing $10\,\text{dB}$ less loss. This proposed structure can greatly benefit the development of large-scale PICs where space is at a premium and low loss is needed.

Acknowledgments

This work is supported by SUPREME, one of seven centers in JUMP 2.0, a Semiconductor Research Corporation (SRC) program co-sponsored by DARPA.

ORCID

Chengxing He ⊙ https://orcid.org/0000-0003-4716-6078
Mohan Shen ⊙ https://orcid.org/0009-0009-4638-3418
Hong X. Tang ⊙ https://orcid.org/0000-0001-5374-2137

References

1. R. Urata, H. Liu, L. Verslegers and C. Johnson, *Silicon Photonics Technologies: Gaps Analysis for Datacenter Interconnects* (Springer Berlin Heidelberg, Berlin, Heidelberg, 2016), pp. 473–488.
2. J. Capmany and D. Novak, "Microwave photonics combines two worlds," Nat. Photonics **1**, 319–330 (2007).
3. M. Shen, J. Xie, Y. Xu, S. Wang, R. Cheng, W. Fu, Y. Zhou and H. X. Tang, "Photonic link from single-flux-quantum circuits to room temperature," Nat. Photonics **18**, 371–378 (2024).
4. Y. Zhang, L. Shao, J. Yang, Z. Chen, K. Zhang, K.-M. Shum, D. Zhu, C. H. Chan, M. Lončar and C. Wang, "Systematic investigation of millimeter-wave optic modulation performance in thin-film lithium niobate," Photon. Res. **10**, 2380–2387 (2022).
5. I. L. Gheorma and G. K. Gopalakrishnan, "Flat frequency comb generation with an integrated dual-parallel modulator," IEEE Photonics Technol. Lett. **19**, 1011–1013 (2007).
6. M. Yu, D. Barton III, R. Cheng, C. Reimer, P. Kharel, L. He, L. Shao, D. Zhu, Y. Hu, H. R. Grant, L. Johansson, Y. Okawachi, A. L. Gaeta, M. Zhang and M. Lončar, "Integrated femtosecond pulse generator on thin-film lithium niobate," Nature **612**, 252–258 (2022).
7. R. Soref and B. Bennett, "Electrooptical effects in silicon," IEEE J. Quantum Electron. **23**, 123–129 (1987).
8. G. T. Reed, G. Z. Mashanovich, F. Y. Gardes, M. Nedeljkovic, Y. Hu, D. J. Thomson, K. Li, P. R. Wilson, S.-W. Chen and S. S. Hsu, "Recent breakthroughs in carrier depletion based silicon optical modulators," Nanophotonics **3**, 229–245 (2014).
9. K. Xu, L.-G. Yang, J.-Y. Sung, Y. M. Chen, Z. Z. Cheng, C.-W. Chow, C.-H. Yeh and H. K. Tsang, "Compatibility of silicon Mach-Zehnder modulators for advanced modulation formats," J. Light. Technol. **31**, 2550–2554 (2013).
10. D. Petousi, L. Zimmermann, A. Gajda, M. Kroh, K. Voigt, G. Winzer, B. Tillack and K. Petermann, "Analysis of optical and electrical tradeoffs of traveling-wave depletion-type Si Mach–Zehnder modulators for high-speed operation," IEEE J. Sel. Top. Quantum Electron. **21**, 199–206 (2015).
11. D. Janner, D. Tulli, M. García-Granda, M. Belmonte and V. Pruneri, "Micro-structured integrated electro-optic LiNbO$_3$ modulators," Laser Photonics Rev. **3**, 301–313 (2009).
12. S. Abel, F. Eltes, J. E. Ortmann, A. Messner, P. Castera, T. Wagner, D. Urbonas, A. Rosa, A. M. Gutierrez, D. Tulli, P. Ma, B. Baeuerle, A. Josten, W. Heni, D. Caimi, L. Czornomaz, A. A. Demkov, J. Leuthold, P. Sanchis and J. Fompeyrine, "Large pockels effect in micro- and nanostructured barium titanate integrated on silicon," Nat. Mater. **18**, 42–47 (2019).
13. C. Xiong, W. H. P. Pernice, J. H. Ngai, J. W. Reiner, D. Kumah, F. J. Walker, C. H. Ahn and H. X. Tang, "Active silicon integrated nanophotonics: Ferroelectric BaTiO$_3$ devices," Nano Lett. **14**, 1419–1425 (2014).
14. L. Alloatti, R. Palmer, S. Diebold, K. P. Pahl, B. Chen, R. Dinu, M. Fournier, J.-M. Fedeli, T. Zwick, W. Freude, C. Koos and J. Leuthold, "100 GHz silicon–organic hybrid modulator," Light. Sci. Appl. **3**, e173 (2014).
15. W. Heni, Y. Kutuvantavida, C. Haffner, H. Zwickel, C. Kieninger, S. Wolf, M. Lauermann, Y. Fedoryshyn, A. F. Tillack, L. E. Johnson, D. L. Elder, B. H. Robinson,

W. Freude, C. Koos, J. Leuthold and L. R. Dalton, "Silicon–organic and plasmonic–organic hybrid photonics," ACS Photonics **4**, 1576–1590 (2017).

16. C. Hoessbacher, A. Josten, B. Baeuerle, Y. Fedoryshyn, H. Hettrich, Y. Salamin, W. Heni, C. Haffner, C. Kaiser, R. Schmid, D. L. Elder, D. Hillerkuss, M. Möller, L. R. Dalton and J. Leuthold, "Plasmonic modulator with >170 GHz bandwidth demonstrated at 100 GBd NRZ," Opt. Express **25**, 1762–1768 (2017).

17. C. Haffner, W. Heni, Y. Fedoryshyn, J. Niegemann, A. Melikyan, D. L. Elder, B. Baeuerle, Y. Salamin, A. Josten, U. Koch, C. Hoessbacher, F. Ducry, L. Juchli, A. Emboras, D. Hillerkuss, M. Kohl, L. R. Dalton, C. Hafner and J. Leuthold, "All-plasmonic Mach–Zehnder modulator enabling optical high-speed communication at the microscale," Nat. Photonics **9** , 525–528 (2015).

18. K. Alexander, J. P. George, J. Verbist, K. Neyts, B. Kuyken, D. Van Thourhout and J. Beeckman, "Nanophotonic pockels modulators on a silicon nitride platform," Nat. Commun. **9**, 3444 (2018).

19. P. Castera, D. Tulli, A. M. Gutierrez and P. Sanchis, "Influence of batio3 ferroelectric orientation for electro-optic modulation on silicon," Opt. Express **23**, 15332–15342 (2015).

20. A. Petraru, J. Schubert, M. Schmid and C. Buchal, "Ferroelectric BaTiO3 thin-film optical waveguide modulators," Appl. Phys. Lett. **81**, 1375–1377 (2002).

21. J. W. Lee, K. Eom, T. R. Paudel, B. Wang, H. Lu, H. X. Huyan, S. Lindemann, S. Ryu, H. Lee, T. H. Kim, Y. Yuan, J. A. Zorn, S. Lei, W. P. Gao, T. Tybell, V. Gopalan, X. Q. Pan, A. Gruverman, L. Q. Chen, E. Y. Tsymbal and C. B. Eom, "In-plane quasi-single-domain batio3 via interfacial symmetry engineering," Nat. Commun. **12**, 6784 (2021).

22. F. Eltes, C. Mai, D. Caimi, M. Kroh, Y. Popoff, G. Winzer, D. Petousi, S. Lischke, J. E. Ortmann, L. Czornomaz, L. Zimmermann, J. Fompeyrine and S. Abel, "A batio3-based electro-optic pockels modulator monolithically integrated on an advanced silicon photonics platform," J. Light. Technol. **37**, 1456–1462 (2019).

23. J. Cheng, H. Yang, C. Wang, N. Combs, C. Freeze, O. Shoron, W. Wu, N. K. Kalarickal, H. Chandrasekar, S. Stemmer, S. Rajan and W. Lu, "Nanoscale etching of perovskite oxides for field effect transistor applications," J. Vac. Sci. Technol. B **38**, 012201 (2019).

24. H. B. Lee, Y.-H. Joo, H. Patil, G.-H. Kim, I. Kang, B. Hou, D. kee Kim, D.-S. Um and C.-I. Kim, "Plasma etching and surface characteristics depending on the crystallinity of the batio3 thin film," Mater. Res. Express **10**, 016401 (2023).

25. M. Bruel, "Silicon on insulator material technology," Electron. Lett. **31**, 1201–1202 (1995).

26. E. Butaud, B. Tavel, S. Ballandras, M. Bousquet, A. Drouin, I. Huyet, E. Courjon, A. Ghorbel, A. Reinhardt, A. Clairet, F. Bernard and I. Bertrand, "Smart Cut™ piezo on insulator (POI) substrates for high performances SAW components," in 2020 IEEE Int. Ultrason. Symp. (IUS), (San Francisco, CA, USA, 2020), pp. 1–4.

27. L. Yang, S. Wang, M. Shen, J. Xie and H. X. Tang, "Controlling single rare earth ion emission in an electro-optical nanocavity," Nat. Commun. **14**, 1718 (2023).

28. P. Rabiei and W. H. Steier, "Lithium niobate ridge waveguides and modulators fabricated using smart guide," Appl. Phys. Lett. **86**, 161115 (2005).

29. G. Poberaj, H. Hu, W. Sohler and P. Günter, "Lithium niobate on insulator (lnoi) for micro-photonic devices," Laser & Photonics Rev. **6**, 488–503 (2012).

30. D. Hisamoto, T. Kaga, Y. Kawamoto and E. Takeda, "A fully depleted lean-channel transistor (DELTA) — A novel vertical ultrathin SOI MOSFET," IEEE Electron Device Lett. **11**, 36–38 (1990).

31. S. Wemple, M. Didomenico and I. Camlibel, "Dielectric and optical properties of melt-grown batio3," J. Phys. Chem. Solids **29**, 1797–1803 (1968).
32. I. H. Malitson, "Interspecimen comparison of the refractive index of fused silica," J. Opt. Soc. Am. **55**, 1205–1209 (1965).
33. P. B. Johnson and R. W. Christy, "Optical constants of the noble metals," Phys. Rev. B **6**, 4370–4379 (1972).
34. J. A. Dionne, K. Diest, L. A. Sweatlock and H. A. Atwater, "PlasMOStor: A metal-oxide-Si field effect plasmonic modulator," Nano Lett. **9**, 897–902 (2009).
35. J. Capmany and D. Pérez, "Field programmable photonic gate arrays," in *Programmable Integrated Photonics,* (Oxford University Press, 2020).
36. M. Zgonik, P. Bernasconi, M. Duelli, R. Schlesser, P. Günter, M. H. Garrett, D. Rytz, Y. Zhu and X. Wu, "Dielectric, elastic, piezoelectric, electro-optic, and elasto-optic tensors of $BaTiO_3$ crystals," Phys. Rev. B **50**, 5941–5949 (1994).
37. R. W. Boyd, *Nonlinear Optics* (3rd Edition) (Academic Press, Inc., USA, 2008).
38. J. Nordlander, F. Eltes, M. Reynaud, J. Nürnberg, G. De Luca, D. Caimi, A. A. Demkov, S. Abel, M. Fiebig, J. Fompeyrine and M. Trassin, "Ferroelectric domain architecture and poling of batio3 on Si," Phys. Rev. Mater. **4**, 034406 (2020).

Micro-Nanostructured Polymeric Scaffolds for Bone Tissue Engineering[#]

Sama Abdulmalik [†], Suranji Wijekoon [†], Khadija Basiru Danazumi [*,†],
Sai Sadhananth Srinivasan [*,†], Laxmi Vobbineni [*,†], Elifho Obopilwe [*,†]
and Sangamesh G. Kumbar [*,†,‡]

*Department of Biomedical Engineering,
University of Connecticut Storrs, CT 06269, USA

†Department of Orthopedic Surgery,
University of Connecticut Health Center, Farmington,
CT 06030, USA

‡Department of Materials Science and Engineering,
University of Connecticut Storrs, CT 06269, USA
‡Kumbar@uchc.edu

While bone tissue allograft and autograft are commonly used in bone healing, their application is limited by factors such as availability, donor site morbidity, and immune response to the grafted tissue. Tissue-engineered implants, such as acellular or cellular polymeric structures, offer a promising alternative, and are a current trend in tissue engineering. Leveraging recent advancements in bone tissue engineering (BTE), we utilize 3D printing to develop biodegradable scaffolds that combine mechanical strength and bioactivity to facilitate bone repair and regeneration. This study focuses on the design and fabrication of mechanically competent 3D printed poly (L-lactic acid) (PLLA) micro-structured scaffolds. These scaffolds are enhanced with collagen type I nanofibrils to create bioactive scaffolds that promote tissue regeneration. The performance of these mechanically competent, micro-nanostructured polymeric matrices, in combination with bone marrow stromal cells (BMSCs), is evaluated in PLLA and PLLA-collagen scaffolds. The resulting micro-nanostructured PLLA-Collagen scaffolds mimic trabecular bone architecture, mechanical strength, and the extracellular matrix environment found in native bone tissue. The composite PLLA-collagen scaffolds exhibit mechanical properties in the mid-range of human trabecular bone. Both PLLA and PLLA-Collagen scaffolds support human BMSCs adhesion, proliferation, and osteogenic differentiation. A significantly higher number of implanted host cells are distributed in the PLLA-Collagen scaffolds with greater bone density, more uniform cell distribution, and attachment compared to the PLLA microstructure. Additionally, the biomimetic collagen nanostructure potently induces osteogenic transcription evidenced by increased alkaline phosphatase activity and upregulation of bone markers such as sialoprotein and collagen type I, ultimately guiding

‡Corresponding author.
#This chapter appeared previously on the International Journal of High Speed Electronics and Systems. To cite this chapter, please cite the original article as the following: S. Abdulmalik, S. Wijekoon, K. B. Danazumi, S. S. Srinivasan, L. Vobbineni, E. Obopilwe and S. G. Kumbar, *Int. J. High Speed Electron. Syst.*, **33**, 2440075 (2024), doi: 10.1142/S0129156424400755.

stem cell-mediated formation of a mature, mineralized bone matrix throughout the interconnected scaffold pores. This study underscores the benefits of micro-nanostructured scaffolds in successfully generating the inductive microenvironment of native bone extracellular matrix, triggering the cascade of cellular events required for functional bone regeneration, repairing critical-sized bone defects, and ultimately serving as an alternative material platform for bone regeneration, thereby instilling confidence in the potential of our research.

Keywords: Bone tissue engineering; poly (L-lactic acid); collagen type I nanofibrils; human mesenchymal stem cells; osteogenic differentiation.

1. Introduction

The United States encounters over 34 million musculoskeletal injuries annually, and these numbers are expected to rise due to the aging population [1,2]. Bone is hard tissue that undergoes continuous remodels to withstand mechanical stress, maintain healthy functions, and repair injury [3–6]. Clinical scenarios frequently involve bone defects, ranging from minor fractures that heal spontaneously to larger defects requiring surgical intervention [3–6]. In addressing severe bone deficiencies caused by congenital defects, degenerative diseases, and tumors, procedures such as bone transplantation are necessary [3–6]. However, the current treatments utilizing bone autografts and allografts often lack optimal regenerative properties and are scarce in resource-limited environments [7–9]. Hence, there exists a pressing need for readily available, off-the-shelf bone graft substitutes with superior healing properties.

Although autologous grafting is considered as the 'gold standard' for treating bone loss, it is plagued by limited availability, donor site morbidity, and challenges in matching shape to defect locations. [10]. Tissue engineering strategies employing biodegradable scaffolds in combination with cells and bioactive factors have the potential to overcome some of the challenges associated with bone grafts. [11]. Various bone tissue engineering (BTE) scaffolds have been developed using polymers, ceramics, and their composites, each with varied outcomes in bone healing. [11,12]. However, suboptimal mechanical strength, inadequate nutrient transport, inefficient waste removal, and limited bioactivity often restrict tissue regeneration to the scaffold surface [3,5,13]. Therefore, innovative designs of novel BTE scaffolds are needed to overcome these limitations and achieve successful bone repair and regeneration strategies.

Additive manufacturing (AM), also known as 3D printing, has revolutionized the fabrication of innovative scaffolds for various tissue regeneration applications, featuring diverse scaffold architectures. [14,15]. 3D printing facilitates precise layer-by-layer construction of complex structures from digital models, allowing for customization of scaffold architecture to closely mimic the extracellular matrix of tissues. [14,15]. These efforts to mimic hierarchical structures of bone, and integrate dimensions, pore properties, and bioactivity effectively. Common methods used in 3D printing for scaffold applications include fused deposition modeling (FDM), stereolithography (SLA), and selective laser sintering (SLS), each offering unique advantages in terms of material compatibility, resolution, and mechanical properties suitable for promoting cell growth and tissue regeneration. Fused deposition modeling (FDM), stereolithography (SLA), and selective laser sintering (SLS) are commonly used in scaffold 3D printing. Among these techniques, FDM is preferred for scaffolding due to its simplicity and ability to use a wide range of biomaterials. [14, 15].

Poly(L-lactic acid) (PLLA) is an FDA-approved biodegradable polyester, widely utilized in orthopaedic fixation devices such as screws, plates, and scaffolds, which are currently available for clinical use [16]. Consequently, this polymer has emerged as a popular choice for tissue engineering due to its processability, mechanical properties, biosafety, and biodegradability. Collagen, the primary structural protein found in bones, tendons, ligaments, and skin, is particularly suitable for biomaterial applications due to its distinctive triple-helix structure and self-assembly properties [3, 4, 13]. Collagen molecular self-assembly into fibrils and complex structures has been utilized to create nanostructures and porous sponges for tissue engineering and drug delivery [3, 4, 13].

Bone bears the weight of the entire body and undergoes constant mechanical load. Therefore, scaffolds used for BTE applications must meet the mechanical load-sharing requirements and offer improved transport features. Nanostructures in the form of self-assembled collagen or electrospun nanofibers, while offering bioactivity, often fall short of meeting the necessary mechanical properties [4, 13, 17]. By combining 3D printing with collagen molecular self-assembly, it is possible to fabricate micro-nanostructured scaffolds suitable for load-sharing applications [5, 18–20]. Thus, integrating PLLA structure with collagen fibers enhances mechanical properties, pore structure, and biocompatibility, making it ideal for BTE. This study specifically focuses on the fabrication and characterization of 3D-printed PLLA-Collagen scaffolds for bone tissue engineering.

2. Materials and Methods

2.1. *Materials*

PolyLite™ PLA was purchased from Polymaker US. Human MSCs were purchased from RoosterBio Inc. (Frederick, MD, USA) and Dulbecco's minimum essential medium (DMEM) media was purchased from Lonza Bioscience (Morrisville, NC, USA). Pen-Strep (P/S), Fetal bovine serum (FBS), phosphate-buffered saline (PBS), ultrapure distilled water, 4% paraformaldehyde solution, Triton™ X-100, L-cysteine hydrochloride 98% anhydrous, 0.25% trypsin EDTA, Quant-iT™ PicoGreen™ dsDNA Assay Kit, ProLong™ Diamond Antifade Mountant with DAPI, Cell Titer 96® AQueous One Solution Cell Proliferation Assay, and IGF-1 recombinant human protein were purchased from Fisher Scientific (Fair Lawn, NJ, USA). Bovine serum albumin, papain, anti-GLP1R produced in rabbit (SAB4501220), sodium acetate anhydrous, cystine-HCL, sodium hydroxide pellets, ethyl alcohol, and chloroform were purchased from Sigma-Aldrich (St. Louis, MO). Normal goat serum, anti-Collagen-I antibody (ab34710), anti-tenomodulin antibody (ab203676), antiSCXA antibody (ab58655), goat anti-mouse Alexa Fluor 488 (ab150117), and goat anti-rabbit Texas Red (ab6719) were purchased from Abcam (Cambridge, MA, USA). QIAzol Lysis Reagent was purchased from QIAGEN Inc. (Germantown, MD) while iScrept cDNA Synthesis Kit and iTaq Universal SYBR Green Supermix were purchased from Bio-Rad (Hercules, CA, USA).

2.2. *Fabrication of PLLA scaffolds and collagen molecular self-assembly*

The PLLA scaffolds were manufactured using an FDM setup (LULZBOT TAZ 6, Aleph Objects, Inc.), with a nozzle size of 0.25 mm being used to print the scaffolds. Initially, scaffolds were modeled by the SolidWorks Software, 2019 Student Edition, Dassault Systems in a cylindrical shape of diameter 5 mm inscribed with 3 rings of diameter 3 mm with a total thickness of 5 mm. The Cura LulzBot Edition – 4.13.9-expV3 software was used to set various parameters for printing, including the nozzle temperature (215°C), the infill percentage (20%), and the bed temperature (60°C) [21]. 3D-printed PLLA scaffolds were incubated in a 0.1% (w/v) collagen type I solution at pH 4.2 and 37°C for 7 days to promote collagen self-assembly [3, 13]. The dried scaffolds were then treated with UV light for 30 minutes on each side to stabilize the collagen nanofibers and were repeatedly washed with DI water to remove buffer salts.

2.3. *Scanning electron microscopy imaging*

Scanning electron microscopy (SEM) was used to examine the cross-sectional morphology of the composite scaffolds ($n = 3$). After collection, samples were freeze-dried and coated with a 10 nm layer of Au/PD using a sputter coater machine (Polaron E5100; Quorum Technologies). The porous structures of the samples were observed using an SEM microscope (Hitachi) at various magnifications. ImageJ software was used to analyze SEM images for microstructure and porosity.

2.4. *Mechanical testing*

To assess the mechanical stability of the composite scaffolds, torsion tests were conducted following ASTM F2502 standards. Samples with a 1:2 ratio (diameter: length) were immersed in PBS (pH = 7.4) overnight before testing. An Instron 5869 electromechanical test system with Blue Hill control software applied a compression rate of 1 mm/min. Stress/strain curves were recorded for each sample to calculate Young's modulus, while torque versus rotation graphs were analyzed to study the torsional properties of polymeric screws. Results, based on at least six dry samples per group, were presented as mean ± standard error.

2.5. *In vitro human mesenchymal stem cell culture*

Human bone marrow mesenchymal stem cells (hMSCs) were cultured and expanded in basal media consisting of α MEM, GlutaMAX™ Supplement with 10% FBS, and 1% penicillin/streptomycin. The cells in all experiments were at passage 3. The composite scaffolds were immersed in 70% ethanol for 30 min and exposed to UV light on each side for 30 min under the cell culture hood. Post sterilization, the scaffolds were submerged in basal media overnight, and 100,000 cells in 50 μL cell suspension were seeded on the scaffolds in 48 untreated well plates. The cells were incubated for at least 4 h and then 25 μL basal media was added. The basal media was switched to the osteogenic media following 48 h. The osteogenic media consisted of the basal media supplemented with 50 μg/mL ascorbic acid and 8 mM β-glycerol phosphate and replaced every 2 days for a period of 21 days [3, 5, 13].

2.6. *Cell proliferation*

The proliferation of the hMSCs on the scaffolds during the osteoinduction was calculated using DNA Picogreen assay ($n = 5$). The cells were collected on days 3, 7, 14, and 21 of osteoinduction, lysed in 1% TritonX-100, subjected to three freeze-thaw cycles from $-80°$C to room temperature, and mixed with a pipette to extract the cell lysate. The double-stranded DNA (dsDNA) was calculated based on the assay manufacturer's instructions. Briefly, 20 μL of the DNA lysate was transferred to a new well plate and mixed with 80 μL of TE reagent (component B) and 100 μL Picogreen reagent (component A). The well plates were covered with aluminum foil for 5 min at room temperature (to avoid light exposure). A BioTek plate reader was used to read the fluorescence at 485 nm/535 nm. Using the standard curve, the optical readings were converted into DNA concentration [3, 5, 13].

2.7. *Alkaline phosphatase activity*

Alkaline phosphatase activity of hMSCs on the scaffolds was evaluated as a marker of osteoblast phenotype progression using an ALP substrate kit. 100 μL of cell lysate was transferred into a well plate to which 400 μL of p-NPP (para-nitro phenol phosphate) substrate and buffer solution were added and incubated at 37°C for 30 min. After 30 minutes, 500 μL of 0.4 N of sodium hydroxide was added to stop the reaction. The intensity of the color produced through the reaction is proportional to the ALP activity. The optical density of the solution was measured at 405 nm using a BioTek plate reader. The results for ALP activity were optical density and these were normalized to scaffold volume [3, 5, 13].

$$\text{ALP activity} = \frac{B}{T \times V} \times D$$

where as B-Amount of product; T-Time; V-Volume; D-Dilution factor.

2.8. Mineralized matrix deposition

Mineralized matrix deposition by osteoinduced hMSCs on the scaffolds was evaluated as a marker of mature osteoblast phenotype using an alizarin red staining method for calcium deposition over 28 days of osteoinduction.

2.9. Osteogenic marker immunostaining

The samples were collected after 21 days of culture and washed with PBS (pH 7.4). They were fixed with 4% formaldehyde in PBS (pH 7.4) for 20 min at room temperature. The scaffolds were washed with ice-cold PBS (pH 7.4) for three times. The permeabilization step was performed by incubating the samples in 0.1% Triton X/1% BSA in PBS (pH 7.4) for 10 min. The samples were washed again with ice-cold PBS (pH 7.4) for three times. The blocking step was performed by incubating the samples in 10% Normal Goat Serum in PBST solution (1% Tween20 in PBS) for 30 min at room temperature to block non-specific antibody binding. The samples were incubated overnight at 4°C in primary antibody (Collagen Rabbit anti-human at 1:200 dilution; or bone sialoprotein rabbit anti-human at 1:100) in a blocking buffer (10% Normal Goat Serum in PBST solution). The samples were incubated in secondary antibody Dylight 594-goat-anti-rabbit (1:200) in a blocking buffer for 1 h. They were covered with aluminum foil and placed on a shaker at room temperature. They were washed with PBS (pH 7.4) three times. NucBlue (DAPI) at 2 drops per ml of PBS (pH 7.4) was added to the samples to stain the cellular nuclei. The samples were imaged using a Zeiss 780/Laser Scanning Confocal Microscope [3,5,13].

2.10. Statistical analysis

All quantitative data are presented as mean $\pm$ standard deviation. Sample sizes of $n = 5 - 6$ were used for mechanical testing, cell proliferation assays, alkaline phosphatase activity, and calcium deposition assays, based on previous studies to ensure adequate statistical power. These sample sizes were chosen based on prior bone tissue engineering studies investigating cellular behavior and biochemical assays on 3D scaffolds. All results were first evaluated using one-way analysis of variance (ANOVA) to determine overall statistical significance between the experimental groups. If the ANOVA result was significant ($p < 0.05$), post-hoc t-tests were performed to identify specific differences between group means using GraphPad Prism (GraphPad Software, Inc., La Jolla, CA). A value of $p < 0.05$ was considered statistically significant for all tests.

3. Results

3.1. Scaffold fabrication and characterization

The PLLA scaffolds were successfully 3D printed with a defined macro-porous structure designed to mimic the architecture of trabecular bone (Fig. 1(a)). This microstructure allowed for the mechanical strength of bone to be replicated while

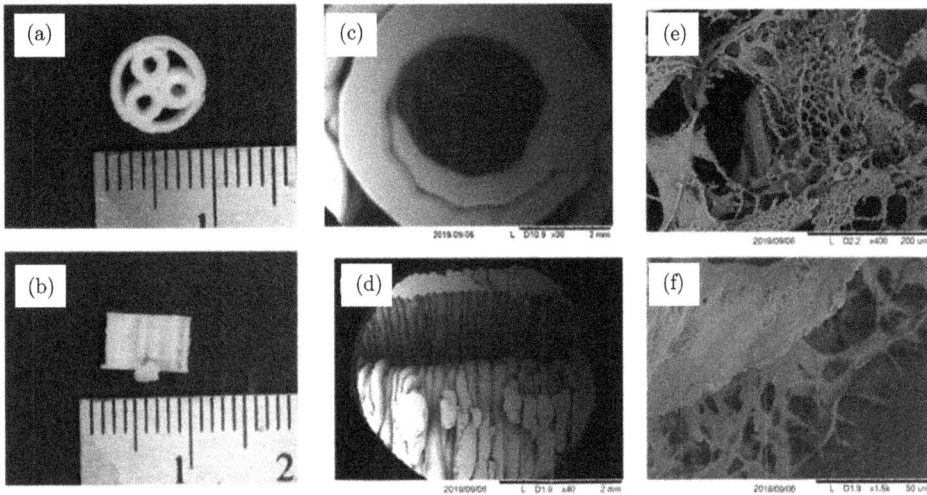

Fig. 1. (a) and (b) Optical images of 3D printed scaffolds PLA, (c) and (d) SEM images of 3D printed scaffolds PLA (scale bar = 2 mm), (e) collagen coated 3D printed scaffolds (top view, scale bar = 200 μm) and (f) collagen coated 3D printed scaffolds (cross-sectional view, scale bar = 50 μm).

supporting nutrient flow and metabolic waste removal, thereby facilitating uniform tissue growth. The incorporation of collagen into the 3D-printed PLLA framework resulted in a hierarchical micro-nanostructure, mimicking the ECM environment found in native bone tissue. SEM analysis confirmed the formation of an interconnected porous collagen network on the PLLA microstructure. The collagen nanostructures were generated by incubating these 3D printed frameworks in a 0.1 wt.% collagen solution under physiological conditions, following laboratory protocols [3, 5, 13]. These nanostructures function as a molecular sieve within the macrostructure, enhancing cell adhesion, proliferation, and osteogenesis by human mesenchymal stem cells (hMSCs) during osteoinduction [3,5,13]. This unique composition combines the mechanical strength provided by the PLLA microstructure with the bioactivity and cell-interactive properties of the collagen nanostructures. Additionally, the torsional mechanical properties of the composite scaffolds (Fig. 2) were evaluated and displayed ideal mechanical properties that align with the midrange of human trabecular bone, making them suitable for load-sharing bone repair applications [22].

3.2. *In vitro cellular behavior*

The unique hierarchical framework of the PLLA-collagen composite scaffolds resulted in significantly enhanced cellular interactions and osteogenic activity compared to the PLLA scaffolds alone. The incorporation of the collagen nanofiber coating introduced biomimetic nanofibrillar features that are known to promote

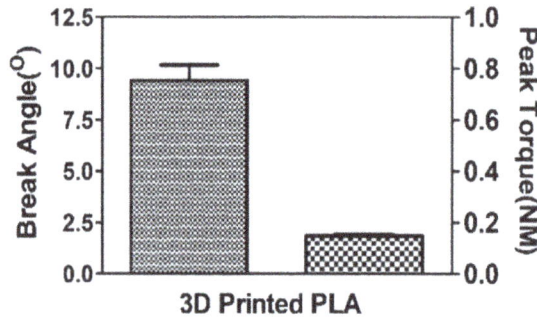

Fig. 2. Torsional analysis of 3 printed scaffolds break angle (o), peak torque (NM).

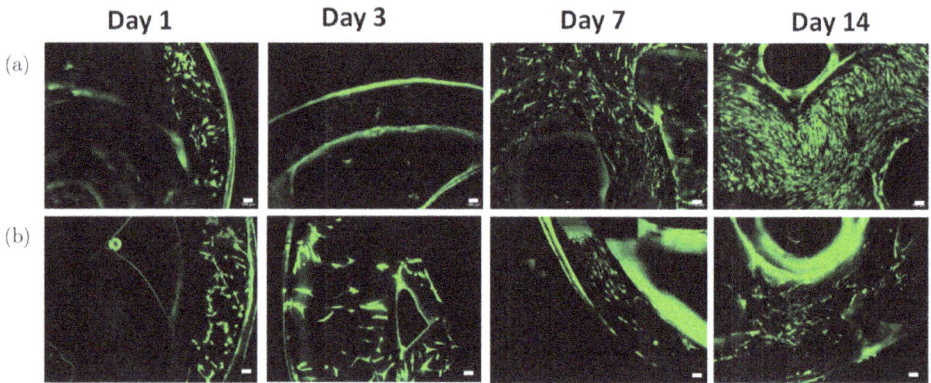

Fig. 3. Live/dead images of 3 printed scaffolds (a) PLA-collagen, (b) PLA (day 1, 3, 7 and 14) scale bar = 100 μm.

protein adsorption and subsequent cell adhesion. As shown in Fig. 3, the PLLA-collagen scaffolds displayed a 3-fold increase in attachment and proliferation of hMSCs over the 14-day culture period. This can be attributed to the collagen's nanofibrillar structure serving as an ECM-mimetic substrate that facilitates robust cell binding and spreading through integrin clustering and cytoskeletal reorganization. In contrast, the relatively flat surface of the uncoated PLLA microstructure lacks these critical cell-interactive cues, resulting in diminished cellular responses [3,5,13].

Beyond simply enhancing initial cell adhesion, the presence of the collagen nanofibers within the 3D microenvironment played a pivotal role in providing the essential biochemical signals to direct stem cell fate commitment toward the osteoblastic lineage. This is evidenced by the quantitative cellular analysis presented in Fig. 4. Over the 28-day culture period in osteogenic media formulations, the PLLA-collagen composite scaffolds supported significantly higher cellular proliferation compared to the uncoated PLLA scaffolds, as measured by total DNA concentration (ng/mL) (Fig. 4(a)) [3, 5, 13]. Aside from day 7 and day 14, the total

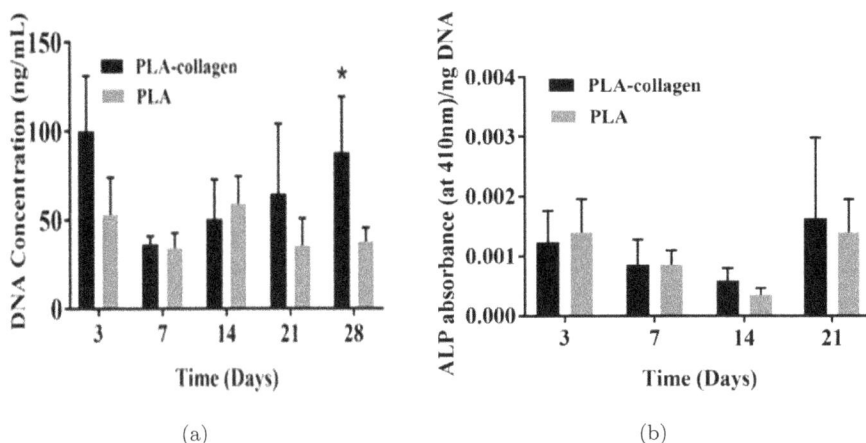

(a)   (b)

Fig. 4. *In vitro* phenotype development by hMSCs seeded onto scaffolds under osteoinduction: (a) DNA content and (b) alkaline phosphatase activity over 3–28 days. Multiple t test statistical significance (*$P < 0.05$).

DNA concentration was doubled with day 3 and 28 having an increase of $50 \, \text{ng/mL}$ and day 21 having a $30 \, \text{ng/mL}$ increase. This enhanced proliferative capacity indicates that the collagen nanofibers created a microenvironment more conducive to sustained stem cell expansion and growth. Additionally, analysis of alkaline phosphatase (ALP) activity (Fig. 4(b)), an early marker of osteoblastic differentiation, revealed a slight upregulation of no more than $0.0015 \, \text{ng}$ of the PLLA-collagen scaffolds compared to PLLA alone. ALP is one of the early phenotypic markers of osteogenesis, playing crucial roles in regulating phosphate homeostasis and matrix mineralization. The slight induction of ALP expression observed in the presence of the collagen nanofibers indicates that the stimulation of the osteogenic transcriptional program within the stem cell population is suboptimal [4].

This osteoinductive effect can be directly attributed to the native role of collagen as the primary protein component of bone ECM. Beyond providing structural support, collagen acts as a bioactive reservoir to sequester and concentrate osteogenic growth factors and morphogens like bone morphogenetic proteins. By incorporating collagen nanofibers that mimic the architecture of this regulatory niche, the PLLA-collagen scaffolds could replicate the inductive microenvironment found in natural bone tissue [3–5, 13]. The high surface area of the collagen fibrils likely facilitated enhanced adsorption and localized presentation of osteoinductive factors to the stem cells [3–5, 13]. This biomimetic recreation of the bone ECM biochemical signaling milieu acted as a powerful trigger to prime the hMSC population for commitment toward the osteoblastic lineage [3–5, 13].

The elevated ALP activity indicates successful initiation of the osteogenic transcriptional cascade in response to these potent collagen-sequestered morphogenic signals. In contrast, the relatively inert PLLA microstructure lacking the collagen nanofiber coating failed to provide the necessary biochemical cues for effective os-teoinduction, as reflected by the minimal ALP upregulation on those scaffolds. This stark difference highlights the indispensable role played by the collagen component in dictating stem cell fate determination within the biomaterial niche. These quan-titative results in Fig. 4 reveal that the micro-nanostructured architecture of the PLLA-collagen composite synergistically combined structural cues for cell adhe-sion/proliferation with the osteoinductive biochemical signals intrinsic to collagen's role in native bone ECM [3–5, 13].

The potent osteoinductive effects of the collagen nanofibers observed in the early stages translated into robust formation of a mineralized ECM by the hMSCs over the 28-day culture period, a critical late-stage marker of mature bone tissue development. Alizarin Red staining in Fig. 5 provides striking visual evidence of progressively increasing calcium deposition on both the PLLA and PLLA-collagen scaffolds. However, the composite scaffolds exhibited considerably higher staining intensity at all timepoints compared to the uncoated PLLA controls, indicative of substantially greater mineralization [3–5, 13].

Immunostaining data in Fig. 6 further substantiated this matrix production and mineralization process. Intense positive staining for the bone-specific proteins bone sialoprotein (BSP) and collagen type I (Coll1) was observed exclusively on the PLLA-collagen composite scaffolds. The presence of these stained proteins confirms osteogenic differentiation of the hMSCs in the presence of the collagen nanofibers, as BSP and Coll1 are key late markers of mature osteoblasts, playing vital roles in regu-lating hydroxyapatite mineralization and providing mechanical strength to the bone matrix. Moreover, the robust expression and deposition of BSP and Coll1 through-out the interconnected pores demonstrate extensive infiltration and osteoblastic differentiation of hMSCs facilitated by the biomimetic collagen nanostructure. In stark contrast, minimal staining for these bone markers was observed on the PLLA scaffolds lacking the collagen signals, indicating reduced osteogenic maturation and incomplete bone matrix development. This difference highlights the critical role

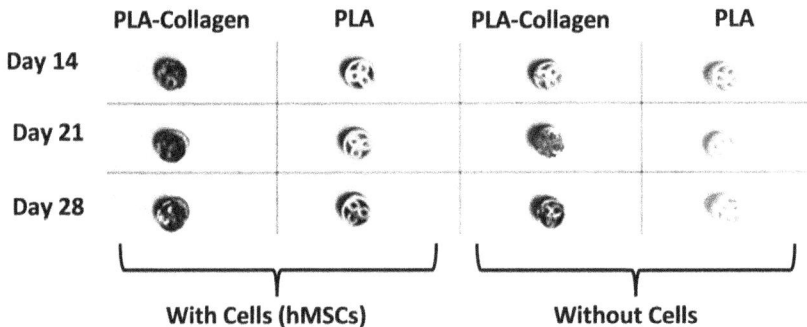

Fig. 5. Alizarin red images for the PlA-collagen coated and PLA 3D printed scaffold for day 14, 21 and 28.

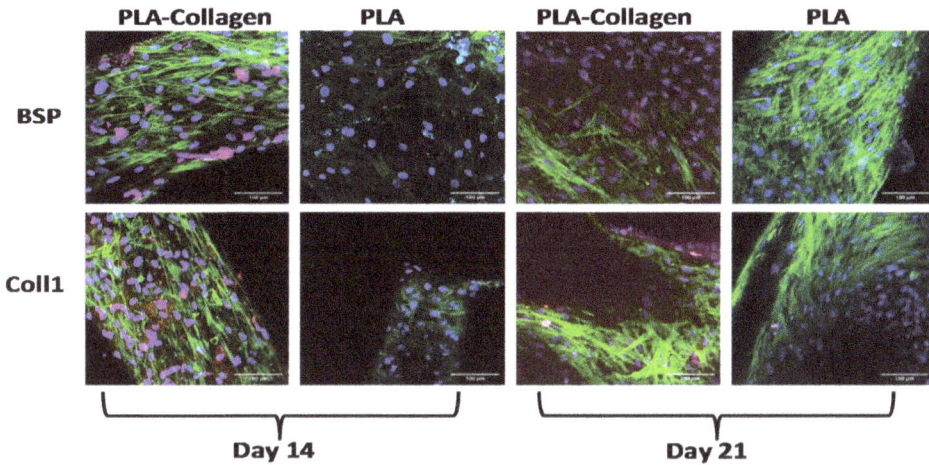

Fig. 6. Immunostaining images for the PlA-collagen coated and PLA 3D printed scaffold for days 14 and 21.

played by the collagen nanofibers in driving the full osteogenic differentiation program of the stem cells to generate a mature, mineralized bone-mimetic matrix. The collagen's hierarchical nano-architecture provided key biochemical and structural signals to guide the progression through each developmental stage from stem cell lineage commitment to terminal matrix mineralization [3–5,13].

4. Discussion

The finding of this study reinforces the crucial significance of recreating the hierarchical micro-nanostructure and composition of native bone ECM to guide stem cell differentiation and functional bone matrix formation. Previous research has highlighted the benefits of integrating biomimetic cues such as collagen nanofibers into 3D scaffolds to augment the osteogenic differentiation of stem cells [3–5,13,23]. However, many of these studies rely solely on collagen nanofibers or gels, lacking the mechanical robustness required for load-bearing bone regeneration applications. The 3D-printed PLLA-collagen composite scaffolds developed here successfully combine the osteoinductive properties of collagen nanofibers with the structural integrity provided by the microporous PLLA framework.

The interconnected porous architecture of the PLLA microstructure mimics key features of trabecular bone reported to facilitate cell infiltration, nutrient transport, vascularization, and mechanical load-sharing [5,22]. Previous studies on 3D printed PLLA and composite scaffolds have similarly highlighted their ability to replicate the anatomical and mechanical properties of bone [24]. The incorporation of the collagen nanofiber coating introduces biomimetic topographical and biochemical signals that synergistically promote every stage of osteogenesis.

The enhanced initial cell attachment, proliferation, and elevated ALP activity observed on the PLLA-collagen scaffolds align with numerous prior reports demonstrating collagen's ability to support robust osteoblastic differentiation of stem cells [3–5,13]. Collagen nanofibers contain integrin-binding motifs and mimic the nanofibrillar architecture of native ECM, making them an ideal substrate for cell adhesion and spreading [25]. Furthermore, collagen acts as a reservoir to sequester and concentrate osteoinductive growth factors like BMPs that initiate transcriptional programs governing osteoblast lineage commitment [3]. The sustained matrix deposition and mineralization reinforce collagen's critical regulatory role in coordinating the intracellular signaling cascades underlying new bone formation [26].

While previous studies have incorporated collagen into 3D scaffolds using techniques like freeze-drying, electrospinning or layer-by-layer assemblies [27], this approach yields a hierarchical multi-scale scaffold that closely mimics the organization of native bone ECM from the nano- to macro-scale. The 3D printed microscale framework integrated with the collagen nanofibril networks more accurately recapitulates the complex biochemical and biophysical environment encountered by cells in vivo. This integrated micro-nanoscale design overcomes the limitations of nanostructures alone, which lack the porosity for cell invasion and limited mechanical properties [17].

In contrast to the PLLA microstructure, which lacks bioactive cues, the hierarchical architecture of the PLLA-collagen composite provided a comprehensive array of signals spanning the initial cell adhesion, stem cell commitment, ECM production, and matrix mineralization stages of bone development. This recapitulation of the native inductive microenvironment was critical for guiding the complete osteogenic program from the undifferentiated mesenchymal stem cell population to the eventual formation of a mature, calcified bone-like matrix throughout the 3D scaffold. These results highlight the potent osteoinductive effects arising from the synergistic combination of the PLLA microstructural scaffold with the collagen nanofibrillar cues. Further investigation in preclinical bone defect models is warranted to evaluate the *in vivo* bone regenerative capacity of these biomimetic scaffolds.

5. Conclusion

This study successfully demonstrated the efficacy of biodegradable 3D printed PLLA-Collagen scaffolds in bone repair and regeneration. The microporous PLLA framework provided the necessary mechanical strength and support, mimicking human trabecular bone. The incorporation of collagen Type I nanofibrils significantly increased the scaffold's surface area, enhancing initial cell attachment during cell seeding. This biomimetic collagen nanostructure also stimulated stem cell proliferation and osteoblastic differentiation, as evidenced by increased alkaline phosphatase activity and the upregulation of bone-specific markers such as bone sialoprotein and collagen type I. The hierarchical micro-nanoarchitecture of the scaffolds successfully replicated the extracellular matrix environment of native bone tissue, guiding stem cells from initial development to forming a mature, mineralized bone matrix within the scaffold pores. By combining the mechanical strength of 3D-printed PLLA with the inherent bioactivity and osteoconductive properties of collagen nanofibers, these composite scaffolds address the primary limitations of current bone graft substitutes by enhancing bone integration and new bone formation.

Acknowledgments

The authors acknowledge the funding support from the National Institutes of Health (#R01NS134604, #R01EB034202, #R01AR078908, #R01EB030060, and #R56NS122753).

ORCID

Sama Abdulmalik ⊕ https://orcid.org/0000-0002-0834-8357

Suranji Wijekoon ⊕ https://orcid.org/0000-0002-7053-5268

Khadija Basiru Danazumi ⊕ https://orcid.org/0009-0008-8421-0537

Sai Sadhananth Srinivasan ⊕ https://orcid.org/0009-0000-7888-6663

Laxmi Vobbineni ⊕ https://orcid.org/0009-0008-8711-1358

Elifho Obopilwe ⊕ https://orcid.org/0000-0001-7515-3990

Sangamesh G. Kumbar ⊕ https://orcid.org/0000-0001-7672-4783

References

1. Cameron, K. L. and B. D. Owens, *The burden and management of sports-related musculoskeletal injuries and conditions within the US military.* Clin Sports Med, 2014. **33**(4): p. 573–589.
2. Fenn, B. P., *et al.*, *Worldwide epidemiology of foot and ankle injuries during military training: A systematic review.* BMJ Mil Health, 2021. **167**(2): p. 131–136.
3. Aravamudhan, A., *et al.*, *Micro-nanostructures of cellulose-collagen for critical sized bone defect healing.* Macromol Biosci, 2018. **18**(2): p. 1–14.
4. Cheng, Y., *et al.*, *Collagen functionalized bioactive nanofiber matrices for osteogenic differentiation of mesenchymal stem cells: bone tissue engineering.* J Biomed Nanotechnol, 2014. **10**(2): p. 287–298.
5. Manoukian, O. S., *et al.*, *Spiral layer-by-layer micro-nanostructured scaffolds for bone tissue engineering.* ACS Biomater Sci Eng, 2018. **4**(6): p. 2181–2192.
6. Brown, J. L. and C. T. Laurencin, *Bone tissue engineering*, in *Biomaterials science*. 2020, Elsevier. p. 1373–1388.
7. Gillman, C. E. and A. C. Jayasuriya, *FDA-approved bone grafts and bone graft substitute devices in bone regeneration.* Mater Sci Eng C Mater Biol Appl, 2021. **130**: p. 112466.
8. Stahl, A. and Y. P. Yang, *Regenerative approaches for the treatment of large bone defects.* Tissue Eng Part B Rev, 2021. **27**(6): p. 539–547.
9. Vidal, L., *et al.*, *Reconstruction of Large Skeletal Defects: Current Clinical Therapeutic Strategies and Future Directions Using 3D Printing.* Front Bioeng Biotechnol, 2020. **12**;**8**(61): p. 1–11.

10. Stevenson, S., *Biology of bone grafts.* Orthopedic Clinics, 1999. **30**(4): p. 543–552.

11. Stratton, S., *et al.*, *Bioactive polymeric scaffolds for tissue engineering.* Bioactive Mater, 2016. **1**(2): p. 93–108.

12. Alonzo, M., *et al.*, *Bone tissue engineering techniques, advances, and scaffolds for treatment of bone defects.* Curr Opin Biomed Eng, 2021. **17**: p. 100248.

13. Aravamudhan, A., *et al.*, *Collagen nanofibril self-assembly on a natural polymeric material for the osteoinduction of stem cells in vitro and biocompatibility in vivo.* RSC Adv, 2016. **6**(84): p. 80851–80866.

14. Zennifer, A., *et al.*, *3D bioprinting and photocrosslinking: emerging strategies & future perspectives.* Biomater Adv, 2022. **134**: p. 112576.

15. Veeman, D., *et al.*, *Additive manufacturing of biopolymers for tissue engineering and regenerative medicine: An overview, potential applications, advancements, and trends.* Int J Polym Sci, 2021. **2021**: p. 1–20.

16. James, R., O. S. Manoukian and S. G. Kumbar, *Poly (lactic acid) for delivery of bioactive macromolecules.* Adv Drug Delivery Rev, 2016. **107**: p. 277–288.

17. Ramos, D. M., *et al.*, *Insulin immobilized PCL-cellulose acetate micro-nanostructured fibrous scaffolds for tendon tissue engineering.* Polym Adv Technol, 2019. **30**(5): p. 1205–1215.

18. Lee, P., *et al.*, *Bioactive polymeric scaffolds for osteochondral tissue engineering: in vitro evaluation of the effect of culture media on bone marrow stromal cells.* Polym Adv Technol, 2015. **26**(12): p. 1476–1485.

19. Lee, P., *et al.*, *Influence of chondroitin sulfate and hyaluronic acid presence in nanofibers and its alignment on the bone marrow stromal cells: cartilage regeneration.* J Biomed Nanotechnol, 2014. **10**(8): p. 1469–1479.

20. Lee, P., *et al.*, *Guided differentiation of bone marrow stromal cells on co-cultured cartilage and bone scaffolds.* Soft Matter, 2015. **11**(38): p. 7648–7655.

21. Dhandapani, R., *et al.*, *Additive manufacturing of biodegradable porous orthopaedic screw.* Bioactive Mater, 2020. **5**(3): p. 458–467.

22. Dhandapani, R., *et al.*, *Additive manufacturing of biodegradable porous orthopaedic screw.* Bioact Mater, 2020. **5**(3): p. 458–467.

23. Aravamudhan, A., *et al.*, *Micro-nanostructures of cellulose-collagen for critical sized bone defect healing.* Macromol Biosci, 2018. **18**(2): p. 1–14. https://pubmed.ncbi.nlm.nih.gov/29178402/.

24. https://www.sciencedirect.com/science/article/pii/S2590006423001205.

25. Dhandapani, R., *et al.*, *Additive manufacturing of biodegradable porous orthopaedic screw.* Bioactive Mater, 2020. p. 458–467. https://www.ncbi.nlm.nih.gov/pmc/articles/PMC7139166/

26. https://www.ncbi.nlm.nih.gov/pmc/articles/PMC7763437/.

27. Vach, A., *et al.*, *Resorbable Biomaterials Used for 3D Scaffolds in Tissue Engineering: A Review.* Materials (Basel), 2023. **8**;**16**(12): p. 4267.

SI/GE Quantum Dot Channel FETs for Multi-Bit Computing[#]

F. Jain [ORCID]*,§, R. H. Gudlavalleti [ORCID]*,‡, J. Chandy [ORCID]* and E. Heller[†]

*Electrical and Computer Engineering,
University of Connecticut Storrs, CT, USA

†Synopsys Inc. Ossining, NY, USA
‡Biorasis Inc., Storrs, CT, USA
§faquir.jain@uconn.edu

This paper presents quantum dot channel (QDC) FETs in quantum wire and coupled quantum dot configurations for cryogenic operation with multi-state operation. It also describes gate-all-around (GAA) quantum dot channel (QDC) FETs that exhibit potential multi-state characteristics at room temperature. FETs with cladded Si and Ge quantum dot layers as a transport channel have been fabricated. The formation of a quantum dot superlattice (QDSL) when SiOx-cladded Si and/or GeOx-cladded Ge quantum dots (QD) are assembled results in mini-energy sub-bands in the conduction and valence band. The intra-mini-energy band transitions results in significant changes in the drain current when gate and/or drain voltages are varied. This novel feature provides a pathway for 16-/32-state logic in CMOS-X configuration. The gate-defined Si quantum dot FETs, comprising of tunnel barrier coupled, have been reported for quantum computing at cryogenic temperatures.

Keywords: Quantum dot channel (QDC) FETs; QDot multi-state; QWire-QDC.

1. Quantum Dot Channel (QDC) FETs with Multi-State Current-Voltage Characteristics

A fabricated QDC-FET hosting two asymmetric layers of SiO_x-Si quantum dot layers serving as the transport channel on the crystalline p-Si substrate is shown in Fig. 1(a). It has exhibited 4-state behavior as shown in the ID-VD characteristics in Fig. 1(b) [1].

Figure 2(a) shows an FET with asymmetric Si quantum dots in the channel region and Ge quantum dots in the gate region over HfO_2-SiOx tunnel oxide layers. The ID-VG characteristic, as shown in Fig. 2(b), exhibited 5 states [2]. Here, a distinct state is defined when ID increases slowly due to the filling of mini-energy bands with a finite density of states (DOS).

‡Corresponding author.
#This chapter appeared previously on the International Journal of High Speed Electronics and Systems. To cite this chapter, please cite the original article as the following: F. Jain, R. H. Gudlavalleti, J. Chandy and E. Heller, *Int. J. High Speed Electron. Syst.*, **33**, 2440076 (2024), doi: 10.1142/S0129156424400767.

(a)

(b)

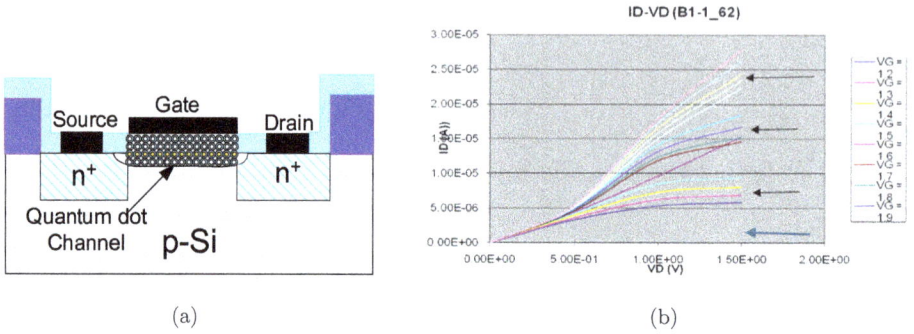

Fig. 1. (a) SiOx-cladded Si quantum dot channel and QD gate FET and (b) Experimental ID-VG showing 4-state ID-VD.

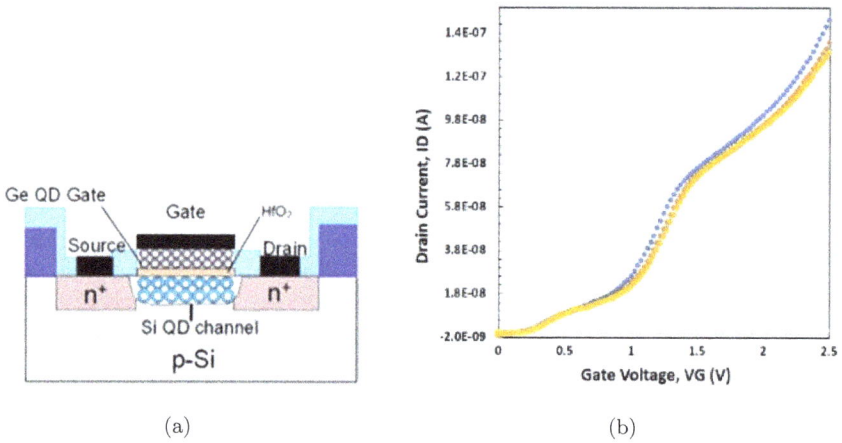

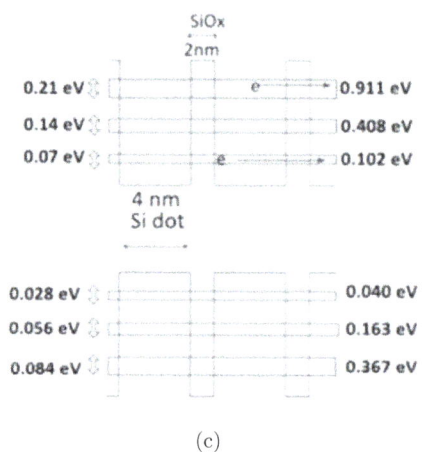

(a)

(b)

(c)

Fig. 2. (a) Cross-sectional schematic of a QDG-QDC-FET with two Si QD layer transport channels and 4 Ge QD layers in the gate. The yellow layer signifies the SiOx cladding of the top pink layer. (b) I-V characteristics showing multi-states in Ge QDSL transport channel. (c) Mini-energy bands in SiOx-Si QDSL.

Figure 2(c) shows QDSL mini-energy bands for SiOx-cladded Si quantum dot superlattice (QDSL). This paper presents multi-state QDC-QDG FETs in a Gate-all-around (GAA) configuration with the potential to design and fabricate sub-2 nm FETs. We have reported a modified Kronig-Penny model for simulating the energy band diagram and density of states (DOS) in GeO_x-Ge and SiO_x-Si quantum dot superlattices (QDSL) [3]. For example, the sub-band is at $0.735\,eV$, $1.61\,eV$, and $2.5\,eV$ for 3 nm Ge dots. By contrast, the mini-energy bands in SiO_x-Si quantum dot superlattices (QDSL) are located from the bottom of the conduction band as $0.102\,eV$, $0.408\,eV$, and $0.911\,eV$ for 4 nm Si dots with 2 nm thin SiOx barrier. The energy width of the mini-energy band is 0.07, 0.14 and 0.21 eV. The energy separation values are an order higher than InGaAs-InP and GaAs-AlGaAs quantum wells and/or quantum well superlattices; thus providing multi-state operation at room temperature. Cryogenic temperature simulations have been performed showing their potential.

2. GAA-QDC-FETs

Figure 3(a) shows a QDC-FET having a SiOx-Cladded Si quantum dot channel with GeOx-cladded Ge quantum dot layers in the gate-all-around configuration [3, 3b]. Here, the n+ drain is realized on a p-Si substrate. Figure 3(b) illustrates a two-QDC stacked as GAA-QDC-FET structure. A variation of Fig. 3(b) using Ge QDs will further enhance the number of bits.

Quantum simulation of QD-FETs incorporating Ge dots in the channel and gate region is shown in Figs. 4(a) and 4(b).

2.1. *Quantum wire QDC-QDG-FET with HfO_2 tunnel oxide*

The interesting scenario is when the channel length L is ~8-9 nm (three QDs between source and drain). Figure 5 shows schematically the cross-section of the

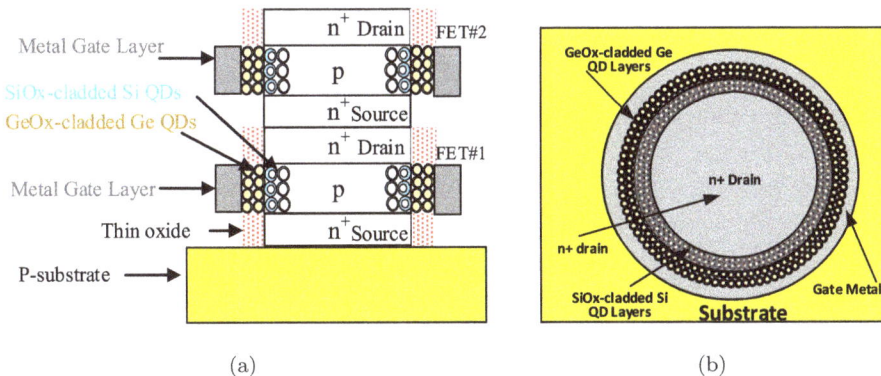

(a)   (b)

Fig. 3. (a) Cross-section of two vertically stacked GAA-QDC FETs. (b) Top view of upper GAA-QDC FET showing Si QDC and Ge QD gate layer [adapted from reference 3(b)].

(a)   (b)

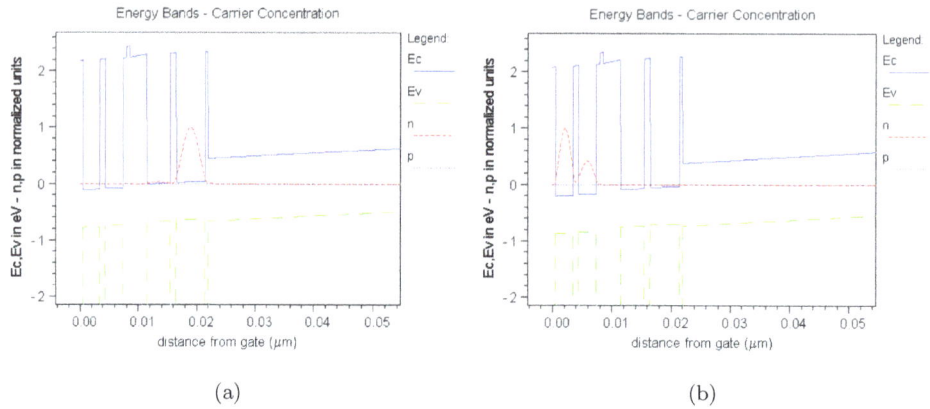

Fig. 4. (a) Electron wavefunctions in the lower QD layer in the channel. (b) Wavefunctions profile in gate GeOx-cladded Ge dot layers at higher Vg.

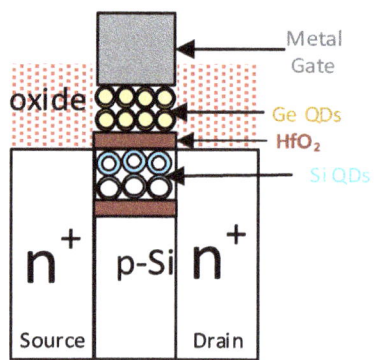

Fig. 5. QDC-QDG-FET in quantum wire configuration.

Quantum Dot Channel with the Quantum dot gate in the QDSL-quantum wire configuration.

Figure 6(a) shows the quantum simulation of Quantum Channel FET in lateral configuration (shown in Fig. 5) at two VG values. Figure 6(b) shows HfO$_2$ under Si QDs in the channel region. Figure 6(c) shows wavefunctions with HfO$_2$ under Si QDs in the channel region and Ge QDs in the gate region.

2.2. Modeling of quantum dot channel (QDC) Si FETs at sub-Kelvin for multi-state logic

Figure 7 compares carrier density at 77, 4.2 and 0.025 K. Here, all steps are shown with Fermi level value up to 1.1 eV [4]. This lays the foundation of QDC-FETs operating at cryogenic temperatures.

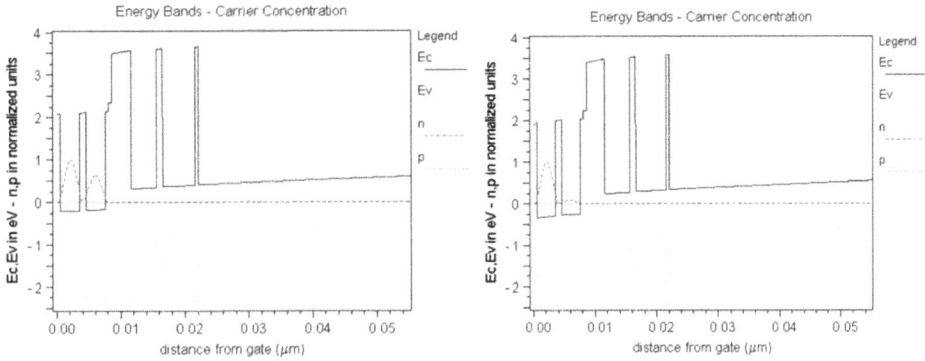

(a)

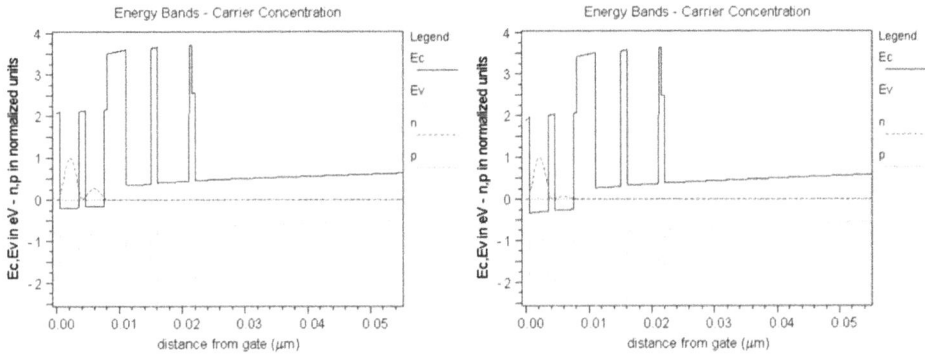

(b)

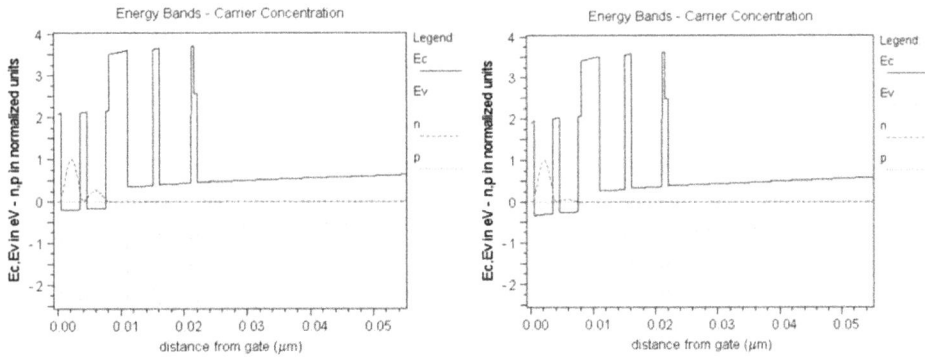

(c)

Fig. 6. (a) Quantum simulation of QDC-FET in lateral configuration (shown in Fig. 5) at VG = −0.2 V (left) and VG = 0 V (right) with HfO2 under Ge QDs in gate region. (b) Quantum simulation of QDC-FET in lateral configuration (shown in Fig. 5) at VG = −0.2 V (left) and VG = 0 V (right) with HfO2 under Si QDs in channel region. (c) Quantum simulation of QDC-FET in lateral configuration (shown in Fig. 5) at VG = −0.2 V (left) and VG = 0 V (right) with HfO2 under Si QDs in channel region and under Ge QDs in the gate region.

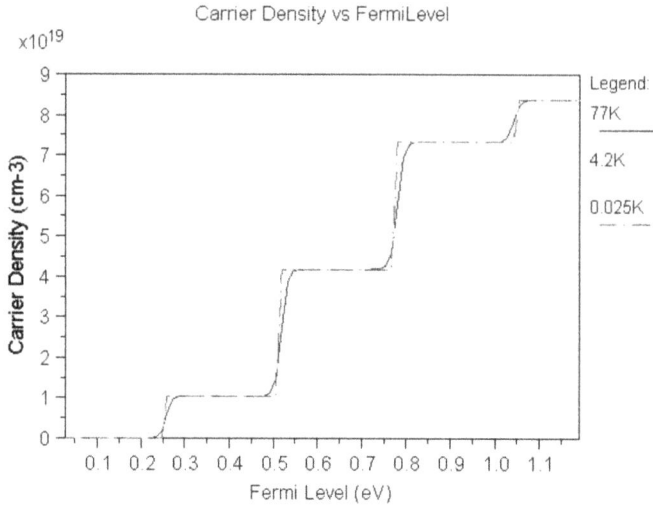

Fig. 7. Modeling of carrier density versus Fermi level at 77 K, 4.2 K and 0.025 K.

2.3. *3-D confined quantum dot superlattice (QDSL) as quantum dot*

In Fig. 7, the electron reservoir (n+ source region) is adjacent to the quantum dot under gate G_1. There is a barrier G_{TB} created between two quantum dots to separate them. Alternately, a barrier can be created between n+ source and drain.

This forms the platform to operate at milli-Kelvin temperatures using electron transfer between two wells creating qubits. This is in contrast to gate defined Si quantum dot devices reported for computing at sub-milli-Kelvin [5].

2.4. *3-D confined quantum dot superlattice (QDSL) as quantum dot*

In Fig. 8, the electron reservoir (n+ source region) is adjacent to the quantum dot under gate G_1. There is a barrier G_{TB} created between two quantum dots to separate them. Alternatively, a barrier can be created between the n+ source and the drain.

This forms the platform to operate at milli-Kelvin temperatures using electron transfer between two wells creating qubits. This is in contrast to gate defined Si quantum dot devices reported for computing at sub-milli-Kelvin [5].

2.5. *Integration of QD-NVRAMs*

Multi-bit QDC-QDG-FETs in regular 2-D, 1-D (wire) and 0-D (dot) configurations have been presented. Quantum simulations show the potential for multi-state operation. The implementation of multi-state QD-NVRAMs has been experimentally

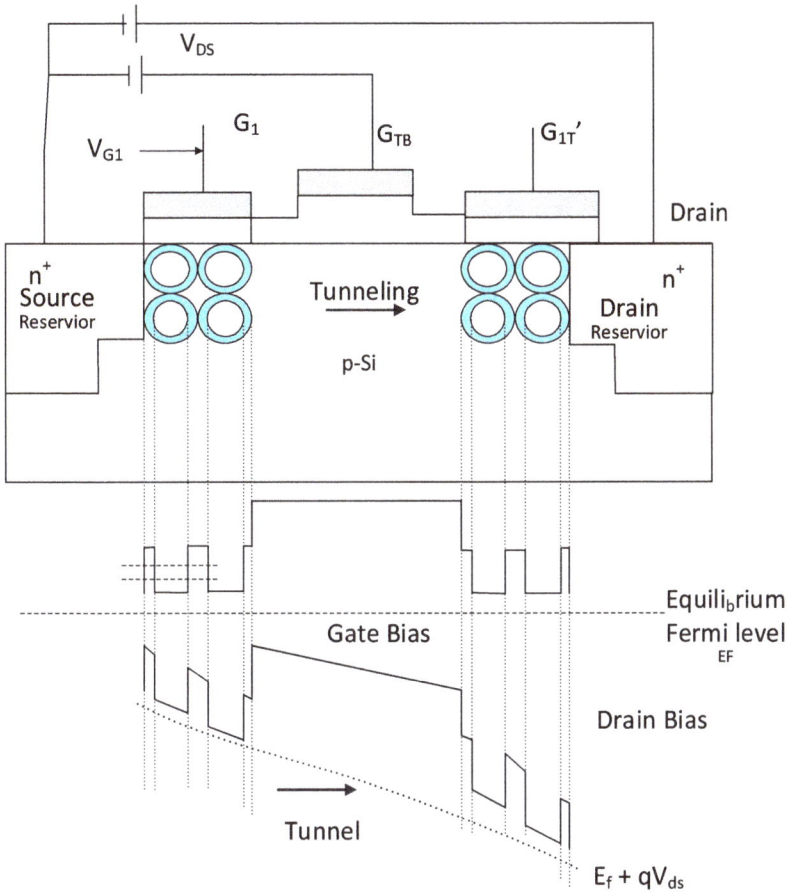

Fig. 8. FET comprising two QDots (QDSL under G_1 & $G_{1T'}$) separated by a tunnel barrier gate G_{TB}.

demonstrated [6]. Finally, the integration of QDC-FET logic with QD-NVRAMs provides a low-power hardware platform for in-memory computing [7] and artificial intelligence applications.

3. Quantum Dots Using Multi-State QDC-QDG-FETs

We also present coupled quantum dots structures (Fig. 8) where QDs are formed under gates G_1 and $G_{1'}$ between source/drain reservoirs, separated by a tunnel barrier GT. Local magnetic fields can be created using FeCo thin films. Unlike recently reported [4] QDots separated by electrostatic gate barriers by Mills *et al.* [8], the quantum dots in Fig. 8 comprise more than one gate-confined quantum dot. In fact, each dot comprises an array of cladded quantum dots forming QDSL-like mini-energy bands. Figure 8 shows that the quantum dots, adjacent to the source region, are cladded with an oxide barrier around the core of QDs to isolate them from the source region and confine them under gate G_1. Similarly, set of quantum dots (coupled to each other due to a thin barrier like finite QDSL) are near the drain region under $G_{1'}$. The two sets of coupled-QDs are separated by tunnel barrier G_{1T}. This structure works on spin-states. Localized magnetic field, using CoFe magnetic line adjacent to gates, is used to control the spin state. Furthermore, the Si quantum dots may be constructed using 28Si isotope to improve coherence time. Finally, control-Z (CZ) and two-qubit gates, CNOT and SWAP gates, are realized [8].

Alternatively, quantum dots are formed using 2-deg electron gas in Al-GaAs/GaAs and Si/SiGe structures by lateral gates confining electron in a quantum dot [9,10]. The authors in Reference 9 describes a multiplexed set of 4 (2×2) double quantum dots.

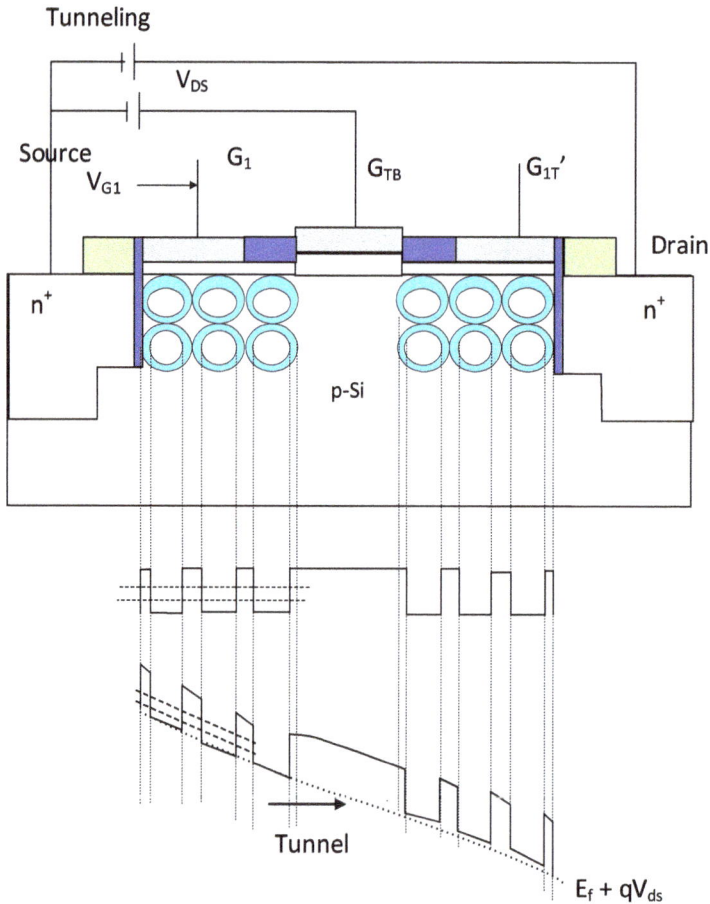

Fig. 9. FET comprising two QDots (QDSL under G_1 & $G_{1T'}$) separated by a tunnel barrier gate G_{TB}.

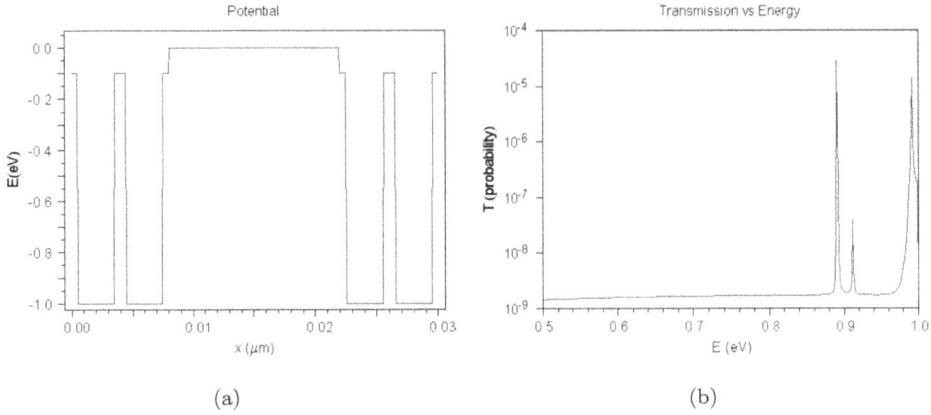

(a)   (b)

Fig. 10. Transmission simulation of Fig. 8 structure, showing channel potential and transmission vs energy for two different V_{ds}: (a) $V_{ds} = 0\,V$ and (b) $V_{ds} = 0.1\,V$.

3.1. *Electrostatic barrier isolated QDs comprising multiple cladded quantum dots*

We present in Fig. 9 a coupled quantum dot structure. We believe that it has the potential to exhibit enhanced multiple current states, separated by pico-amps (or lower) at around 4.2 K, controlled by gate and source-drain voltages. Each quantum dot has its own gate voltage. Both QDs are separated from source and drain regions by a thin tunnel barrier. The barrier could be a resonant tunnel structure.

A simulation of tunneling probabilities is presented in Fig. 10. Figure 10 shows the channel potentials and transmission vs energy for $V_{ds} = 0\,V$ and 0.1 V, respectively. The structure of Fig. 8 was approximated at each V_{ds} as a 1D potential between source and drain and then simulated using a Transfer Matrix Method solution of the Schrodinger Equation [13]. The transmission results show a voltage-dependent resonant tunneling effect.

4. Conclusion

QDC-QDG-FETs are presented in 2-D, 1-D, and 0-D configurations. The important difference with conventional QDot is that each QD is comprised of $2 \times 2 \times 2$ or more arrays forming a QDSL. QDSL comprises thin SiO_x or GeO_x cladding around Si or Ge dot, respectively, and mini-energy band characteristics could be designed based on the application. In addition, quantum interference transistors, QUITs, mimicking GAA-FETs, can be realized [11, 12]. The I-V characteristics are dependent on mini-energy bands, barriers between source and drain, coupling tunnel barrier, and screening effects.

Finally, we anticipate 16-32 states of QD channel FET-based logic integrating QD-NVRAMs for the implementation of hardware platforms for in-memory and quantum computing at cryogenic temperatures.

Acknowledgments

The authors dedicate this work to the memory of Prof. T.-P. Ma of Yale University. We gratefully acknowledge the feedback of Prof. F. Xia (Yale).

ORCID

F. Jain ⊚ https://orcid.org/0000-0003-3961-6665
R. H. Gudlavalleti ⊚ https://orcid.org/0000-0002-7727-8030
J. Chandy ⊚ https://orcid.org/0000-0003-3449-3205

References

1. F. Jain, S. Karmakar, P.-Y. Chan, E. Suarez, M. Gogna, J. Chandy and E. Heller, "Quantum dot channel (QDC) field-effect transistors (FETs) using II-VI barrier layers," *Journal of Electronic Materials*, 41, 2775, 2012.
2. F. Jain, R. H. Gudlavalleti, J. Chandy and E. Heller, "Novel multi-state QDC-QDG FETs and gate all around (GAA) FETs for integrated logic and QD-NVRAMs," *International Journal of High Speed Electronics & Systems*, 32(2–4), 2350026, 2023. DOI: 10-1142/S012915642350026X
3. F. Jain, R. Gudlavalleti, R. Mays, B. Saman, P.-Y. Chan, J. Chandy, M. Lingalugari and E. Heller, Multi-state quantum dot channel (QDC) FETs for multi-bit computing, *52nd IEEE SISC*, San Diego, December 8–11, 2021; also (3b) F. Jain, R. H. Gudlavalleti, J. Chandy and E. Heller, Vertically stacked cladded Si/Ge quantum dot GAA-FETs for multi-bit computing potentially integrating logic, SRAMs, and NVRAMs, *54th IEEE SISC*, San Diego, December 8–11, 2023.
4. F. Jain, R. H. Gudlavalleti, R. Mays, B. Saman, J. Chandy and E. Heller, Modeling of quantum dot channel (QDC) Si FETs at sub-Kelvin for multi-state logic, *International Journal of High Speed Electronics and Systems* 29(01n04), 2040017, 2018.
5. A. J. Sigillito, J. C. Loy, D. M. Zajac, M. J. Gullans, L. F. Edge and J. R. Petta, "Site-selective quantum control in an isotopically enriched ^{28}Si/Si$_{0.7}$Ge$_{0.3}$ quadruple quantum dot," *Physics Review Applied*, 11, 061006, 2019.
6. M. Lingalugari, P.-Y. Chan, E. K. Heller, J. Chandy and F. C. Jain, "Quantum dot floating gate nonvolatile random access memory using quantum dot channel for faster erasing," *Electronic Letters*, 54, 36, 2018.
7. D. Lelmini and H.-S. P. Wong, "In-memory computing with resistive switching devices," *Nature Electronics*, 1, 333–337, 2018.
8. A. Mills, C. Guinn, M. Gullans, A. Sigillito, M. Feldman, E. Nielsen and J. Petta, *Science Advances* 8, eabn5135, 2022.
9. M. A. Eriksson, S. N. Coppersmith and M. G. Lagally, *MRS Bulletin*, 38, 794–801, 2013.
10. D. R. Ward, D. E. Savage, M. G. Lagally, S. N. Coopersmith and M. A. Eriksoon, *Applied Physics Letters* 102, 213107, 2013.
11. E. Heller and F. Jain, "Simulation of one-dimensional ring quantum interference transistors using the time-dependent finite-difference beam propagation method," *Journal of Applied Physics*, 87, 8080–8087, 2000.
12. E. Heller, S K. Islam, G. Zhao and F. Jain, "Analysis of In0.52Al0.48As/In0.53Ga0.47As/InP quantum wire MODFETs employing coupled well channels," *Solid-State Electronics*, 43, 901–914, 1999.
13. Y. Ando and T. Itoh, "Calculation of transmission tunneling current across arbitrary potential barriers," *Journal of Applied Physics*, 61(4), 1497–1502, 1987.

8-State SRAMS Based on Cladded GE Quantum Dot Gate FETS[#]

B. Saman ©[*], A. Almalki ©[†], J. Chandy ©[†], E. Heller ©[‡] and F. C. Jain ©[†,§]

*Department of Electrical Engineering,
Taif University, P. O. Box 888 - 21974 Al-Hawiyah,
Taif, Kingdom of Saudi Arabia

†Department of Electrical and Computer Engineering,
University of Connecticut, Connecticut 06269, USA

‡Synopsys Inc., Ossining, New York 10562, USA
§faquir.jain@uconn.edu

This paper describes the fabrication of quantum dot gate (QDG) n-FETs using GeOx-cladded Ge quantum dot self-assembled on tunnel gate oxide. Experimental I–V characteristics exhibiting 4-states are presented. Simulations are presented for the operation of a viable 8-state SRAM using QDG-FETs.

Keywords: QDGFET; SRAM; ABM model.

1. Introduction

Our group has fabricated quantum dot gate (QDG) FETs and inverters [1–3]. Experimental ID-VG characteristics of a FET, fabricated via site-specific self-assembly of GeO_x -cladded Ge quantum dots, exhibited multiple thresholds and intermediate states, as shown in Fig. 1 [4]. In parallel, Jain *et al.* [5] demonstrated an inversion charge transferring from the lower quantum dot channel to the upper quantum dot channel in the quantum spatial wavefunction switched (SWS) FETs. 8-state SRAMs using SWS-FETs have been reported by Saman *et al.* [6].

2. Simulation of 8-State QDG-FET-Based SRAMs

Figure 2 shows the schematic of a QDG-FET-based SRAM memory cell utilizing conventional six-transistor (6T) architecture. However, the pull-down transistors

§Corresponding author.
#This chapter appeared previously on the International Journal of High Speed Electronics and Systems. To cite this chapter, please cite the original article as the following: B. Saman, A. Almalki, J. Chandy, E. Heller and F. C. Jain, *Int. J. High Speed Electron. Syst.*, **33**, 2440077 (2024), doi: 10.1142/S0129156424400779.

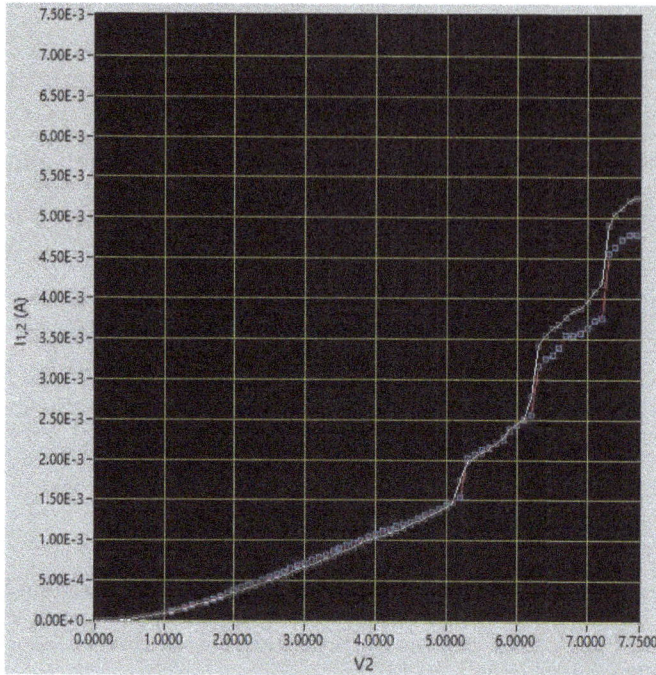

Fig. 1. QDG n-FET with self-assembled GeOx-cladded Ge quantum dot on tunnel gate oxide.

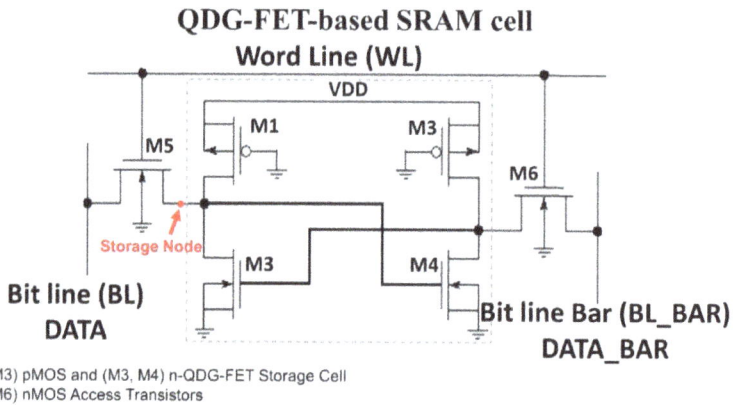

Fig. 2. Schematic of 8-state SRAM integrating QDG-FET as pull-down device and P-MOS as pull-up FET.

in the cross-coupled inverter circuits are replaced by QDG-FETs to be used as the driver transistors and to give the 8-state outputs for the provided DC input voltage. The two inverters in the SRAM cell use P-MOS as pull-up FETs. These 8-state outputs can be thought of as 3-bit binary output or as 8 digital states.

Approximately, 66% of cell area saving can be achieved by using QDG-FETs in conjunction with the conventional 6T SRAM architecture.

2.1. *Analog behavioral model*

The QDG-FET was modeled using an analog behavioral model (ABM) in conjunction with the Berkeley Short-channel IGFET BSIM3 version 3V3 of 1 μm technology node with capacitances and other device parameters.

A QDG-FET behaves as a MOSFET. The drain current for an n-channel QDG-FET is given in the following equation:

$$I_D = \frac{W}{L} c_{ox} \mu_n \left[(V_G - V_{\text{THQDGFET}}) V_D - \frac{V_D^2}{2} \right]. \tag{1}$$

Here, the threshold VTH_QDGFET is defined in the following equation:

$$V_{\text{THQDGFET}} = V_{TH} - \frac{q}{c_{ox}} \left[\sum \frac{x_{QD1} n_1 N_{QD1}}{x_g} + \sum \frac{x_{QD2} n_2 N_{QD2}}{x_g} \right]. \tag{2}$$

The ABM libraries is used to establish an n-channel QDG-FET model that can address the change in threshold voltage V_{THQDGFET}.

Figure 3 shows the input–output characteristics of QDG-FET-based inverters and was obtained using Cadence Simulator. Simulations of the inverter transient behavior are shown in Fig. 4.

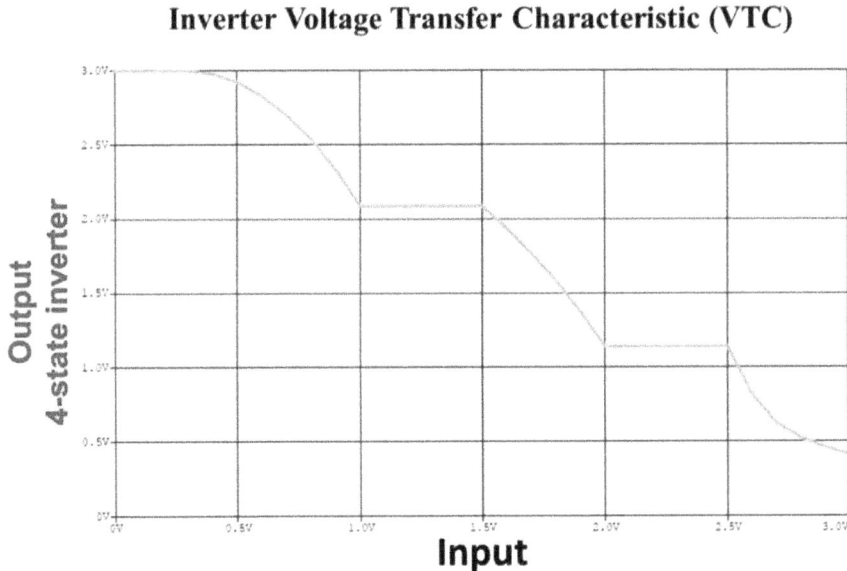

Fig. 3. Simulation of 4-state inverter voltage transfer characteristic (VTC).

Inverter transient behavior

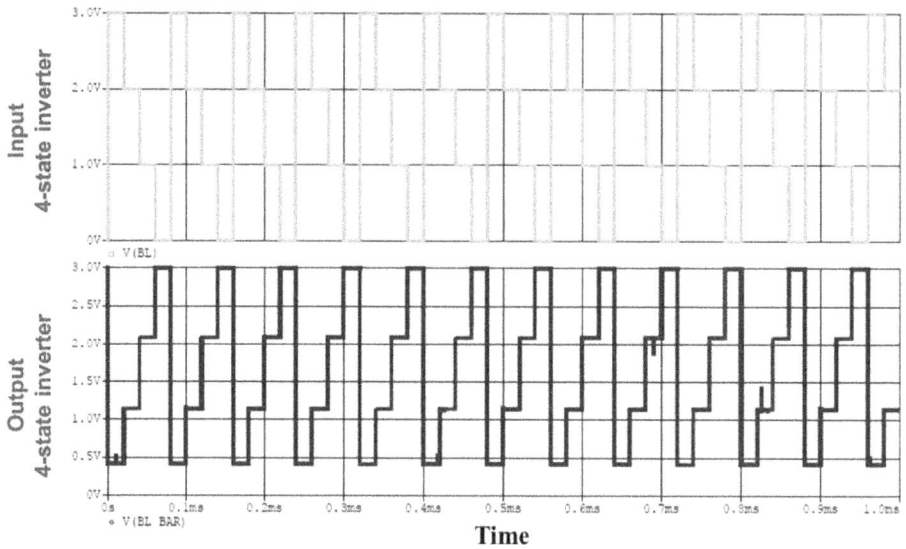

Fig. 4. Simulation of 4-state inverter transient behavior: Input upper green panel and output lower blue panel.

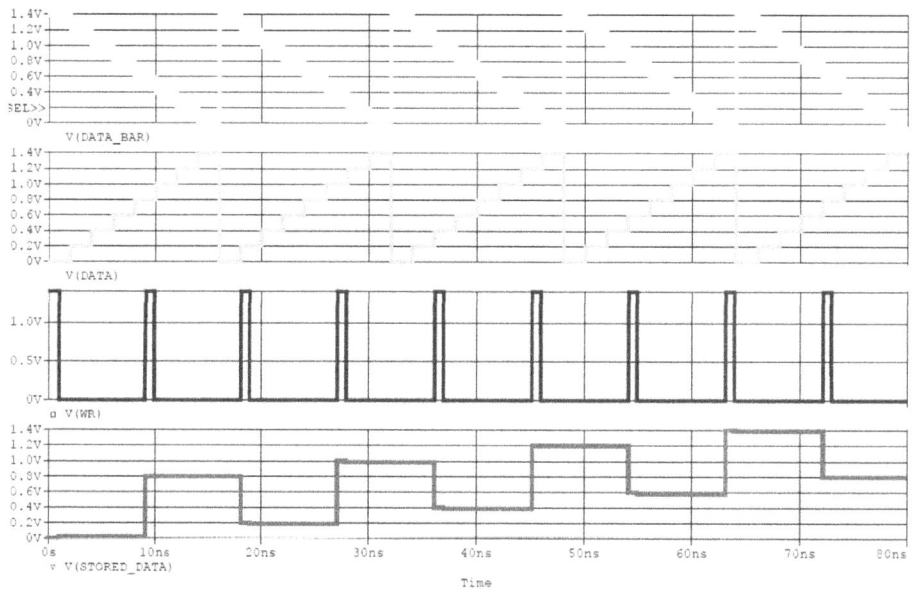

Fig. 5. SRAM Simulation results of Data (green), Word line (blue) and data stored in the memory cell (red).

2.2. *Simulation of 8-state SRAM*

Figure 5 shows the simulation result for the write operation of a 3-bit SRAM cell based on 6T QDG-FET. The top panel shows 8-state input data (green) connected to the bit line (BL). The middle panel shows the word line's pulse voltages at different states (blue). Finally, the QDG-FET-based SRAM's stored data are illustrated in the bottom panel (red) showing the storage of each state in the SRAM cell.

3. Conclusion

Fabrication of QDG n-FETs using self-assembled GeOx-cladded Ge quantum dot on tunnel gate oxide exhibited 5/6-state operation (Fig. 1). Simulation of 4-state inverters is presented. Simulations are also presented for a viable operation of 8-state SRAMs using QDG-FETs.

ORCID

B. Saman ⊙ https://orcid.org/0000-0001-6917-5763

A. Almalki   https://orcid.org/0009-0001-1954-4644

J. Chandy   https://orcid.org/0000-0003-3449-3205

E. Heller   https://orcid.org/0009-0005-4405-7089

F. C. Jain   https://orcid.org/0000-0003-3961-6665

References

1. F. C. Jain, E. Heller, S. Karmakar and J. Chandy, "Device and circuit modeling using novel 3-state quantum dot gate FETs," *2007 ISDRS*, 2007, pp. 1–2, doi: 10.1109/IS-DRS.2007.4422254.
2. M. Lingalugari et al., "Novel multi-state quantum dot gate FETs using SiO_2 and lattice-matched ZnS-ZnMgS-ZnS as gate insulators," Journal of Electronic Materials, 42, pp. 3156–3163, 2013.
3. B. Khan, R. Mays, R. Gudlavalleti and F. C. Jain, "Fabrication and characterization of nMOS inverters utilizing quantum dot gate field effect transistor (QDGFET) for SRAM device," Nanostructures for Electronics, Photonics, Biosensors and Emerging Systems, pp. 95–108, 2022.
4. A. Almalki and F. Jain (private communication).
5. F. Jain, M. Lingalugari, B. Saman, P.-Y. Chan, P. Gogna, E.-S. Hasaneen, J. Chandy and E. Heller, "Multi-state sub-9 nm QDC-SWS FETs for compact memory circuits," *46th IEEE Semiconductor Interface Specialists Conference (SISC)*, Atlanta (VA), December 2–5, 2015.
6. F. Jain, R. Gudlavalleti, R. Mays, B. Saman, P.-Y. Chan, J. Chandy, M. Lingalugari and E. Heller, "Quantum dot channel FETs harnessing mini-energy band transitions in GeOx-Ge and Si QDSL for multi-bit computing," In Selected Topics in Electronics and Systems, pp. 133–142, 2022.

Threshold Inverter Quantizer-Based 2-Bit Comparator using Spatial Wavefunction Switched (SWS) FET Inverters: Power Dissipation Analysis#

W. Alamoudi ⊙*, B. Saman ⊙†, R. H. Gudlavalleti ⊙*, A. Almalki ⊙*,†,
J. Chandy ⊙*, E. Heller ⊙‡ and F. Jain ⊙*,§

*Department of Electrical and Computer Engineering,
University of Connecticut, CT 06269, USA

†Department of Electrical Engineering, College of Engineering,
Taif University, Al-Hawiyah 21944, Saudi Arabia

‡Synopsys Corporation, Ossining, New York 10562, USA
§faquir.jain@uconn.edu

This paper aims to assess the power dissipation of a threshold quantizer (TIQ) 2-bit-based comparator using a SWS-FET-based inverter [1, 2, 4, 5]. Unlike conventional comparators, SWS-based comparator functionalized with TIQ comprises two or more vertically stacked quantum dots or well channels [1–3, 5, 6]. Herein, power dissipation analysis of the simulated circuit is carried out using Cadence by integrating the Berkeley Short-Channel IGFET Model (BSIM) and the analog behavioral model (ABM) [1,3,4,9]. The transient behavior of the inverter circuit is evaluated using 180 nm technology node. Our results demonstrated a significant reduction in power dissipation which overcome the limitation of previous 4-state logic implementations.

Keywords: Threshold inverter quantizer; 2-bit SWS-CMOS inverter; SWS-FETs power dissipation analysis.

1. Introduction

As the field of very large-scale integration (VLSI) is emerging, it is associated with increasing the number of used transistors. Increasing the transistor number within the chip leads to an increase in the power dissipation and the number of interconnects. Therefore, reducing the power consumption increase the reliability of tool load [7,9]. CMOS technology is one of the most common devices that is essential to lower power consumption [9].

§Corresponding author.
#This chapter appeared previously on the International Journal of High Speed Electronics and Systems. To cite this chapter, please cite the original article as the following: W. Alamoudi, B. Saman, R. H. Gudlavalleti, A. Almalki, J. Chandy, E. Heller and F. Jain, *Int. J. High Speed Electron. Syst.*, **33**, 2440078 (2024), doi: 10.1142/S0129156424400780.

A spatial wavefunction switched field effect transistor (SWSFET) is one such device, which can be utilized to develop multiple state logic circuits. It works based on the electron wavefunction switching between two channels depending on the voltage applied to the gate. A SWSFET has two or more vertically stacked quantum dot or quantum well channels, where the spatial location two of carriers within these channels is used to encode the logic states (00), (01), (10) and (11) [1]. The application of two strained layers of Si/Si0.5Ge0.5 in SWSFET improves performance and device integration [1, 4].

2. Threshold Inverter Quantizer (TIQ) Flash ADC

A flash analog-to-digital comparator (ADC), mainly known for its high-speed conversion is compared with other ADC architectures such as pipelined, successive-approximation register (SAR) ADCs. However, flash ADC consumes more power compared to other ADCs. A threshold inverter quantizer (TIQ) flash ADC is one approach to reducing power consumption in flash ADC while maintaining the conversion speed of the flash ADC. In this work, a TIQ-based voltage comparator is used to quantize analog input signals in flash ADC designs.

This work aims to assess the power dissipation of a threshold quantizer (TIQ) 2-bit-based comparator using a SWS-FET-based inverter. Herein, the power dissipation analysis of the simulated circuit is carried out using cadence by integrating the Berkeley Short-Channel IGFET Model (BSIM) and the analog behavioral model (ABM) [1, 3, 4, 9]. The transient behavior of the inverter circuit is evaluated using

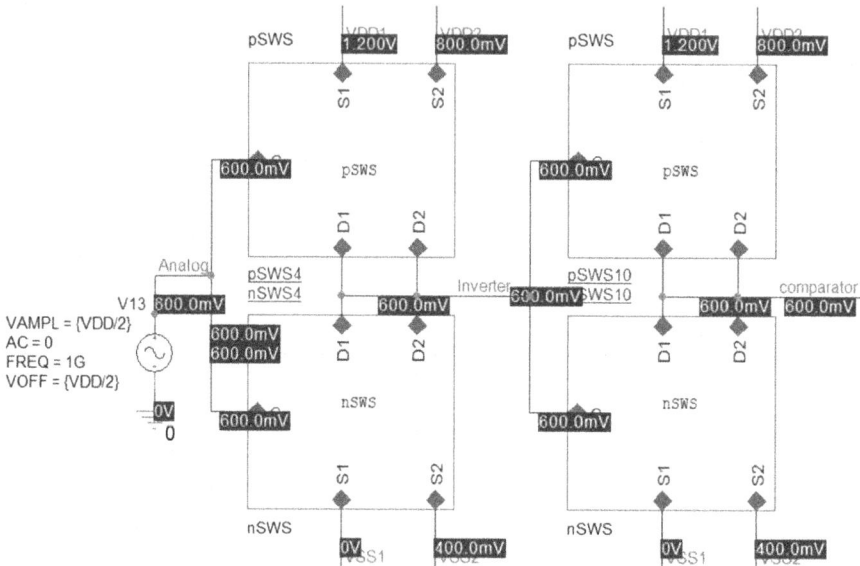

Fig. 1. 2-bit ADC comparator using SWS-FET-based inverters in CMOS-X.

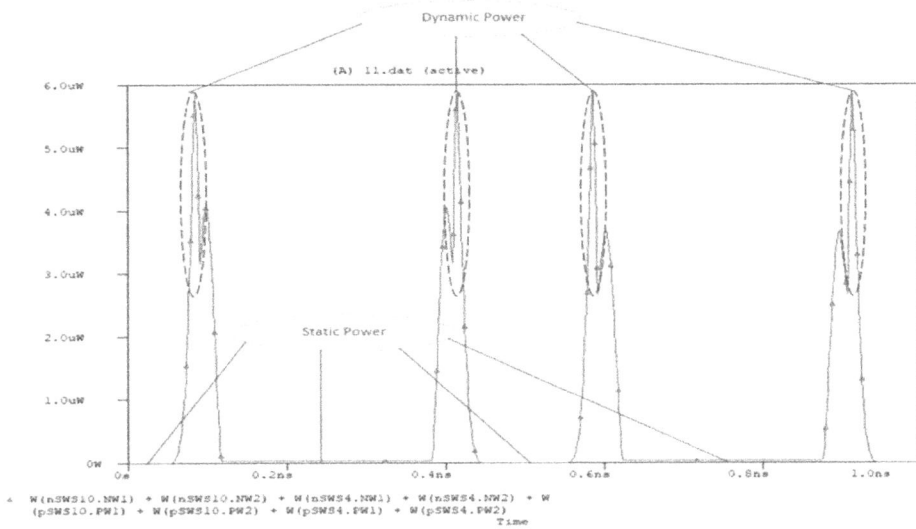

Fig. 2. Power dissipation in the TIQ-based SWS-FET inverter.

Table 1. Average power consumption.

	[8]	This work
Voltage supply	1.8 V	1.2 V
Power dissipation	1.33 nW	0.547 nW
Average dynamic power dissipation	0.46 mW	0.23 nW

180 nm technology node. Our results demonstrated a significant reduction in power dissipation, which overcame the limitation of previous 4-state logic implementations.

The circuit schematic is shown in Fig. 1. The transient behavior of the inverter circuit is evaluated using 180 nm technology node.

- Figure 2 demonstrated the power dissipation graph where red curve represents dynamic power and the blue line pointed to the static power. Results demonstrated reduction in power dissipation (0.547 μW) with an average dynamic power of 0.239 μW as mentioned in Table 1.
- The implementation of TIQ within SWS-FET inverter displayed around 97.45% improvement in power consumption [7].
- It also demonstrated about 98.24% improvement when TIQ was implemented within conventional CMOS [8].
- Therefore, the presented design overcomes the limitation of previous 4-state logic implementations.

3. SWSFET-Based Flash Analog-to-Digital Converter

TIQ comparators compare the input voltage with internal threshold voltage (V_T) of the inverter, which is determined by the transistor sizes in the inverters. Hence, we do not need the resistor ladder circuit used in a conventional flash ADC. The comparator outputs a binary code in two steps through an encoder. The comparator's role is to convert an input voltage (Vin) into a logic '1' or '0' by comparing an internal reference voltage (Vref) with Vin. If Vin is greater than Vref, the output of the comparator is '1', otherwise '0'. The TIQ comparator uses two cascading CMOS inverters as a comparator for high-speed and low-power consumption [4].

To realize SWSFET-based Flash ADC, the threshold of the inverter can be varied by connecting the upper quantum -dot/well channel of the P-SWSFET to the upper quantum -dot/well channel of the N-SWSFET, upper quantum-dot/well channel of the P-SWSFET to the lower quantum-dot/well channel of the N-SWSFET, the lower quantum-dot/well channel of the P-SWSFET to the upper quantum-dot/well channel of the N-SWSFET, lower quantum-dot/well channel of the P-SWSFET to the lower quantum-dot/well channel of the N-SWSFET. These four configurations provide a 2-bit ADC quantizer. To further increase the resolution of the ADC, the size of the P-SWSFET and N-SWSFET can be varied to generate different inverter threshold voltages. The design and simulation of this SWSFET-based Flash ADC will be realized in our future work.

4. Conclusions

The power dissipation in a 2-bit TIQ comparator design using a SWS-CMOS-X inverter was evaluated and compared to other logic designs. The four-state logic is achieved using four voltage levels. The 2-bit TIQ comparator design using SWS-CMOS-X simulations was carried out in Cadence to analyze the power dissipation. The model was made by combining an analog behavioral model (ABM) and the Berkeley Short-channel IGFET Model (BSIM4.6). The implementation of TIQ within the SWS-FET inverter displayed around 97.45% improvement in power consumption. Therefore, the presented design overcomes the limitation of previous 4-state logic implementations.

ORCID

W. Alamoudi ◎ https://orcid.org/0009-0007-4146-3676

B. Saman ◎ https://orcid.org/0000-0001-6917-5763

R. H. Gudlavalleti ◎ https://orcid.org/0000-0002-7727-8030

A. Almalki ◎ https://orcid.org/0009-0001-1954-4644

J. Chandy ◎ https://orcid.org/0000-0003-3449-3205

E. Heller ◎ https://orcid.org/0009-0005-4405-7089

F. Jain ◎ https://orcid.org/0000-0003-3961-6665

References

1. Jain, F. C., Chandy, J., Miller, B., Hasaneen, E. S. and Heller, E., 2011. Spatial wavefunction-switched (SWS)-FET: A novel device to process multiple bits simultaneously with sub-picosecond delays. International Journal of High Speed Electronics and Systems, 20(3), pp. 641–652.
2. Jain, F., Lingalugari, M., Saman, B., Chan, P. Y., Gogna, P., Hasaneen, E. S., Chandy, J. and Heller, E., 2015. Multi-state sub-9 nm QDC-SWS FETs for compact memory circuits. In 46th IEEE Semiconductor Interface Specialists Conference (SISC) (pp. 2–5).
3. Saman, B., Gogna, P., Hasaneen, E. S., Chandy, J., Heller, E. and Jain, F. C., 2017. Spatial wavefunction switched (SWS) FET SRAM circuits and simulation. International Journal of High Speed Electronics and Systems, 26(3), p. 1740009.
4. Alamoudi, W., Saman, B., Gudlavalleti, R. H., Almalki, A., Chandy, J., Heller, E. and Jain, F., 2023. Threshold inverter quantizer (TIQ)-based 2-Bit comparator using spatial wavefunction switched (SWS) FET inverters. International Journal of High Speed Electronics and Systems, 32(02n04), p. 2350025.
5. Almalki, A., Saman, B., Chandy, J., Heller, E. and Jain, F. C., 2022. Propagation delay evaluation for spatial wavefunction switched (SWS) FET-based inverter. International Journal of High Speed Electronics and Systems, 31(01n04), p. 2240008.
6. Gudlavalleti, R. H., Saman, B., Mays, R., Heller, E., Chandy, J. and Jain, F., 2020. A novel peripheral circuit for SWSFET based multivalued static random-access memory. International Journal of High Speed Electronics and Systems, 29(01n04), p. 2040010.
7. Husawi, A., Saman, B., Almalki, A., Gudlavalleti, R. and Jain, F. C., 2022. Power dissipation and cell area: Quaternary logic CMOS inverter vs. four-state SWS-FET inverter. International Journal of High Speed Electronics and Systems, 31(01n04), p. 2240009.
8. Halim, I. S. A. and Abidin, N. A. N. B. Z., 2011. Low power CMOS charge sharing dynamic latch comparator using 0.18 μm technology. In 2011 IEEE Regional Symposium on Micro and Nano Electronics (pp. 156–160). IEEE.
9. Chaudhary, A. and Rana, A., 2020. Ultra low power SRAM cell for high speed applications using 90 nm CMOS technology. In 2020 8th International Conference on Reliability, Infocom Technologies and Optimization (Trends and Future Directions) (ICRITO) (pp. 1107–1109). IEEE.

Study of Soft Errors in Spiking Neural Network Hardware[#]

Zongming Li [ORCID]* and Lei Wang [ORCID]†

*Department of Electrical and Computer Engineering,
University of Connecticut, Storrs, CT 06269, USA*
*zongming.li@uconn.edu
†lei.3.wang@uconn.edu*

The problem of soft errors in the hardware implementation of Spiking Neural Networks (SNN) has always been a challenge. SNNs, unlike traditional deep learning networks, simulate the temporal dynamic behavior of biological neurons and emit spikes when a specific threshold is reached. In recent years, the hardware implementation of SNNs has shown great potential in performing efficient and low-power tasks, but as technology nodes shrink and integrated circuit complexity increases, soft errors become a key challenge. To this end, we propose a novel approach based on input and weight analysis, through specific algorithms at the hardware level, to significantly reduce the probability of soft errors affecting the results. In terms of training accuracy, our method maintains a high accuracy under various voltage fluctuation conditions, demonstrating its superior robustness.

Keywords: Spiking neuron network; soft error; leaky integrate-and-fire.

1. Introduction

With the rapid development of artificial intelligence on mobile and edge devices, spiking neural networks (SNNs) have attracted much attention from the research community as a computational model that efficiently mimics biological neural systems. SNNs are the third generation of artificial neural networks (ANNs) [1]. Their core feature is to exploit biological neuron process and transmit information through discrete electrical signals (spikes). Compared with conventional neural networks (CNNs), SNNs show great potential for achieving embedded intelligence with better energy efficiency [2].

Soft errors, which is a class of non-permanent data corruption due to external radiation or internal disturbances [3], are a critical issue in SNN hardware devices [4,9].

---

†Corresponding author.
#This chapter appeared previously on the International Journal of High Speed Electronics and Systems. To cite this chapter, please cite the original article as the following: Z. Li and L. Wang, *Int. J. High Speed Electron. Syst.*, **33**, 2440112 (2024), doi: 10.1142/S0129156424401128.

Such errors may seriously affect the performance and reliability of SNNs, especially in embedded systems deployed in the field that require high accuracy and long-term stability. Researchers are seeking effective techniques to mitigate the problem of soft errors, including improved training algorithms, introducing new hardware architectures, and exploring new neuron and synapse models.

Some existing works have been developed targeting the mitigation of soft errors on SNN hardware devices. Rachmad *et al.* proposed SoftSNN [4], an approach aimed at reducing soft errors in SNN accelerators without re-execution, thus maintaining accuracy with low latency and energy overhead. A detection method [5] is developed to detect synaptic weight errors in Resistive RAM (RRAM) based SNN accelerators and a correction method is proposed to address soft errors. Some researchers exploit AI-based hardware detectors to monitor soft errors [6]. Most of these works use conventional methods that are previously applied to CNNs without considering the unique features of SNNs.

In this paper, we perform a comprehensive study on the impact of soft errors that may be encountered in the hardware implementations of SNNs. Our contributions are summarized in the following points:

- For SNNs based on the LIF model, we propose to monitor the LIF neurons and study the impact of soft errors on the membrane potential.
- Our approach considers the unique temporal characteristics of input spikes as well as neuron weights to prevent scenarios that may lead to soft errors.
- A low-cost dynamic protection strategy is proposed to ensure the accurate results of the SNNs in the presence of soft errors.

2. LIF Model and Soft Errors

2.1. *SNNs with LIF model*

Leaky Integrate-and-Fire (LIF) is a neuron model of SNNs, a biologically inspired neural network in which each neuron accumulates its input until it reaches a threshold [7], then fires a spike and resets its status. The neuron's membrane potential "leaks", or gradually decays, back to its resting potential. The neuron then accumulates (or integrates) its inputs until it reaches a threshold and then fires a spike. Although the LIF model is a simplified neuronal model, it has been widely used in neuroscience and neuromorphic computing due to its computational simplicity and reasonable approximation of biological neuron activities [8]. The mathematical representation of the LIF neuron model is given by the following differential equation:

$$\tau_m \frac{dV(t)}{dt} = -V(t) + RI(t) \tag{1}$$

where τ_m is the membrane time constant representing the rate of voltage decay, R is the membrane resistance, and $I(t)$ is the current input to the neuron.

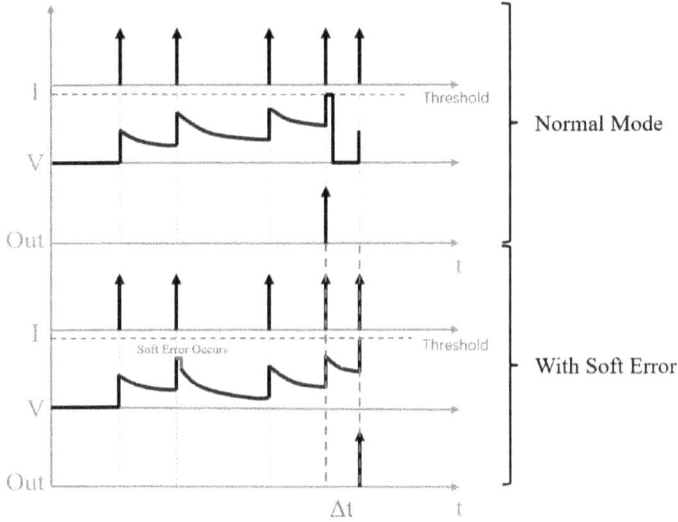

Fig. 1. Neuronal dynamics in normal mode and under soft errors.

When the membrane potential $V(t)$ reaches a threshold, the neuron fires a spike. Figure 1 illustrates the potential accumulation and spike firing processes from the LIF model.

2.2. *Soft errors*

Soft errors refer to the errors or failures that occur temporarily in electronic components or computer systems, usually caused by external radiation, electromagnetic interference, voltage noise, temperature changes and other factors [8]. They are unpredictable, temporary, and can cause system performance degradation or data corruption. Soft errors are usually not permanent problems caused by hardware failures, but transient problems caused by external interference or random events. In very large-scale integrated circuit (VLSI) systems, soft error-induced bit flips are a well-known problem. These bit flips can propagate along the data paths in deep learning architectures, impacting multiple computational operations. This can cause significant performance degradation because the predictive ability of deep learning models is highly dependent on these numerical operations. Bit flips due to soft errors may trigger a series of chain effects during model operation, leading to erroneous calculation results.

To mitigate the impact of soft errors [10], hardware and software measures are usually taken, including hardware fault tolerance techniques, error detection and correction codes, backup systems, data integrity checks, temperature and voltage monitoring, etc. These measures help to improve system reliability and reduce the impact of soft errors on electronic components and computer systems.

3. Error-Tolerant SNNs

In this section, we propose a method to model the soft errors in SNN hardware systems. We also present a technique to minimize the system performance degradation caused by soft errors.

3.1. *Soft errors in SNNs*

In SNNs with the LIF neuron model, soft errors can cause multiple effects. Soft errors may lead to random changes in neuron weights and thresholds, which in turn affect the learning ability and stability of a neural network. In addition, for neuronal circuits with long-term memory functions, soft errors may cause memory loss or errors, weakening the network's prediction capabilities. Specially in SNNs, a sudden soft error will cause the voltage to deflect to a certain extent, causing the neuron that is about to fire to miss the scheduled time. This will affect the output of the neuron and the input of the subsequent neurons.

As shown in Fig. 1, under the influence of soft errors, we can see several effects:

- The membrane potential might be affected. The magnitude and speed of its accumulation are no longer as predictable as under ideal conditions. At some points, the accumulation in membrane potential might be prematurely truncated, meaning that soft errors cause delays in firing time.
- The timing of the output spike might be shifted, indicating that the temporal characteristics [11] (which contain the important information in SNNs [12]) of the neuron are disturbed. In some cases, soft errors can cause neurons to miss firing or fire when they should not.

3.2. *Membrane potential analysis*

When the membrane potential reaches the threshold, the neuron fires a spike and resets its potential to V. Upon V hitting the threshold, the neuron initiates a spike. However, the exact computation of firing probability is complicated, as it depends on various factors such as input patterns, weight distributions and the neuron's initial state. Different from CNNs, SNNs have inputs in the form of discrete spike trains, whose arriving time and rate vary for different neurons and significantly affect a neuron's firing probability. On the other hand, the neuron weight also affects the firing probability. As a result, the membrane potential should be the main factor to be considered when determining the fire probability, as its value is determined by both input spike temporal characteristics and neuron weights.

From the LIF model described in Eq. (1), to deduce the relationship between a neuron's firing probability and its inputs and weights, it is needed to compute the input current I. In an SNN setup, a spike from a presynaptic neuron introduces an input current to the current neuron scaled by a weight w_i, such as

$$I = \sum w_i \cdot s_i \qquad (2)$$

where s_i is the input spike signal, whose value is 1 when the ith presynaptic neuron fires and 0 otherwise. For a short interval Δt, the change in the membrane potential Δv can be expressed as:

$$\Delta v = \frac{-(v - v_{\text{rest}}) + RI}{\tau_m} \Delta t. \tag{3}$$

From Eq. (3), we can calculate how many consecutive time intervals of such input current would bring the membrane potential to the threshold. Let n be the number of these time intervals, such as

$$n\Delta v \geq v_{\text{threshold}} - v_{\text{rest}}. \tag{4}$$

We can solve for n as

$$n \geq \frac{(v_{\text{threshold}} - v_{\text{rest}})\tau_m}{-(v - v_{\text{rest}}) + R_m I} \times \frac{1}{\Delta t}. \tag{5}$$

Assume that T is the full operation time of the neuron. The expected number of spikes within T is $T/(n\Delta t)$. Then, the firing probability can be expressed as

$$P_{\text{fire}} = \frac{\Delta t}{n} = \frac{\Delta t^2(-R_m \sum_{i=1}^{n} s_i \cdot w_i + v - v_{\text{rest}})}{\tau_m(v_{\text{rest}} - v_{\text{threshold}})}. \tag{6}$$

Figure 2 visualizes the combined impact of input spike rate and neuron weight. Under the condition of small input spike rate and synaptic weight, the firing probability of neurons is low. As these two parameters increase, the firing probability also increases significantly, indicating that neurons with higher input spike rates and stronger synaptic connections (i.e., higher neuron weights) have high neural activities and thus are in critical need of soft error protection.

3.3. *Soft error protection*

Soft errors are difficult to detect and cannot be eliminated [13] completely. Thus, our idea is to reduce the probability of soft-error induced misbehaviors occurring in SNN hardware. To achieve this, we propose a technique based on the model developed in Sec. 3.2. The proposed technique is quite effective without adding large hardware and power overheads.

The potential of a neuron is determined by the input temporal characteristics and the weight of the neuron. A more active neuron is likely to have a higher input rate and larger weight, and its potential can reach the firing threshold more frequently. As a result, this neuron should be dynamically identified and protected from soft errors. By monitoring the neuron potential at runtime, we can identify a neuron that is vulnerable to soft errors. The proposed technique then isolates this neuron from its neighboring neurons while preserving the current potential and direct connection weights to the preceding and following neurons. After that, a soft-error protection scheme such as triple modular redundancy [14] can be applied to this neuron for network computation. Note that this scheme is used dynamically

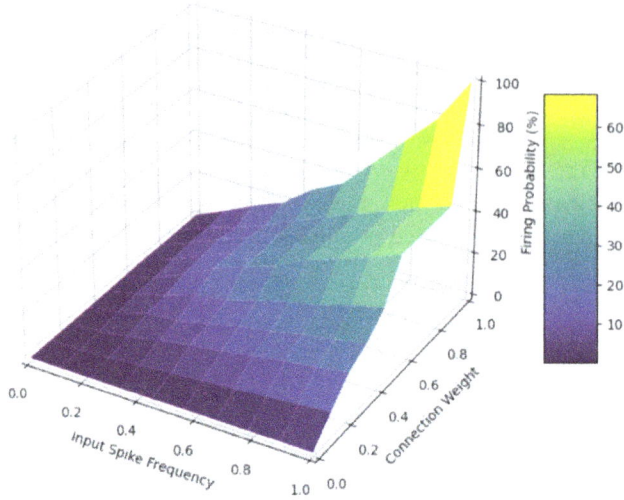

Fig. 2. Effects of input spike frequency and connection weight on firing probability.

only for a few neurons that need to be protected. Once the protected neuron finishes it computation (either firing a spike or its potential falling below a threshold due to leakage), it will be returned to the network. As a result, the proposed technique does not introduce large hardware and power overheads.

4. Simulation Results

We evaluate the performance of SNNs in the presence of soft errors. These transient errors are assumed to occur randomly with a rate ranging from 10^{-9} to 10^{-6} in the neurons. We simulated an SNN with 500 LIF neurons. The input data are composed of high-dimensional spike arrays. These values are used to explore their responses and behaviors under different soft error conditions. The occurrence of soft errors is simulated by randomly introducing bit upsets with a probability equal to that of the soft errors. This will reduce or reset the membrane potential of randomly selected neurons with a certain probability at each time step. In SNNs, the performance is measured by the accuracy during inference operations. A high accuracy indicates that the SNN can withstand disturbances caused by soft errors and maintain the accuracy of its computational tasks. From the results shown in Fig. 3, we observe that when the probability of soft error is low, the accuracy of SNN remains at a high level. This shows that the SNN itself has certain error-tolerant capabilities. This is not surprising as SNN is essentially an approximate computing scheme, and its performance is qualified by statistical measures such as classification rate. However, as the probability of soft errors increases, the accuracy of the network begins to decrease sharply. This indicates that when soft errors reach a certain level, the SNN is beyond its inherent capability to deal with soft errors.

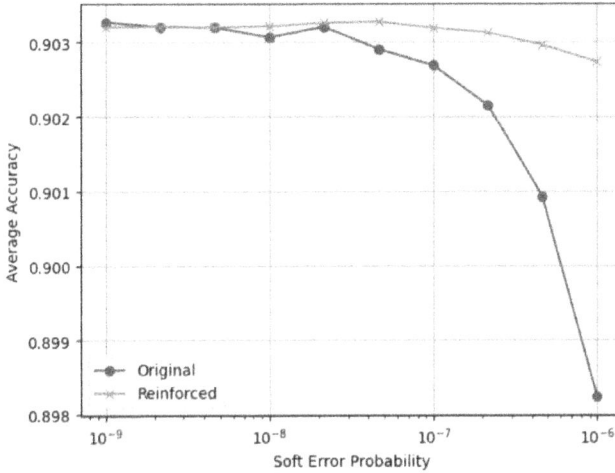

Fig. 3. Comparison of accuracy of spiking neural network after using our strategy.

Using the proposed error-protection technique, the performance of the SNN maintains a relatively stable level even under a high soft error probability. Under a high soft error probability, the average accuracy of the original network dropped sharply, while the proposed technique was able to maintain a stable accuracy. This validates the effectiveness of the proposed technique. The hardware and power overheads associated with the proposed soft error protection technique are minimal, as only 2% of neurons at the maximum need to be dynamically protected at run times.

5. Conclusion

In this paper, we studied the soft error problem in spiking neural and proposed a low-complexity error protection technique. By analyzing the input data and network weights, we can dynamically detect the neurons that are more vulnerable to soft errors in the SNN hardware. In addition, we introduced a mechanism that when the membrane potential of a neuron exceeds a set threshold, it is immediately isolated and apply a triple modular redundancy network computation, and then reconnected to the network after restoring its initial potential. Through simulations, we found that the proposed spiking neural network can maintain a high level of training accuracy in the presence of soft errors, enhancing the robustness of the network. Future work is being directed toward power and performance analysis on hardware implementation of the proposed technique.

ORCID

Zongming Li ⊙ https://orcid.org/0009-0000-4527-4038
Lei Wang ⊙ https://orcid.org/0009-0009-6653-9950

References

1. Ghosh-Dastidar, Samanwoy, and Hojjat Adeli. "Spiking neural networks." International Journal of Neural Systems 19.04 (2009): 295–308.
2. Cao, Yongqiang, Yang Chen and Deepak Khosla. "Spiking deep convolutional neural networks for energy-efficient object recognition." International Journal of Computer Vision 113 (2015): 54–66.
3. Mukherjee, Shubhendu S., Joel Emer and Steven K. Reinhardt. "The soft error problem: An architectural perspective." 11th International Symposium on High-Performance Computer Architecture. IEEE, (2005).
4. Putra, Rachmad Vidya Wicaksana, Muhammad Abdullah Hanif and Muhammad Shafique. "SoftSNN: Low-cost fault tolerance for spiking neural network accelerators under soft errors." Proceedings of the 59th ACM/IEEE Design Automation Conference, IEEE 2022.
5. Saha, Anurup, Chandramouli Amarnath and Abhijit Chatterjee. "A resilience framework for synapse weight errors and firing threshold perturbations in RRAM spiking neural networks." 2023 IEEE European Test Symposium (ETS). IEEE, 2023.
6. Kasap D, Carpegna A, Savino A, *et al.* "Micro-Architectural features as soft-error markers in embedded safety-critical systems: preliminary study." IEEE European Test Symposium (ETS). (2023): 1-5.
7. Maass, Wolfgang, and Christopher M. Bishop, eds. Pulsed neural networks. MIT press, 2001.
8. Nahmias, Mitchell A., *et al.* "A leaky integrate-and-fire laser neuron for ultrafast cognitive computing." IEEE Journal of Selected Topics in Quantum Electronics 19.5 (2013): 1–12.
9. Pham, Quoc Trung, *et al.* "A review of SNN implementation on FPGA." 2021 International Conference on Multimedia Analysis and Pattern Recognition (MAPR). IEEE, 2021.
10. Chandra, Vikas and Robert Aitken. "Impact of technology and voltage scaling on the soft error susceptibility in nanoscale CMOS." 2008 IEEE International Symposium on Defect and Fault Tolerance of VLSI Systems. IEEE, 2008.
11. Baumann, Robert. "Soft errors in advanced computer systems." IEEE Design & Test of Computers 22.3 (2005): 258–266.
12. Camuñas-Mesa, Luis A., Bernabé Linares-Barranco and Teresa Serrano-Gotarredona. "Neuromorphic spiking neural networks and their memristor-CMOS hardware implementations." Materials 12.17 (2019): 2745.
13. Shivakumar, Premkishore, *et al.* "Modeling the effect of technology trends on the soft error rate of combinational logic." Proceedings International Conference on Dependable Systems and Networks. IEEE, 2002.
14. Lyons, Robert E. and Wouter Vanderkulk. "The use of triple-modular redundancy to improve computer reliability." IBM Journal of Research and Development 6.2 (1962): 200–209.

Author Index